T0185298

B

Progress in Physics
No. 6

Edited by
A. Jaffe and
D. Ruelle

Birkhäuser
Boston · Basel · Stuttgart

Third Workshop on Grand Unification

University of North Carolina,
Chapel Hill
April 15–17, 1982

Paul H. Frampton,
Sheldon L. Glashow, and
Hendrik van Dam, editors

1982

Birkhäuser
Boston • Basel • Stuttgart

Editors:

Paul H. Frampton and Hendrik van Dam
Institute of Field Physics
Department of Physics and Astronomy
University of North Carolina
Chapel Hill, NC 27514

Sheldon L. Glashow
Lyman Laboratory of Physics
Harvard University
Cambridge, MA 02138

Library of Congress Cataloging in Publication Data

Workshop of Grand Unification (3rd : 1982 :
    University of North Carolina, Chapel Hill)
    Third Workshop on Grand Unification, University of
North Carolina, Chapel Hill, April 15-17, 1982.

    (Progress in physics ; no. 6)
    1. Grand unified theories (Nuclear physics)--
Congresses. I. Frampton, Paul H. II. Van Dam, H.
III. Glashow, Sheldon L. IV. Title. V. Series:
Progress in physics (Boston, Mass.) ; v. 6.
QC794.6.G7W67 1982   539.7'54   82-17717
ISBN 3-7643-3105-4 (Switzerland)

CIP - Kurztitelaufnahme der Deutschen Bibliothek
Workshop on Grand Unification:
... Workshop on Grand Unification. - Boston; Basel;
Stuttgart ; Birkhäuser
3. University of North Carolina, Chapel Hill, April
15 - 17, 1982. - 1982.
    (Progress in Physics: No. 6)
    ISBN 3-7643-3105-4
NE University of North Carolina (Chapel Hill): GT

© Birkhäuser Boston, Inc., 1982

ISBN 3-7643-3105-4

Printed in USA

TABLE OF CONTENTS

EDITORS' INTRODUCTION............................ vii
    Paul H. Frampton, Sheldon L. Glashow,
    Hendrik van Dam

MAGNETIC MONOPOLES ABOUT US....................... 1
    Sheldon L. Glashow

GUD: GIANT UNDERGROUND TRACK-DETECTOR,
A PROJECT FOR THE GRAN SASSO LABORATORY........... 7
    Guido De Zorzi

GAUGE SYMMETRY BREAKING WHICH PRESERVES
SUPERSYMMETRY.................................... 22
    Paul H. Frampton

THE UCI MOBILE NEUTRINO OSCILLATION EXPERIMENT.... 30
    Henry W. Sobel

SUPERSYMMETRY AT ORDINARY ENERGY................. 42
    Burt A. Ovrut

THE KAMIOKA PROTON DECAY EXPERIMENT.............. 56
-PRESENT STATUS-
    Kasuke Takahashi

GEOMETRIC HIERARCHY............................. 72
    Savas Dimopoulos

THE HOMESTAKE SPECTROMETER
A ONE-MILE DEEP 1400-TON LIQUID SCINTILLATION
NUCLEON DECAY DETECTOR.......................... 89
    Richard I. Steinberg

COMPLEX ANOMALY-FREE REPRESENTATIONS
FOR GRAND UNIFICATION.......................... 105
    Kyungsik Kang

NUCLEON DECAY EXPERIMENT AT KOLAR GOLD FIELDS..... 119
    S. Miyake and V. S. Narasimham

FROM FLUX QUANTIZATION TO MAGNETIC MONOPOLES...... 131
    Blas Cabrera

COMPOSITE/FUNDAMENTAL HIGGS MESON................. 158
    Howard Georgi

SEARCH FOR PROTON DECAY - THE HPW
DEEP UNDERGROUND WATER CERENKOV DETECTOR.......... 174
    Robert Morse

SUPERSYMMETRY, GRAND UNIFICATION
AND PROTON DECAY.................................... 191
   Serge Rudaz

NEUTRINO OSCILLATION EXPERIMENTS
AT ACCELERATORS.................................... 206
   Herbert Chen

COSMOLOGICAL CONSTRAINTS ON WITTEN'S
HIERARCHY MECHANISM................................ 222
   So-Young Pi

THE LEPTON ASYMMETRY OF THE UNIVERSE.............. 231
   Paul Langacker

STATUS OF THE IRVINE-MICHIGAN-BROOKHAVEN
NUCLEON DECAY SEARCH............................... 236
   William R. Kropp

FRACTIONALLY-CHARGED COLOR SINGLETS.............. 249
   Michael T. Vaughn

NEUTRINO MASSES FROM β
END-POINT MEASUREMENTS............................ 258
   John J. Simpson

NEUTRINO WEIGHT WATCHING.......................... 269
   Alvaro de Rujula

PLANNING FOR THE NEXT GENERATION OF PROTON DECAY
EXPERIMENTS IN THE UNITED STATES.................. 289
   David S. Ayres

RECENT DEVELOPMENTS OF THE INVISIBLE AXION........ 305
   Jihn E. Kim

NEUTRON ANTINEUTRON CONVERSION EXPERIMENTS........ 322
   Herbert L. Anderson

MASS OF THE t-QUARK............................... 348
   Sandip Pakvasa

THE MONT-BLANC FINE GRAIN EXPERIMENT ON NUCLEON
STABILITY... 356
   Donald Cundy

WHERE IS SUPERSYMMETRY BROKEN?.................... 359
   Steven Weinberg

ORGANIZING COMMITTEE 369

LIST OF PARTICIPANTS 370

PROGRAM 374

EDITORS' INTRODUCTION

This workshop held at the University of North Carolina was in the series which started with meetings in the University of New Hampshire (April 1980) and the University of Michigan (April 1981). More than one hundred participants congregated in the Carolina Inn in April 1982 to discuss the status of grand unified theories and their connection to experiment. The spring foliage of Chapel Hill provided a beautiful back-drop to this Third Workshop on Grand Unification. As mentioned in the first talk herein, these three workshops have heard of indications, respectively, of neutrino oscillations, proton decay and a magnetic monopole. Since all three experimental reports remain unconfirmed, grand unifiers must wait expectantly and patiently.

These proceedings preserve faithfully the ordering of the workshop talks which followed the tradition of alternation between theory and experiment. Only our introduction will segregate them.

The experimental presentations mainly concerned proton decay and massive neutrinos.

Four U.S. proton decay experiments were reported: the Brookhaven-Irvine-Michigan experiment in the Morton Salt Mine at Fairport Harbor, Ohio was described by WILLIAM KROPP, and ROBERT MORSE represented the Harvard-Purdue-Wisconsin group in the Silver King Mine, Utah. The Homestake Mine, South Dakota and Soudan Mine, Minnesota, experiments were reported respectively by RICHARD STEINBERG and DAVID AYRES, the latter providing also a survey of future U.S. experiments.

Equally intensive are the proton decay searches abroad. European experiments were discussed by GUIDO DE ZORZI from the Gran Sasso Laboratory and by DONALD CUNDY of the Mont Blanc tunnel collaboration. The Japanese Kamioka proposal was covered in a talk by KASUKE TAKAHASHI. The only experiment with actual candidates for proton decay events is that in the Kolar Gold Mine, India, a Japanese-Indian collaboration which was represented in Chapel Hill by SABURO MIYAKE and V.S. NARASIMHAM.

On massive neutrinos, ALVARO DE RUJULA described the progress at CERN in measuring radiative electron capture; JOHN SIMPSON surveyed critically other end-point determinations of neutrino mass; HERBERT CHEN reviewed accelerator searches for neutrino oscillations; and HENRY SOBEL reported on a new movable detector installation at the Savannah River reactor facility.

In addition, there were two experimental talks concerning respectively neutron-antineutron conversion and a magnetic monopole search. The talk on neutron-antineutron oscillation this year was by HERBERT ANDERSON who emphasized the efforts at Los Alamos National Laboratory.

Finally, on experiment, BLAS CABRERA described a magnetic monopole search at Stanford, including one impressive event suggestive of a monopole with one Dirac unit of charge.

We turn now to the theoretical contributions to the workshop.

Possible origins for the monopole flux indicated by the Stanford result were provided by SHELDON GLASHOW who favored a cloud of monopoles in solar orbit fed by slow evaporation of monopoles from the Sun.

A novel approach to the gauge hierarchy problem was provided by HOWARD GEORGI. Recent work on the invisible axion solution of the strong CP puzzle was surveyed by JIHN KIM. A prediction for the top quark mass was made by SANDIP PAKVASA.

MICHAEL VAUGHN constructed grand unified theories containing fractionally-charged leptons or hadrons. KYUNGSIK KANG built complex anomaly-free SU(N) representations. The cosmological lepton number was discussed by PAUL LANGACKER.

The most popular theoretical topic this year was undoubtedly super-symmetry which paradoxically combines strong theoretical appeal with, as yet, absolutely no empirical basis. PAUL FRAMPTON analyzed the group theory of gauge group breaking while retaining supersymmetry unbroken. A review of supersymmetric electroweak models was made by BURT OVRUT. An ambitious supersymmetric attempt to derive both the Planck scale and the electroweak scale from one input mass was made by SAVAS DIMOPOULOS. Implications of supersymmetry for the lifetime and decay branching ratios of the proton were explained by SERGE RUDAZ. SO-YOUNG PI studied the cosmology of a mass hierarchy generated in a supersymmetric gauge theory.

The general question of at what mass scale supersymmetry is broken was addressed by STEVEN WEINBERG. He argued against preserving supersymmetry down to the electroweak scale, and imposed cosmological bounds on the allowed higher energy scales of supersymmetry breaking. He also suggested how gravity can distinguish between degenerate supersymmetric vacua having different residual gauge symmetries, and pointed to cosmological difficulties that would arise if supersymmetry were broken below about 10^{13} GeV.

The experimental and theoretical speakers provided the participants with a clear overall picture of the current status of grand unification. Quite apart from the speakers' program, many other participants made valuable contributions to informal discussions.

We acknowledge financial support by grants from the U.S. Department of Energy and the National Science Foundation, as well as from the University of North Carolina.

Thanks are due to MAURICE GOLDHABER for his after-dinner speech, to TOM KEPHART for help with the workshop and these proceedings, and to DEBBIE WILSON who acted as workshop secretary.

May 1982   P.H. Frampton
                                                   S.L. Glashow
                                                   H. van Dam

MAGNETIC MONOPOLES ABOUT US

S.L. Glashow

Lyman Laboratory of Physics
Harvard University

In the first of these workshops, indications of neutrino oscillations were reported. In the second, candidate nucleon-decay events
were shown. In this third workshop, we have heard of the Valentine's
Day monopole. What is left for future workshops? Perhaps we will have
axions next year, and neutron-antineutron oscillations the year after.

The magnetic monopole was invented by Paul Dirac half a century
ago, and it is one of his only unverified predictions. Today, its
existence is one of the few predictions of grand unified theories,
along with proton decay and the great desert. Perhaps, by the time
these notes are published, the existence of magnetic monopoles will be
confirmed.

Blas Cabrera reports the observation of one monopole candidate
(with the Dirac value of magnetic charge $2eg = \hbar c$) passing through a
20 cm^2 superconducting loop during half a year. The most likely value
for the flux of monopoles upon the Earth is $0.1/\text{cm}^2$ yr $(2\pi \text{ sr})$. This is
a very large flux. If monopoles were to accumulate on the Earth's
surface, we would expect to find many of them per gram of terrestrial
material. We do not. But today's monopole is supposed to be very
massive, $\sim 10^{16}$ GeV. The force of surface gravity amounts to ~ 0.1 eV/Å,
and is sufficient to propel a heavy monopole through the solid Earth.
If all the ambient monopoles have fallen into the Earth's core, then
there are now 3×10^{27} monopoles inside the Earth. This is impossible.
A far smaller number of monopoles is sufficient to poison the geomagnetic dynamo. Either the Cabrera flux is wrong, or the ambient
monopoles must pass through the Earth so readily as to almost never be
trapped within it.

Incident monopoles have been searched for with particle detectors.

1

No heavily ionizing monopoles have been seen. Thus the monopole velocity must be smaller than the velocity of atomic electrons. Negative searches for lightly ionizing slow monopoles assure us that the energy loss of a monopole due to ionization is less than 0.2 MeV/cm. Slow monopoles may lose some energy (and produce photons) by the process of collisional excitation of atoms. They will lose even more energy due to ohmic dissipation as they pass through conductive media. Taking these various energy-loss mechanisms into account, we may imagine that monopoles passing through the Earth's core should lose as much kinetic energy as 10^6 GeV. Unless the mean monopole energy is considerably larger than this, the geomagnetic field is put in jeopardy. Indeed, ambient monopoles with velocities less than Earth escape velocity of 3×10^{-5} c, corresponding to 10^7 GeV for GUT monopoles, are certainly captured. Thus, ambient monopoles must have energies in excess of 10^8 GeV and velocity in excess of 30 km/sec.

Truly cosmic monopoles satisfy the preceding criteria. With typical "peculiar" gravitational velocities of 300 km/sec, they are fast enough to avoid being trapped in the Earth, yet slow enough to avoid detection by copious ionization. Can the Cabrera flux be cosmic? With these velocities, the mean monopole density is $n = 10^{-16}$/cm^3. This corresponds to an energy density of 1 GeV/cm^3. Such an energy density cannot be universal--it is some five orders too great, as Preskill observed years ago. Can the monopoles be concentrated within the galaxy? Parker has shown that this, too, is impossible for the monopoles would poison the galactic magnetic field. The "observed" monopole density exceeds the Parker bound by about six orders of magnitude. The Cabrera flux can neither be typical of the Universe, nor typical of the Galaxy. If it is right, it must correspond to a local flux, and it must be due to the fact that Cabrera's laboratory is situated in a very special part of the Universe, on the cool green hills of Earth. This is the conclusion I have come to with my colleagues, Savas Dimopoulos, Edward Purcell and Frank Wilczek.

The local monopole flux cannot originate from within the Earth, for reasons already presented. Can it originate from the Sun? Sunlight is, after all, eight orders of magnitude more intense than starlight for earthbound observers. If the observed monopole flux is a direct solar emanation, then the Sun must emit 10^{19} monopoles/sec, and there must be 10^{36} monopoles within the Sun. However, the Sun is a magnetically active body, with a characteristic magnetic growth time

of 22 years, the period of sunspot activity. Parker's arguments limit
the monopole content of the Sun to 10^{26} monopoles, ten orders of
magnitude too few. Moreover, 10^{36} monopoles in the Sun would lead to
the ohmic dissipation of far too much power--five orders of magnitude
more than the solar luminosity. The Sun is not the direct source of
the Cabrera flux.

And yet, the Sun can emit some monopoles. Suppose that there are
in fact ~10^{26} monopoles in the Sun, a number which is consistent with
the existence of an internal solar magnetic field. In order to be
kept from falling into the center of the Sun, the solar magnetic field
must compensate the ohmic losses

$\quad$ gB > dE/dx ~ 10 MeV/cm,

where 10 MeV/cm is our estimate for monopole energy loss due to induced
eddy currents in solar material. It follows that interior solar mean
fields of ~10^{3} Gauss are required. The characteristic dissipation
time of the solar magnetic field due to energy transfer to the
monopoles is given by Parker to be

$\quad$ T = B/8πgvn.

Taking B = 10^{3} Gauss, n = 10^{-7}/cm^{3} in accordance with a monopole number
of 10^{26}, and v = 400 km/sec corresponding to a gravitationally bound
monopole, we find T ~ 20 years. This successful estimate of the
characteristic time of the solar magnetic field is to be contrasted
with the conventional and unsuccessful estimate

$\quad$ T = $(4\pi\sigma/c^{2})R^{2}$ ~ 10^{7} years

where σ is the electrical conductivity in the convective zone and R is
the solar radius. Monopoles in the Sun may be good for something.

Can the solar monopoles escape from the Sun? Escape velocity from
the center of the Sun is 1400 km/sec, and from the surface it is
600 km/sec. Solar fields of order 10^{5} Gauss extending over the Solar
radius are necessary to extricate monopoles from the solar core.
However, only a few kiloGauss, such as is observed during solar flares,
suffices to eject monopoles from the outer reaches of the Sun. Let us
imagine that the Sun manages to excrete something like a billion
monopoles per second, corresponding to the number of solar monopoles
divided by the age of the Sun. What is the fate of these monopoles
once they leave the Sun?

One question we shall not address is how the monopoles got into
the Sun in the first place. Furthermore, we shall assume that the net
magnetic charge of the Sun is nearly zero, with roughly half of the

monopoles being North poles and half being South. This should be true
of the monopole flux on Earth as well. Monopole-antimonopole annihilation is expected to be a very rare occurrence, since its cross
section is controlled by the very high monopole mass.

The evaporation of monopoles from the Sun during solar flares can
be expected to produce a cloud of monopoles about the Sun in low solar
orbit. The principal force acting on the monopoles will be gravity,
once the solar flare has subsided. Perturbations due to sunlight are
negligible, since photon-monopole scattering is small. Magnetic
perturbations will change the monopole energies, so that some of the
monopoles will be led to higher solar orbits. Parker, in 1964,
estimated the effects of magnetic perturbations upon the Newtonian
orbits of charged micrometeorites. His calculations are readily
adapted to deal with monopoles in solar orbit. The time over which
the magnetic perturbation will significantly alter the energy of a
monopole in solar orbit is given by

$$t = (v/c)^2 \, T^{-1} \omega^{-2}$$

where v is the monopole velocity, T is the characteristic time of the
external solar magnetic field, and $\omega = gB/Mc$ is the cyclotron frequency
of the monopole. For an orbiting monopole, T is given by half the solar
rotation period, or, $T = 10^6$ sec. At a distance from the Sun of one
astronomical unit, $B = 3 \times 10^{-5}$ Gauss so that $\omega = 2 \times 10^{-15}$ Hz. The resi-
dence time for a monopole in the vicinity of Earth is estimated to be
100 My. It follows that the number of monopoles within Earth's orbit
is $\sim 10^{24}$ and that their local density is $\sim 10^{-16}/cm^3$.

We have demonstrated that a weak source of monopoles from the Sun
can be sufficient to produce a monopole density near the Earth which
explains the Cabrera flux. This hypothesis is subject to experimental
verification. The velocity distribution of monopoles incident upon
the Earth should be rather like the velocity distribution of meteorites,
being bounded above by 70 km/sec. This is certainly slow enough to
suppress ionization effects of monopoles to known experimental limits
yet fast enough to evade capture by the Earth. There should be
characteristic variations of the monopole flux with time of day, with
season of the year, and with latitude on Earth, as there are for the
flux of meteorites.

Eventually, monopoles in solar orbit will escape the solar system,
injecting 10^9 monopoles/sec into the Galaxy. If all the stars do it,

10^{28} monopoles are released annually. Galactic fields then accelerate them to escape velocity, whereupon they leave the Galaxy. The required magnetic power of 10^{38} GeV/year is easily compatible with the Galactic magnetic energy of 10^{57} GeV and its restoration time of 10^8 years.

Wilczek and Callan have suggested that magnetic monopoles may catalyze nucleon decay with a significant cross section σ. In the laboratory, monopole induced nucleon decay is distinguishable from spontaneous nucleon decay. In either case, the energy release is 1 GeV, but in catalytic decay visible momentum is not conserved since the recoiling monopole is not seen. A limit may be placed on σ from known limits on proton decay: $f\sigma < \tau_p^{-1}$. Taking $\tau_p > 10^{29}$ years as a plausible experimental limit on proton decay when momentum conservation is not imposed and putting the monopole flux f equal to .1 cm^2 yr, we obtain $\sigma < 10^{-28}$ cm^2. It follows from this limit that catalytic proton decay can be responsible for no more than a millionth of the solar luminosity. Thus, the process can have little effect on the solar neutrino problem. Of course, monopoles could be more effective in the catalysis of nuclear fusion, or in promoting heat transfer from the solar interior. Either of these effects could reduce the central solar temperature. However, catalytic nucleon decay could produce an observable flux of energetic solar antineutrinos. On Earth, this flux could be as much as $10^3/cm^2$ sec of GeV electron antineutrinos. Of course, we are now dealing with a _third_ order speculation, which depends upon the correctness of the Cabrera flux, upon our explanation of it, and upon the existence of a significant cross section for monopole-induced nucleon decay.

It is appropriate to conclude with a note of caution. Ambulance chasers must remember the unconfirmed status of previous reports to Grand Unification Workshops. Cabrera only claims an upper limit to the flux of monopoles: he does not claim to have discovered one. His monopole candidate is promising, but is very far from a decisive existence proof. Whatever be the wonderful cosmological and astro-physical consequences of the existence of magnetic charges, Paul Dirac must maintain his patient vigil for at least a little while longer.

*
This research is supported in part by the National Science Foundation under Grant No. PHY77-22864.

REFERENCES

B. Cabrera, Phys. Rev. Lett. 48, 1378 (1982).

C. Callan, "Dyon-Fermion Dynamics", Princeton University Preprint (May, 1982).

S. Dimopoulos, S.L. Glashow and E.M. Purcell, Harvard Preprint HUTP-82/A016.

P.A.M. Dirac, Proc. Roy. Soc., A133, 60 (1931); Phys. Rev. 74, 817 (1948).

H. Georgi and S.L. Glashow, Phys. Rev. Lett. 32, 438 (1974).

G. Lazarides, Q. Shafi, and T.F. Walsh, Phys. Lett. 100B, 21 (1981).

E.N. Parker, Astrophys. Journal 139, 951 (1964).

E.N. Parker, Astrophys. Journal 160, 383 (1970).

A.M. Polyakov, Pis'ma Ekxp. Teor. Fiz. 20, 430 (1974) [JETP Lett. 20, 194 (1974).]

J. Preskill, Phys. Rev. Lett. 19, 1365 (1979).

G. 't Hooft, Nucl. Phys. B79, 276 (1974).

F. Wilczek, Morris Loeb Lectures at Harvard (April, 1982).

GUD: GIANT UNDERGROUND TRACK-DETECTOR, A PROJECT FOR THE GRAN SASSO LABORATORY

Guido De Zorzi

Istituto di Fisica "G.Marconi" Università di Roma, Rome Italy.

Istituto Nazionale di Fisica Nucleare, Sezione di Roma, Rome, Italy.

- ABSTRACT -

A project is presented for a Giant Underground track-Detector (GUD) to be installed deep underground in the Gran Sasso Laboratory in Italy. With a mass of the order of 12 Ktons and a fine granularity, this calorimeter detector should allow one to study the nucleon stability, to test a number of "GUT" and "SUSY" predictions ($p \rightarrow K^{+}\bar{\nu}$, $n \rightarrow K^{\circ}\bar{\nu}$, etc.), and to attack several problems of interest for Astrophysics, opening up a new field of medium and high energy "Neutrino Astronomy".

1. - THE GRAN SASSO LABORATORY -

A large underground laboratory will be constructed in Italy, under the Gran Sasso mountain, at about 100 miles from Rome. Its depth will be around 4 km water equivalent and the total volume should be of the order of 10^4 m^3. Inside this huge laboratory the installation of more than one large detector will be possible. The final shape of the laboratory, at present, is still to be defined. Two auto-tunnels under the mountain are already existing and open to car trafic. In February 1982 the Italian Authorities have approved the financial support, so that the Gran Sasso Laboratory might be completed in about three years. International Collaborations of many Institutions are expected for experiments.

7

2. - GUD : PROJECT STATUS AND GENERAL FEATURES -

The GUD-project is under study, at present, by several groups following a "letter of intent" of June 1980 from a Frascati, Milan, Rome and Turin Collaboration. Other groups have shown interest and are expected to participate to this project.

The main features of GUD, plus some relevant technical solution, have been already presented by M.Conversi[1] at the "GUD Workshop", held in Rome in October 29-31, 1981. It should be mentioned that a Pavia-Rome Collaboration is preparing, at the Pavia University reactor, an experiment on neutron-antineutron oscillations[2] which uses essentially the same techniques proposed for the GUD-project. (The Pavia-Rome n - n̄ experiment is scheduled to take data in 1983).

The main features of present version of the GUD-project are given below:

a) GUD should have a modular structure expandable into the 10 Kt region, made of calorimeter modules;

b) each module is an active-passive fine-grain calorimeter made of flash-chambers, for purposes of tracking and of energy measurement, and of "resistive plate counters"[3], (RPC), for purposes of flash-chamber triggering and accurate time measurements (~1 ns resolution), allowing time-of-flights to be measured;

c) GUD's "over all granularity" will imply a good resolution both in event reconstruction and in energy measurements[4]: iron absorbers of 3 mm thickness and flash-chamber cells of $(4 \text{ mm})^2$ cross section are proposed;

d) a special feature of GUD (new for calorimeter devices) is the sense of motion determination of tracking particles by RPC time-of-flight measurements;

e) a horizontal disposition of flash chambers, trigger counters and absorbers is presently assumed.

The underground experiments which are now in progress or near to start, should provide relevant information for the conclusive decisions on the final design of the present GUD project.

3. - GUD PHYSICS -

GUD should be able to attack many fundamental problems of Physics and Astrophysics such as:

a) nucleon lifetime measurements, with possible identification of various expected decay modes;
b) nucleon decay branching ratios;
c) search for $\Delta B = 2$ transitions $(NN \rightarrow pions)$[5];
d) search for neutrino oscillations by exploiting the path difference of upward and downward atmospheric neutrinos[6];
e) detection of galactic and/or extragalactic neutrino sources[7];
f) detection of bursts of (low energy) neutrinos from stellar collapses;
g) other (unexpected) phenomena.

4. - MODULAR STRUCTURE AND DETECTION TECHNIQUES PROPOSED FOR GUD -

In its present proposal GUD is basically a cube of 18 m side, made of eight cubic flash calorimeters of average density ~2.1 g/cm^3, each of 9 m side (~1.5 Kt) as shown in Fig. 1. Alternative solutions, involving a much larger number of smaller flash-calorimeter modules (e.g. 40 modules of $(5\ m)^3$ volume) are also possible, still keeping a GUD fiducial mass close to its total mass for most of the forseen nucleon decay modes.

In Fig. 2 the basic structure of a calorimeter is shown. Four double planes of flash chambers alternate with four absorbers (3 mm thick iron). Every 1.2 cm iron (about 7 cm in space) there is one trigger plane made of a double layer "resistive plate counter". Suitable coincidences among the RPC pulses are used to get a flexible trigger system to sensitize the chambers.

5. - FLASH CHAMBERS -

Flash chambers of plastic material[8] are used today adopting extruded planes with honeycomb structure[9] in many large multiton apparata for neutrino physics. The capability of large flash calorimeters in reconstructing a great number of tracks with nearly any track inclination is illustrated in Fig. 3, where a 100 GeV charged current neutrino interaction is shown (Fermilab, experiment E594).

In case of the GUD calorimeter modules of the present proposal (Fig. 1), the chamber area is $(9 \text{ m})^2$ and the flash cell cross-section $(4 \text{ mm})^2$, so as to provide a good track resolution. Then the total number of biplane chambers turns out to be 3440, and the corresponding total number of cells is of the order of 10^7. Due to this huge number of cells, cost and reliability must be carefully satisfied. In particular the trigger and read-out systems are under study. There are several possible read-out techniques:

a) optical read-out[9] (e.g. by vidicon/plumbicon systems);

b) magnetostrictive[10] (an example is shown in Fig. 3);

c) electric signal pick-up by means of electrodes directly in touch with the plasma inside the cells[8,11];

d) electric signal pick-up by means of capacitive strips[12].

In order to reduce the number of read-out channels, a possibility is to read-out the "OR" at hardware level (pick-up) of any two adjacent cells in each biplane. This allow one to increase the detection efficiency, which is about 80%/plane for present operating flash chambers.

In a flash calorimeter the energy E of e.m. showers originated from electrons, positrons or photons impinging perpendicularly on the chamber planes, can be measured over a wide energy range with a relative resolution

$$\frac{\sigma}{E} = R \sqrt{\frac{t(r.l.)}{E(GeV)}}$$

merely by counting the number of flashes[9]. In this formula t is the plate thickness and R is a coefficient[13] depending on the plate material (12% for lead; 17% for iron). Then an energy resolution ~20% should be achieved with GUD for energies E ≳ 200 MeV. Fig. 4 shows the linear relation between incident energy and average number of "fired" cells as found[14] for a flash calorimeter (made of glass flash tubes) exposed to electrons of energies up to 0.5 GeV. The energy of a muon emitted and stopping in GUD can be measured accurately by range, due to the small thickness of the iron plates. Should the flash chamber be efficiently operated with a few μsec memory time (possibility now being investigated using especially treated polypropilene sheets) the charge of the muon would be identified from the e^+ track from the stopping μ^+, since μ^-'s at rest in iron are captured. For charged pions the energy resolution is limited by the nuclear absorption.

6. - RESISTIVE PLATE COUNTERS (RPC) -

The resistive plate counter recently developed in Rome by R.Santonico and R.Cardarelli[3] is a wireless, d.c. operated, fast and low-cost particle detector. It makes use of high resistivity parallel plate electrodes, filled with a gas mixture at atmospheric pressure. Differently from the high-pressure counter developed by Yu.N.Pestov and G.V.Fedotovich[15] the RPC can be built with large area at low cost.

A sketch of a double layer RPC is given in Fig. 5, where the built-in "chip electronics" is also shown. The counter is made of two layers with a common ground electrode, in double coincidence, so that the accidental counting is strongly reduced. The overall thickness is about 2 cm. A strong electric field (~ 50 Kvolt/cm) is present in the 2 mm gap between the resistive plate electrodes (bakelite, $\rho = 10^{10} - 10^{11} \Omega$ cm). The gap is filled with a gas mixture (72% Ar, 25% Butane, 3% Freon 13B1) which does not allow the discharge to propagate via photoionization processes after a particle traversal as in the case of Geiger counters, flash tubes, etc. Under the effect of the strong electric field, the ionization electrons freed by the primary particle develop into an avalanche, and then into a local streamer which, however, cannot evolve further into a spark because of the large resistivity of the electrodes. The "local" discharge induces a signal onto the external pick-up strip lines. No amplification of output pulses is needed.

The RPC main working conditions are summarized below:

high voltage	9.5 Kvolt
output pulse rise-time	3 nsec
pulse height	$\geq$ 300 mvolt/ 25 Ω
efficiency	98%
time resolution	~1 nsec

A large number of these counters are in construction at present in Rome for use in the neutron-antineutron oscillation experiment, at the Pavia reactor[2]. Namely an RPC of 36 m^2 area will complement the scintillation trigger system used in this experiment to sensitize the flash chambers, while RPC's of 250 m^2 area will be utilized as a veto system to reject cosmic rays.

In GUD, the flash calorimeters will be sensitized by high voltage pulses triggered by logical signals derived from the trigger counters. The overall number of RPC planes (double layer RPC) is about 880. Since two consecutive counter planes are separated by only ~ 12 g/cm^2 of material (~7 cm in space), the energy trigger threshold is very low, corresponding to some 10 MeV for minimum ionizing particles.

Tests have been made with RPC inside the existing Gran Sasso tunnel[16,17], where the counting rate turns out to be about 0.6 Hz/m^2 (Fig. 6). Then the expected accidental rate of twofold coincidences between any two trigger planes in GUD is of order of 10^{-4} Hz. This figure, however, should be lower under the final GUD conditions. In fact: (a) tests were made under worse radiation background conditions (concrete walls, instead of rock, in test tunnel, and furthermore no screening of the counters against the radiation coming from the walls); (b) electronic coincidences can be realized between "sectors" of contiguous trigger planes, rather than between two whole (9x9 m^2) planes.

As pointed out already, an appealing feature of RPC's is in the possibility of accurate time-of-flight measurements, at the nsec level. Fig. 7 shows the distribution of the relative delay between the pulses of a single RPC layer and a fast scintillation counter (NE110 scintillator viewed by a XP2020 photomultiplier). The sense of motion of the detected particles can be determined also on track going through a few counter planes.

7. - PROSPECTS OF PRESENT PROJECT -

In order to gain confidence in the proposed technique, a fine-grain flash-calorimeter sample of a few m^3 volume triggered by RPC's is planned to be built in 1982 for exposure to photons, electrons, charged pions and possibly neutrinos of various energies.

Nevertheless, the preliminary project presented here might undergo substantial changes before reaching the final design. Contributions of new groups are welcome from now.

- REFERENCES -

/1/ M.Conversi: Talk delivered at the Workshop on "Physics and Astrophysics with a Multikiloton Modular Underground track-Detector", Rome 29-31 October 1981 ("GUD Workshop"); CERN Int. Rep. EP 81-13; (to be published in the "GUD Workshop" Proceedings).

/2/ S.Ratti: Talk delivered at ICOBAN-Bombay 11-15 January 1982; Pavia University Int. Rep. IFNUP/AE 04/82; (to be published in the ICOBAN Proceedings).

/3/ R.Santonico and R.Cardarelli: Nucl. Instr. & Meth. 187 (1981) 377.

/4/ A.Grant: Proceedings of "GUD Workshop".

/5/ See e.g. R.N.Mohapatra and R.E.Marshak: Phys. Rev. Lett. 44 (1980) 1316;
Phys. Lett. 94B (1980) 183.

/6/ M.Dardo: Proceedings of "GUD Workshop".

/7/ G.Auriemma: Proceedings of "GUD Workshop".

/8/ M.Conversi: Nature 241 (1973) 160;
M.Conversi and L.Federici: Nucl. Instr. & Meth. 151 (1978) 93.

/9/ L.Federici et al.: Nucl. Instr. & Meth. 151 (1978) 103.

/10/ F.E.Taylor et al.: IEEE Trans. Nucl. Sci. NS-25 (1978).

/11/ H.Meyer: Proceedings of "GUD Workshop".

/12/ R.C.Allen, G.A.Brooks and H.H.Chen: IEEE Trans. Nucl. Sci. NS-28 (1981).

/13/ U.Amaldi Jr.: Physica Scripta 23 (1981) 409.

/14/ G.Brosco: Thesis, University of Rome 1972 (unpublished).

/15/ Yu. N.Pestov and G.V.Fedotovich: Preprint IYAFT 77-78 SLAC Translation n. 184 (1978).

/16/ M.Bertino et al.: Rome University Int. Rep. n. 788 (1982).

/17/ It is with deep regret that I recall the death of Lorenzo Federici, who first carried out systematic measurements on the cosmic ray flux in the Gran Sasso tunnel.

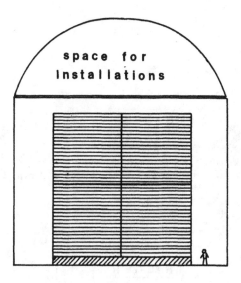

Fig. 1   Front view of 12 Ktons detector in the Gran Sasso Laborato-
          ry. Only the four frontal of the eight cubic modules are
          shown.

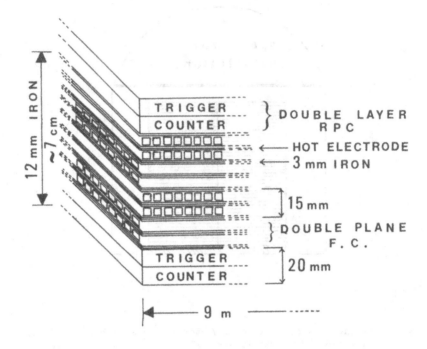

Fig. 2   The $(9\ m)^3$ module structure. The horizontal disposition of
four flash chamber biplanes (F.C.) alternated with four
iron plates is shown. Every 1.2 cm of iron one double layer
"resistive plate counter" (RPC) is inserted for triggering
purposes.

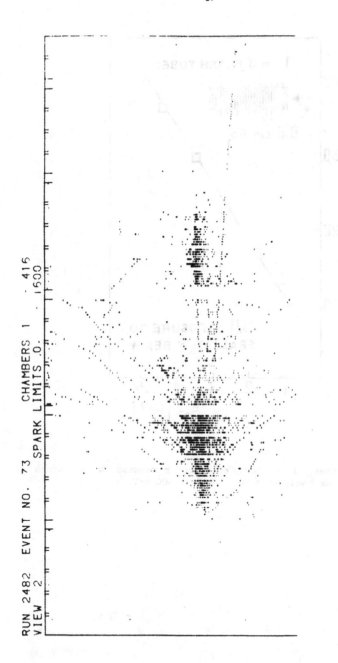

Fig. 3   A 100 GeV charged current neutrino event is shown in an expanded view (experiment E594 at Fermilab). Magnetostrictive read-out is used.

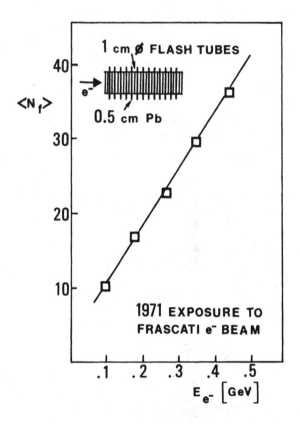

Fig. 4   Response of a flash calorimeter exposed to electrons of
various energies from 50 MeV to 500 MeV.

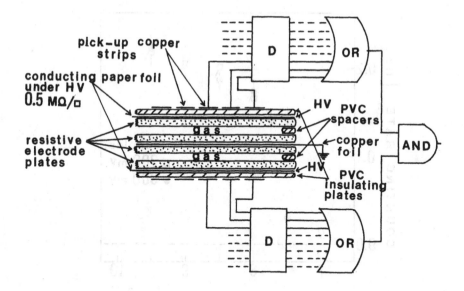

Fig. 5 Sketch of a RPC (double layer) with the read-out electronic. In this example, coincidences of opposite groups of eight strips are realized. The output signals were collected by 3 cm wide pick-up strips of 50 Ω characteristic impedance.

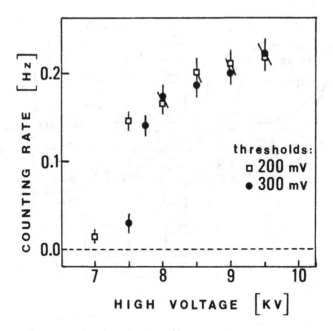

Fig. 6   Test done inside the Gran Sasso tunnel. Counting rate of a
          RPC of 0.15 x 2.2 m² useful area versus the high voltage
          applied to both gaps for different values of the discri-
          mination threshold.

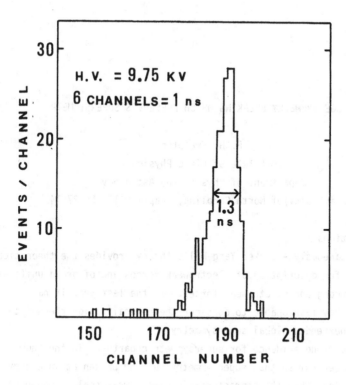

Fig. 7   Distribution of the relative delay between the pulses of a
         single layer RPC and a NE110 scintillator with XP2020
         photomultiplier.

GAUGE SYMMETRY BREAKING WHICH PRESERVES SUPERSYMMETRY

P. H. Frampton

Institute of Field Physics
Department of Physics and Astronomy
University of North Carolina, Chapel Hill, NC 27514

I. Motivations

Spontaneously - broken Yang-Mills theory provides the theoretical framework for description of electroweak forces and of grand unification of strong and electroweak forces. In the last year it has become highly fashionable to consider the generalization thereof to include unextended global supersymmetry.

There is no evidence for supersymmetric partners in the known particle spectrum so that supersymmetry must be broken at an energy scale not less than the characteristic electroweak scale, though it could be broken at a much higher energy [1].

In the last year, global supersymmetry has been applied in two ways - to the electroweak sector [2-7] and to the grand unification sector [8-17].

The application to grand unification may ameliorate both (i) the gauge hierarchy problem, since masslessness of scalar fields may be understood as a consequence of the chiral symmetry, and (ii) family unification for SU(5) embedded in SU(N), since the fermion and scalar representations of SU(N) are the same.

In order to resolve the gauge hierarchy we assume that scalars are kept massless at the superheavy scale by supersymmetry plus chiral symmetry for the fermion partners. Supersymmetry is preserved down to $\lesssim v/\alpha$, say $\lesssim 10$ TeV. In particular, supersymmetry is preserved at the grand unification mass and this constrains the gauge symmetry breaking. For example, given a grand unification gauge group G can we have

$$G \rightarrow SU(3) \times SU(2) \times U(1) \times \tilde{U}(1)?$$

Family unification without supersymmetry has 3-family examples such as [18]

$$SU(9) \supset SU(5)$$

$$9^3 + 9(\overline{9}) \supset 3(10 + \overline{5})$$

With supersymmetry the fermion-boson linkage changes the rules of this game. Supersymmetric **SU**(9) has a quite different 3-family solution, as given below.

II.   Group Theory of Superheavy Breaking in SusyGUTS

Let me first recall the structure of the lagrangian for super-symmetrized Yang-Mills theory [19]. Without any matter the pure gauge piece in Wess-Zumino gauge is

$$\mathcal{L} = -\frac{1}{4} F^a_{\mu\nu} F^a_{\mu\nu} + \frac{1}{2} \chi i \not{D} \chi + \frac{1}{2} D^5_\alpha D^5_\alpha \tag{1}$$

The auxiliary field satisfies $D^5_\alpha = 0$. Introducing chiral super-fields $\Phi_a = (\psi_a, \phi_a)$ the supercovariant derivatives give

$$-g_\alpha D^5_\alpha \, \overline{\phi}_a (T^\alpha)^a_b \phi^b - \sqrt{2} \chi_\alpha \overline{\gamma}^0 \overline{\phi}_a (T^\alpha)^a_b \psi^b \tag{2}$$

so that now $\overline{D}^5_\alpha = g_\alpha \overline{\phi}_a (T^\alpha)^a_b \phi^b \ (= K_\alpha)$. The kinetic energy terms give

$$D_\mu \overline{\phi}_a \, D_\mu \phi^a + \frac{1}{2} \overline{\psi}_a \, i \not{D} \psi^a + F^{5*}_a F^{5a} \tag{3}$$

where $F^{5a} = 0$.

Finally, one may add a superpotential $V(\Phi^a)$ which is a general gauge-invariant polynomial up to cubic order. The F term is then $F^5_a = V_a = \partial V/\partial \phi^a$ and the Higgs potential has the two terms

$$V(\phi) = \overline{V}_a V^a + \frac{1}{2} K_\alpha K_\alpha \tag{4}$$

In such a supersymmetric theory the vacuum energy cannot be nega-tive, and must be zero if supersymmetry is unbroken. Both terms in $V(\phi)$ are ≥ 0, hence unbroken supersymmetry requires

$$V_a(\langle \phi \rangle) = 0 \quad \text{for all a} \tag{5}$$

$$\langle K_\alpha \rangle = \langle \bar{\phi}_a \rangle \, (T^\alpha)^a_b \langle \phi^b \rangle = 0 \quad \text{for all } \alpha \tag{6}$$

These are the two basic conditions.

As an example, consider an adjoint chiral superfield Σ^b_a in SU(N). The most general superpotential is then

$$V = \frac{m}{2} \, \text{Tr}\Sigma^2 + \frac{\lambda}{3} \, \text{Tr}\Sigma^3 \tag{7}$$

Putting $\langle \Sigma^a_b \rangle = v_a \delta^a_b$ (still general) gives a quadratic equation for v_a with only two roots. Hence either SU(N) is broken or

$$SU(N) \rightarrow SU(k) \times SU(N-k) \times U(1) \tag{8}$$

for all $k = 1, 2, \ldots$ [N/2 - 1]. We have degenerate zero-energy supersymmetric vacua each with different residual gauge symmetry.

We shall consider asymmetric vacua appropriate to SusyGUTS and not discuss how the degeneracy is lifted. It may be by adding extra fields, or, more likely, by adding gravity (curved space-time) -- only one vacuum e.g. SU(3) × SU(2) × U(1) may have an acceptably flat space-time [20].

For the adjoint, $K_\alpha = 0$ is automatic if Σ^b_a is real. More generally, this condition has a simple interpretation in Dynkin weight space.

Note that $V_a = 0$ may be satisfied supernaturally (exploiting the non-renormalization property) but not $K_\alpha = 0$ since it depends on the gauge coupling constant g_α. Thus consider now

$$K_\alpha = g_\alpha \bar{\phi}_a (T_\alpha)^a_b \phi^b$$

In particular, consider first [k] of SU(N) -- the antisymmetric rank-k tensor. Now write

$$(T_\alpha)^a_b = (T_\alpha)^{m_1 m_2 \cdots m_k}_{n_1 n_2 \cdots n_k} \tag{9}$$

$$= \sum (-1)^P (\lambda_\alpha)^{m_1}_{n_1} \delta^{m_2}_{n_2} \cdots \delta^{m_k}_{n_k} \tag{10}$$

where the λ_α are $N \times N$ generators and $a, b = 1, 2, \ldots \binom{N}{k}$. Put $\alpha = (\hat{\alpha}, \alpha')$ where $\hat{\alpha} = 1, 2, \ldots (N - 1)$ designates the Cartan subalgebra and $\alpha' = N, N + 1, \ldots (N^2 - 1)$. Then choose the σ_3 - form for the Cartan subalgebra

$$\left(\lambda_{\hat{\alpha}}\right)^m_n = \left[\delta^m_{\hat{\alpha}} \delta^{\hat{\alpha}}_n - \delta^m_{\hat{\alpha} + 1} \delta^{\hat{\alpha} + 1}_n\right] \tag{11}$$

For example, with $k = 1$ we find

$$\left|V_{\hat{m}}\right|^2 - \left|V_{\hat{m} + 1}\right|^2 = 0 \tag{12}$$

for $\hat{m} = 1, 2, \ldots, (N - 1)$. corresponding to

$$(V_1^* V_2^* \ldots V_N^*) \begin{pmatrix} 0 & & & & \\ & 0 & & & \\ & & 0 & & \\ & & & 1 & \\ & & & & -1 \\ & & & & & \ddots \end{pmatrix} \begin{pmatrix} V_1 \\ V_2 \\ \vdots \\ \vdots \\ V_N \end{pmatrix} = 0 \tag{13}$$

Taking the more general case

$$\Phi = \Phi_m \oplus \Phi_{mn} \oplus \Phi_{mn\ell} \oplus \ldots \tag{14}$$

gives

$$\left(|V_{\hat{m}}|^2 - |V_{\hat{m} + 1}|^2\right) + \left(\sum_{n \neq \hat{m}+1} |V_{\hat{m}n}|^2 - \sum_{n \neq \hat{m}} |V_{\hat{m}+1,n}|^2\right) +$$

$$+ \left(\sum_{n, \ell \neq \hat{m}+1} |V_{\hat{m}n\ell}|^2 - \sum_{n, \ell \neq \hat{m}} |V_{\hat{m}+1,n\ell}|^2\right) + \ldots = 0 \tag{15}$$

The remaining $(N^2 - 1) - (N - 1) = N(N - 1)$ conditions are: $(\sigma_1$ - type)

$$\sum_{m \neq n} V_m^* V_n + \sum_{m \neq n} V_{m\ell}^* V_{n\ell} + \sum_{m \neq n} V_{m\ell p}^* V_{n\ell p} + \ldots = 0 \tag{16}$$

which is automatic provided the VEVs have at least two non-overlapping indices. Also $(\sigma_2$ - type)

$$\sum_{m \neq n} (V_m^* V_n - V_n^* V_m) + \sum_{m \neq n} (V_{m\ell}^* V_{n\ell} - V_{n\ell}^* V_{m\ell}) + \ldots = 0 \tag{17}$$

which is automatic for real VEVs.

Returning to the Cartan subalgebra and taking k = 1 note that the coefficients in Equation (12) are

$\hat{m}$ = 1   (+1 -1   0 0... 0 0)

$\hat{m}$ = 2   (0 +1   -1 0 ...0 0)

$\hat{m}$ = (N - 1) (0 0 0 0 ... +1 -1)

and this coincides with the Dynkin weights for the defining representation of SU(N).

We can prove that [12] for an arbitrary component of any irreducible representation of SU(N) the coefficients of $|V|^2$ in $K_{\hat{\alpha}}$ = 0 equations are precisely the Dynkin weights.

Hence the VEV must have Dynkin weight zero and mimic a singlet of the gauge group. This observation facilitates the solution greatly.

Consider one adjoint $\sum$ plus antisymmetry tensors. The Dynkin weight can be derived as (for [k])

$$a_\alpha(V^{m_1 m_2 \cdot \cdot m_k}) = \sum_{\{m_i\}} (\delta_{a_i,\alpha} - \delta_{a_i,\alpha+1}) \tag{18}$$

Such a simple formula may itself be new since it has been <u>tabulated</u> by some authors in the literature. The complex conjugate has

$$a_\alpha(\bar{V}_{m_1 m_2 \cdot \cdot m_k}) = - a_\alpha(V^{m_1 m_2 \cdot \cdot m_k}) \tag{19}$$

It is trivial then to find that in terms of the defining representation k = 1

$$a_\alpha(V^{m_1 m_2 \cdot \cdot m_k}) = a_\alpha(V^{m_1} \oplus V^{m_2} \oplus ... \oplus V^{m_k}) \tag{20}$$

Thus VEVs which "contract" together such as

$$V_{m_1 m_2} \bar{V}_{m_3 m_4 m_5 m_6} V^{m_1 m_2 m_3} V^{m_4 m_5 m_6} \tag{21}$$

(indices unsummed) must have vanishing Dynkin weight a_α = 0. Note that the use of the antisymmetric tensor $\varepsilon_{m_1 m_2 \cdots m_n}$ can always be suitably re-written.

Now the simplest case $\bar{V}_m V^m$ allows a mass term so, in general, naturalness must disallow quadratic and cubic "contractions". The superpotential itself may still have cubic terms provided not more than one of the three fields has a VEV.

The procedure for building a model with gauge group G is to decide the residual symmetry

$$G(= SU(N)?) \rightarrow SU(3) \times SU(2) \times U(1)(\times \tilde{U}(1)?)$$

then

(i)   Write down the independent quartic and higher contractions respecting the residual symmetry required, or a higher symmetry.

(ii) Choose a set from (i) breaking to the residual symmetry only.

(iii) Write the superpotential up to cubic terms. It is guaranteed that the most general V will leave supersymmetry unbroken since now both $K_\alpha = 0$ and $V_a = 0$.

III. Naturalness and Model Building.

Naturalness in its strongest form dictates that one keeps the most general renormalizable (cubic) superpotential. A weaker form is supernaturalness where some F - terms are set equal to zero; they stay zero to all orders of perturbation theory due to non-renormalization theorems.

Supersymmetry changes the old rules for family unification. We now keep only those chiral superfields which have VEVs. For simplicity and economy no additional superfields (without VEVs) are included. We then want a fermion structure sufficiently chiral to yield e.g. three light families of quarks and leptons. Characteristic of supersymmetry is that there are not separate choices for fermion scalars.

One other new question arises with supersymmetry: should one include a chiral superfield in the adjoint representation? The answer [21] without fractional charges is: yes.

The results are that in SU(N) for N = 5,6,7,8 there is no natural 3- family model. In SU(9) there are two with matter fields transforming as

(i) $\overline{36}$ + 2(126) + 5($\overline{9}$) + 80

(ii) 2($\overline{36}$) + 84 + 126 + 4 ($\overline{9}$) + 80

In both cases the residual symmetry is SU(3) × SU(2) × U(1). To obtain an additional $\tilde{U}(1)$ is surprisingly difficult; even in SU(9) the supernaturalness criterion is essential. This may suggest that the $\tilde{U}(1)$ is not there at low energy?

Aside from model building, let us summarize the general procedure.

Given a supersymmetric gauge theory with gauge group G and chiral superfields in representation R of G, the question is how many ways may G be broken to a given little group G → G' while keeping supersymmetry exact and full naturalness? The quickest procedure is then:

(i) The VEVs must have vanishing Dynkin weight, hence the contraction rule.

(ii) Keep only quartic and higher independent contractions ensuring naturalness of the superpotential.

(iii) Select from the (short) list, solution(s) with the desired little group G'.

This method proves superior to looking first at little groups then checking supersymmetry [c.f. Ref. 15]. And it does give the most general solution. Nothing is missed.

References

[1] It has recently been pointed out by S. Weinberg, Phys. Rev. Letters 48, 1303 (1982) that cosmological considerations limit the allowed energy scale for supersymmetry breaking.

[2] P. Fayet, Nucl. Phys. B90, 104 (1975).
     P. Fayet, Phys. Letters 64B, 159 (1976); ibid 69B, 489 (1977); ibid 70B, 461 (1977).
     The phenomenology was studied in G. Farrar and P. Fayet, Phys. Letters 89B, 191 (1980).

[3] C. R. Nappi and B. Ovrut, IAS Princeton preprint.

[4] L. J. Hall and I. Hinchliffe, Berkeley preprint.

[5] L. Alvarez - Gaumé, M. Claudson and M. B. Wise, Harvard University preprint.

[6] M. Dine and W. Fischler, preprint.

[7] S. Weinberg, Harvard University preprint.

[8] S. Dimopoulos, S. Raby and F. Wilczek, Phys. Rev. D24, 1681 (1981)

[9] S. Dimopoulos and H. Georgi, Nucl. Phys. B193, 150 (1981)

[10] E. Witten, Nucl. Phys. B188, 513 (1981).

[11] N. Sakai, Z. Phys. C11, 153 (1981).
      N. Sakai and T. Yanagida, Max Planck Institute preprint.

[12] P. H. Frampton and T. W. Kephart, Phys. Rev. Letters 48, 1237 (1982); and UNC-Chapel Hill preprint IFP 170-UNC (Jan 1982).

[13] J. Ellis, D. V. Nanopoulos and S. Rudaz, CERN preprint.

[14] W. J. Marciano and G. Senjanovic, preprint.

[15] F. Buccella, J. P. Derendinger, C. A. Savoy and S. Ferrara, CERN preprint.

[16] P. Ginsparg, Phys. Letters B (to be published).

[17] S. Y. Pi, Phys. Letters B (to be published).

[18] P. H. Frampton, Phys. Letters 89B, 352 (1980).

[19] An elegant presentation of supersymmetrized Yang-Mills theory was provided in A. Salam and J. Strathdee, Phys. Rev. D11, 1521 (1972).

[20] S. Weinberg, University of Texas preprint, and these proceedings.

[21] P. H. Frampton and T. W. Kephart, UNC-Chapel Hill preprint IFP 172-UNC (March 1982).

THE UCI MOBILE NEUTRINO OSCILLATION EXPERIMENT

N. Bauman,[*] H. Gurr[†], W. R. Kropp, M. Mandelkern,
E. Pasierb, L. Price, F. Reines, H. W. Sobel

1. Introduction

We have installed a detector at our Savannah River Project
neutrino laboratory to investigate the possibility of neutrino
oscillations. This detector is designed to be movable and to be
positioned at any distance from the reactor core between 15 and 50
meters. A single detector, looking at a particular reactor at various
distances, is expected to avoid most of the systematic errors which
were possible in previous experiments. In particular, the results
will be independent of any predicted reactor neutrino spectrum.

2. The Reactor as a Neutrino Source

A power reactor produces neutrinos from the fission of ^{235}U, ^{238}U
and ^{239}Pu. The neutrino spectrum of each component is different (Fig.
1). At the Savannah Project (SRP) reactor where we work, ^{239}Pu
produces less than 8% of the fissions and ^{238}U produces less than
4%. As a result, for neutrino energies between 2 and 8 MeV, the
difference from a pure ^{235}U neutrino spectrum is less than 1.5%.

Neutron activation of the reactor surroundingss can also
produce $\bar{\nu}_e$'s but they are less than 1.3 MeV. [1] Neutrino
production in this mode is also possible with $\nu_e/\bar{\nu}_e$ .0005
and $E_{\nu_{max}}$ = 0.8 Mev. [2] Experimentally, the ratio $\nu_e/\bar{\nu}_e$ has been
determined to be < 0.02. [3]

---

[*] Savannah River Laboratory, Aiken, SC
[†] University of South Carolina, Aiken, SC

3. Disappearance Experiment

The detector is designed to observe the inverse beta decay reaction:

$$\bar{v}_e + P \rightarrow n + e^+$$

Since the neutrino energy is less than 10 MeV, the reaction is nonrelativistic and the positron kinetic energy is:

$$E_{e^+} = E_v - (m_n - m_p + m_e)$$
$$= E_v - 1.8 \text{ MeV}$$

A measurement of the positron spectrum therefore, is a measurement of the neutrino spectrum.

If the $\bar{v}_e$ changes to another neutrino type, it will be below prodcution threshold and thus unobservable at reactor energies. A reactor experiment is therefore a disappearance experiment, where we look for a paucity of neutrinos.

There have been three categories of experiment discussed:

a. Experimental results can be compared with theoretical prediction: This technique suffers from the uncertainty in the predicted neutrino spectrum in that any observed discrepancies can be attributed to that source.

b. We can compare the results of different detectors at various source to detector distances: This has been done by Silverman and Soni [4] and they conclude that if neutrino oscillations, rather than experimental errors, are the cause of an observed distance dependence exhibited by the data, then neutrino oscillation fits are better than no oscillation fits by factors of from 5 to 15.

c. We can compare the results of the same detector taken at different distances: This clearly is the experiment of choice since analysis of this type of data does not depend on a predicted neutrino spectrum, and since first order systematic effects are removed.

4. Detector

A schematic of the detector is shown in Figure 2. The target is 300 liters of NE313, a xylene based scintillator loaded with 1/2% gadolinium which can provide pulse shape information distinguishing lightly (e^+) from heavily (p) ionizing particles. It is surrounded by 1100 liters of mineral oil based scintillator divided into two volumes. Each volume is viewed by a large number of photomultiplier tubes. The detector is surrounded by two inches of low background lead, a three-inch thick plastic scintillator anticoincidence and finally eight inches of lead. There is room for additional neutron shielding (Cd or B) if required.

An electron antineutrino ($\bar{\nu}_e$) from the reactor will interact with a target proton. The positron stops and promptly annihilates. Each .5 MeV annihilation γ ray has high probability of depositing its energy in the liquid scintillator. The neutron with kinetic energy of tens of KeV, quickly comes to rest and is captured by a Gd nucleus with a characteristic time of 10μs giving rise to ~ 8 MeV in gamma rays which are detected with high efficiency. The signature of an event is then, Fig. 3.
a) A pulse in the target with suitable shape and 2 MeV < E < 10 MeV.
b) A second pulse within ~ 20μs for which the sum of deposited energies exceeds ~ 5 MeV.

5. Detector Characteristics

a. Energy Resolution
We have measured the target detector and surrounding or "blanket" liquid detector resolution by inserting various sources. They are
$$\sigma_{target} = .12 \sqrt{E(MeV)}$$
and
$$\sigma_{blanket} = .2 \sqrt{E(MeV)}$$

A typical calibration spectrum (^{60}Co) is shown in Figure 4. In this case, σ = 7.6 % at 2.5 MeV.
b. A Monte Carlo simulation has been made and gives the e^+ detection efficiency as a function of neutrino energy (Fig. 5), and the delayed neutron detection efficiency as a function of gamma ray threshold (Fig. 6). The overall e^+ efficiency, integrating over the expected spectrum, is about 0.7 for E_{e^+} > 2 MeV and the delayed neutron efficiency is about 0.7 for E_γ > 5 MeV.
c. We have measured the capture time for neutrons in the target to be ~ 10μs, in agreement with the Monte Carlo calculation, using an Am - Be neutron source and cosmic ray neutrons. A time distribution for source neutrons is shown in Fig. 7.

6. Expected Rates

a. Signal

At 15 meters from the core, the $\bar{\nu}_e$ flux is about
$1.2 \times 10^{13} cm^{-2} sec^{-1}$. If we assume an overall efficiency of 0.5 for
$E_{e^+} > 2$ MeV and $E_\gamma > 5$ MeV then we anticipate an event rate of 550
day^{-1}. At 50 meters, this scales to 50 day^{-1}. We expect to run at a
minimum of three distances, accumulating > 5000 events at each, and
expect to measure the ratio of rates for any two points to better than
3%.

b. Backgrounds

There are two principle backgrounds:

i)   Uncorrelated (accidental) counts due to the ambient gamma
flux from the reactor and the materials in the detector (principally U
and Th isotopes and ^{40}K): We have employed relatively clean materials
and shielded heavily to minimize this effect. The gamma backgrounds
in the experimental area are measured and range from several mr/hour
at near points to .01 mr/hour at most distant points. The accidental
rate is estimated to be ~ 10/day for the worst case, falling to
negligible rates at 30 meters. The accidental background will be
continuously monitored.

ii)   Correlated backgrounds due to cosmic rays.

Figure 8 shows a neutron from a penetrating cosmic ray which
recoils from a proton in the target giving a prompt pulse, then is
captured by Gd giving a delayed pulse. From various data we estimate
this background to give ~ 75 events/day.

Pulse Shape Discrimination in the target scintillator is valuable
in distinguishing e^+ pulses from such proton recoils. We show a
typical separation spectrum in Figure 9 demonstrating gamma rays and
neutrons from a source in a particular energy band. For 98%
acceptance of gamma pulses, neutron rejection is about 100:1. This
discrimination reduces the rate to ~ 1 day^{-1}.

Figure 10 shows a cosmic ray muon producing a bremstrahlung gamma
and a neutron from a nuclear collision, giving an "e^+ like" prompt
pulse and a neutron capture. An estimate from other data gives
~ 30/day.

In general, cosmic ray background can be established by taking
data with the reactor off. The SRP reactor has frequent reactor off

periods.

7. Sensitivity

We consider a two neutrino oscillation model. In this case a physical neutrino can be written as a super-position of two pure states i.e.:

$$\bar{\nu}_e = \nu_1 \cos\theta + \nu_2 \sin\theta$$

and

$$\bar{\nu}_{\mu/\tau} = -\nu_1 \sin\theta + \nu_2 \cos\theta$$

where θ is the mixing angle. Then, the probability of finding a $\bar{\nu}_\mu$ (or $\bar{\nu}_\tau$) at time t, given that we started with a $\bar{\nu}_e$ is:

$$\text{Prob.} = |a(t)|^2 = \tfrac{1}{2} \sin^2 2\theta \left[1 - \cos\frac{(\Delta m)^2 (t)}{2E_\nu} \right]$$

where $\Delta m^2 = |m_1^2 - m_2^2| c^4$, and the oscilation length

$$\lambda(m) = \frac{2.5 E_\nu (\text{MeV})}{\Delta m^2 (\text{eV}^2)}$$

For $\bar{\nu}_e$ sensitive experiments, the ratio of counting rates with and without oscillations is given by:

$$R(d) = 1 - \tfrac{1}{2}\sin^2 2\theta \left[1 - \frac{\int N(E_\nu)\sigma(E_\nu)\cos(\frac{\Delta m^2 d}{2c\hbar E_\nu})dE_\nu}{\int N(E_\nu)\sigma(E_\nu)dE_\nu} \right]$$

where d= distance from source to detector

$N(E_\nu)$ = Neutrino flux

and

$\sigma(E_\nu)$ = Reaction cross section.

Figure 11 gives the 68% limits on Δm^2 and $\sin^2 2\theta$ which this experiment can achieve for a change in distance of 35 meters. The results are given for different values of the experimental precision.

The three solutions ofSilverman and Soni [4] have rather large values of Δm^2 and fairly small mixing angles. The parameters are (.9 eV^2, .29), (2.2 eV^2, .17) (3.7 eV^2, .23) where we give (Δm^2, $\sin^2 2\theta$). The corresponding oscillation lengths are comparable to the reactor dimensions. Due to the short oscillation lengths, a comparison of integrated rates at two points will not necessarily demonstrate these oscillations because of the intrinsic averaging over detector and reactor size and neutrino energy. We consider R_e defined in Ref. 4,

$$R_e \equiv \frac{\text{\# of } e^+ \; (2.2 < E_e < 6.7 \text{ MeV})}{\text{\# of } e^+ \; (4.4 < E_e < 6.7 \text{ MeV})}$$

Figure 12 gives the distance variation of R_e for each of the solutions of Ref. 4, and includes the averaging due to reactor size. Measurements at two points with ~ 2000 events/point provides a determination of R_e with 5% error.

We observe that confrontation of the $\Delta m^2 > .9$ eV^2 solution is quite feasible for this experiment. The higher mass solutions are not accessible via simply considering the distance dependence of a quantity such as R_e.

8. Schedule

The detector is installed at SRP. The external lead shield is partially complete. Data taking could begin within two months.

References

[1] S. Blankenship - UCI internal report, UCI-10P19-102 (1976).

[2] S. Blankenship - UCI internal report, UCI-10P19-104 (1976).

[3] R. Davis, Jr., and D. S. Harmer, Bull. Amer. Phys. Soc. (2) 4, 219 (1959).

[4] A. Soni and D. Silverman, Phys. Rev. Lett. 46, 467 (1981).

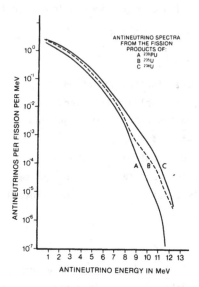

Figure 1. Antineutrino spectra from fission products.

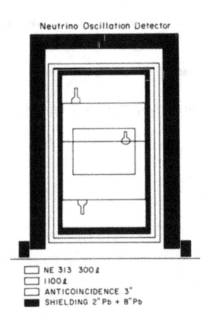

Figure 2. Schematic of detector.

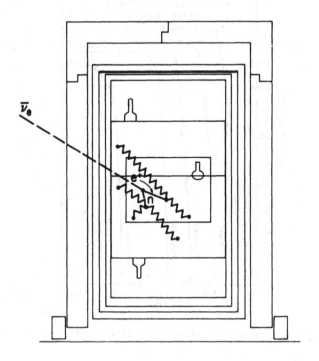

Figure 3. An inverse beta decay event. $\bar{\nu}_e + p \rightarrow e^+ + n$

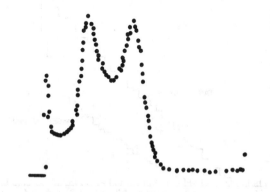

Figure 4. A ^{60}Co spectrum in the target detector.
$\sigma = 7.6\%$ at 2.5 MeV.

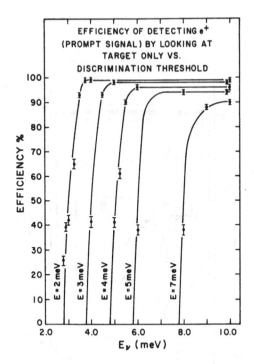

Figure 5. Monte Carlo calculation of positron detection efficiency.

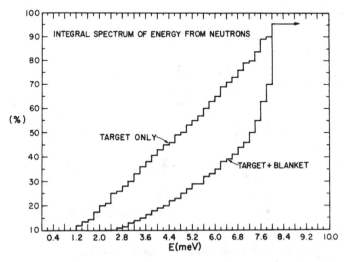

Figure 6. Monte Carlo calculation of delayed neutron detection efficiency.

Figure 7. Time distribution of neutron captures from an Am-Be
source. τ ≏ 9.4 μs.

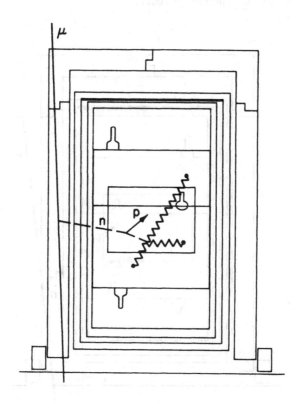

Figure 8. A background event due to a cosmic ray neutron.

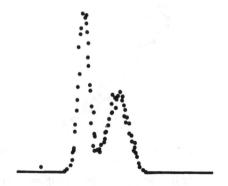

Figure 9. Neutron-gamma separation in the target detector using pulse shape discrimination. For a γ efficiency of ≃ .98, n rejection ≃ 100:1.

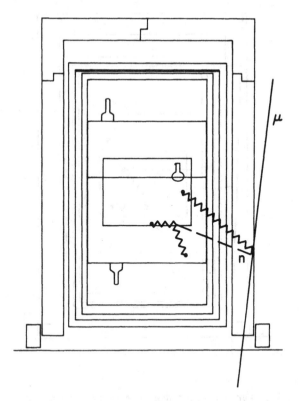

Figure 10. A background event due to a neutron and a muon-electron bremstrahlung gamma ray.

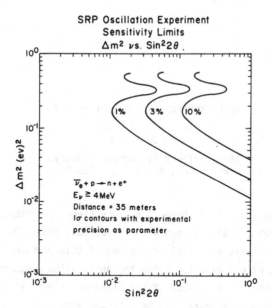

Figure 11. Detector sensitivity limits.

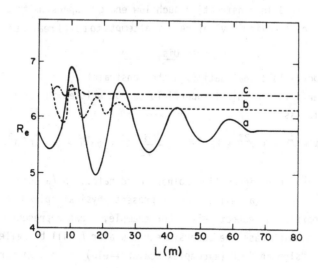

Figure 12. R_e for the three solutions of reference 4 as a function of distance. The SRP reactor dimensions are taken into account;

$a(\Delta m^2 = .9 eV^2)$, $b(\Delta m^2 = 2.2 eV^2)$, $c(\Delta m^2 = 3.7 eV^2)$.

SUPERSYMMETRY AT ORDINARY ENERGY

Burt A. Ovrut

The Institute for Advanced Study
Princeton, New Jersey 08540

Everybody knows that there is a gauge hierarchy problem in Grand Unified Theories (GUTS). By now almost everybody knows that supersymmetry offers at least a partial resolution of this problem. A supersymmetric GUT, at energies far below the unification mass, must behave like a supersymmetric extension of the standard color × electro-weak model where the supersymmetry has been broken either explicitly or spontaneously. In this lecture I would like to first discuss the general problems involved in constructing such low energy supersymmetric theories and then to review several recent attempts to build realistic models.

General Discussion

A superfield ϕ that satisfies the constraint $\bar{D}_{\dot\alpha}\phi = 0$ is called a chiral superfield. A chiral superfield can be written in terms of component fields as

$$\phi = A + \sqrt{2}\theta\psi + \theta^2 F + i\theta\sigma^m\bar\theta\partial_m A - \frac{i}{\sqrt{2}}\theta^2\partial_m\psi\sigma^m\bar\theta + \frac{1}{4}\theta^2\bar\theta^2\partial^2 A \quad (1)$$

where θ and $\bar\theta$ are anticommuting spinor coordinates. A (a complex scalar field) and ψ (a Weyl spinor) represent physical spin-0 and spin-1/2 particles respectively. For example, ψ can represent a quark or lepton in which case the supersymmetric scalar A will be called a "squark" or "slepton." F (a complex scalar field) is an auxiliary field and does not represent a physical particle. However, a non-vanishing vacuum expectation value (VEV) of F is associated with spontaneous breaking of supersymmetry. Hence, F is of fundamental importance.

A superfield V that satisfies the constraint $V = V^\dagger$ is called a vector superfield. In Wess-Zumino gauge a vector superfield can be written in terms of component fields as

$$V = -\theta\sigma^m\bar{\theta}V_m + i\theta^2\bar{\theta}\bar{\lambda} - i\bar{\theta}^2\theta\lambda + \frac{1}{2}\theta^2\bar{\theta}^2 D \tag{2}$$

Fields V_m (a real vector field) and λ (a Weyl spinor) represent physical spin-1 and spin-1/2 particles, respectively. Generally V_m represents a gauge field such as a photon or a gluon in which case the supersymmetric fermionic partner λ will be called a photino or gluino. D (a real scalar field) is an auxiliary field and does not represent a physical particle. However, as with F , a non-vanishing VEV of D is associated with spontaneous breakdown of supersymmetry. Hence, D is of fundamental importance.

Let G be an unspecified gauge group and ϕ_i a set of chiral superfields transforming under G. The most general, locally G-invariant, supersymmetric Lagrangian is given by (suppressing all group indices)

$$L = \frac{1}{2g^2}\int d^2\theta\ W^\alpha W_\alpha + 2\kappa\int d^4\theta\ V$$

$$+ \int d^4\theta\ \phi^\dagger\ e^{gV}\phi - \left[\int d^2\theta\ W(\phi) + h.c.\right] \tag{3}$$

where

$$W_\alpha = -\frac{1}{4}\bar{D}^2(e^{-gV}\ D_\alpha\ e^{gV}) \tag{4}$$

$$W(\phi) = f\phi + \frac{m}{2}\phi^2 + \frac{\lambda}{3}\phi^3 \tag{5}$$

are the covariant curl of V and the superpotential, respectively. There is one $2\kappa\int d^4\theta\ V$ term associated with each U(1) factor and one $f\phi$ term associated with each G-singlet chiral superfield. Using (1) and (2), Lagrangian (3) can be written in terms of component fields. The equations of motion for the auxiliary fields are

$$F^\dagger = f + m A + \lambda A^2 \tag{6}$$

$$D = -\kappa - \frac{g}{2}A^\dagger A \tag{7}$$

Eliminating F and D from L using (6) and (7), we find that

$$L = \ldots - V_{P.E.} \tag{8}$$

where ... stands for kinetic energy terms or terms involving at least one fermion, and

$$V_{P.E.} = F^{\dagger} F + \frac{1}{2} D^2 \qquad (9)$$

$V_{P.E.}$ is the potential energy of the system. Note that $V_{P.E.}$ is non-negative.

The Hamiltonian associated with L can be written in the form

$$H = \frac{1}{4} \sum_{\alpha=1}^{2} (\bar{Q}_\alpha Q_\alpha + Q_\alpha \bar{Q}_\alpha) \qquad (10)$$

where Q_α and $\bar{Q}_\alpha$ are the spinor generators of the superalgebra. Since H is a sum of squares, it follows that $<0|H|0> \geq 0$. Supersymmetry will be spontaneously broken if and only if (iff) $Q_\alpha |0> \neq 0$ for some Q_α and, hence, iff $<0|H|0> > 0$. Translating this result to quantum field theory, we conclude that supersymmetry will be spontaneously broken iff

$$V_{P.E.}(<A>) > 0 \qquad (11)$$

where $<A>$ are the fields at the absolute minimum of $V_{P.E.}$. Using (11) we can ask which field theories exhibit this phenomenon. This is determined by parameters κ and f. If both vanish, then spontaneous supersymmetry breakdown does not occur. There may, however, be spontaneous breakdown of G. If $\kappa \neq 0$ (at least one U(1) factor in G), but $f = 0$ (no G-singlet chiral superfields), then spontaneous supersymmetry breakdown usually occurs. There may, or may not, be spontaneous breakdown of G. We call such theories U(1) models, depending upon the values of the parameters. If $\kappa = 0$ (no U(1) factors in G), but $f \neq 0$ (at least one G-singlet chiral superfield), then, in most cases, spontaneous breakdown of supersymmetry does not occur ($f\phi$ can be transformed away by shifting the superfields). There is, however, a unique class of theories in which $f\phi$ cannot be transformed away. In these theories supersymmetry is spontaneously broken. Again, there may, or may not, be spontaneous breakdown of G, depending upon the values of the parameters. We call such theories O'Raifeartaigh models after their inventor. Finally, it is possible to combine U(1) and O'Raifeartaigh models.

Whether or not supersymmetry is spontaneously broken there is a mass sum rule associated with each of the extrema of the potential energy. For example, let G = U(1) and ϕ_+, ϕ_- be chiral superfields with charge +1 , -1 respectively. Then, for a certain range of parameters, supersymmetry, but not G, is spontaneously broken. If we let $A_\pm = \frac{1}{\sqrt{2}}(a_{r\pm} + ia_{i\pm})$, then the mass matrices for the r and i scalars and the fermions are given by

$$
m_r^2 = \begin{pmatrix} m^2 + \frac{\kappa g}{2} & \\ & m^2 - \frac{\kappa g}{2} \end{pmatrix} , \quad
m_i^2 = \begin{pmatrix} m^2 + \frac{\kappa g}{2} & \\ & m^2 - \frac{\kappa g}{2} \end{pmatrix} , \quad
m_\psi^2 = \begin{pmatrix} m^2 & \\ & m^2 \end{pmatrix} \tag{12}
$$

respectively. These matrices satisfy the sum rule

$$
\mathrm{Tr}\, m_r^2 + \mathrm{Tr}\, m_i^2 - 2\, \mathrm{Tr}\, m_\psi^2 + 3\, \mathrm{Tr}\, m_\nu^2 = 0 \tag{13}
$$

where m_ν is the mass of the U(1) gauge field ($m_\nu = 0$, since U(1) is unbroken). Note that two scalars are lighter than the fermions. This follows from the vanishing of the right side of (13). The general form of the sum rule is

$$
\sum_J (-1)^{2J} (2J+1)\, \mathrm{Tr}\, m_J^2 = \sum_i \frac{g_i^2}{2} \langle D_i \rangle\, \mathrm{Tr}\, Y_i \tag{14}
$$

where $\sum_J$ is a sum over all spin states, $\sum_i$ is a sum over all U(1) factors in G and g_i , $\langle D_i \rangle$ and Y_i are the gauge coupling constant, D VEV, and hypercharge matrix associated with the i th such factor. The right side of (14) does not necessarily vanish.

The sum rule (14) imposes strong constraints on model building. First, we examine its implications at the tree level. Take the gauge group to be

1)　　$G = SU(3) \times SU(2) \times U(1)$

The usual anomaly free quantum number assignments lead to the condition that Tr Y = 0 and, hence, the right side of (14) vanishes. We expect, therefore, that some scalar will be lighter than all the fermions as indicated above. This expectation has been sharpened into a theorem by Dimopoulos and Georgi [1] which states that

$$
\begin{pmatrix} \text{mass of some charge } \frac{2}{3}(-\frac{1}{3}) \\ \text{scalar triplet} \end{pmatrix} \leq \begin{pmatrix} \text{mass of lightest charge } \frac{2}{3}(-\frac{1}{3}) \\ \text{quark triplet} \end{pmatrix} \tag{15}
$$

This result is phenomenologically unacceptable. The only alternative is to enlarge the gauge group in such a way that the right side of (14) does not vanish. It follows that we must take

2) $G = SU(3) \times SU(2) \times U(1) \times \widetilde{U(1)} \times \ldots$

where $<D_i> \text{Tr } Y_i \neq 0$ for at least one of the new $U(1)$ factors. This approach is due to Fayet [2] and has been extensively studied by S. Weinberg [3]. The addition of new $U(1)$ factors to the gauge group and the constraints of phenomenology lead to anomalies. These anomalies can be cancelled by introducing new chiral superfields into the model but then, as a rule, the scalar component fields develop VEVs which can break color and lepton number and even restore supersymmetry. This approach is interesting but, for the above reasons, it is difficult to write down an acceptable model. We therefore reject it and go back to the gauge group $G = SU(3) \times SU(2) \times U(1)$.

We have shown that the tree level mass spectrum is unacceptable in models based on this group. Our hope is that radiative corrections may alter the mass spectrum. There are two statements that can be made in this regard.

  A) The mass sum rule is maintained by radiative corrections which are first order in the supersymmetry breaking parameter (to any order of perturbation theory in the coupling constants).

  B) The sum rule is modified by radiative corrections which are higher order in the supersymmetry breaking parameter.

Statement A) was proven by Girardello and Iliopoulos [4] and statement B) shown explicitly in the papers discussed in the second part of this lecture. If fermion-scalar mass splitting occurs at tree level, then corrections due to B) are too small to alter the tree level conclusion. It follows that we don't want quark and lepton masses to be split from squark and slepton masses at tree level. This immediately rules out trying to break supersymmetry through the $U(1)$ factor (since squarks and sleptons carry non-vanishing hypercharge and therefore couple to $<D>$). Luckily, in O'Raifeartaigh models, quarks, leptons and their scalar partners can have degenerate mass at tree level (they need not couple directly to the non-vanishing $<F>s$). If radiative mass corrections are of type B) only then the spectrum might be acceptable. Henceforth, we consider O'Raifeartaigh models.

We introduce these models within the context of an $SU(3) \times SU(2) \times U(1)$ gauge group. Let E, H, $\bar{F}$, $\bar{G}$, and Y be five chiral superfields which transform as shown in Table 1. R is the quantum number of the R-transformation, which acts on scalars and fermions with differing phases. The most general superpotential invariant under these transformations is

$$W = \int d^2\theta [m_1 \bar{G}E + m_2 \bar{F}H + Y(\gamma_1 \bar{G}H - M^2)] \tag{16}$$

Some of the F equations of motion are

$$F_{EA}^+ = m_1 A_{GA}$$

$$F_F^{+A} = m_2 A_H^{\ A} \tag{17}$$

$$F_Y^+ = \gamma_1 A_{GA} A_H^{\ A} - M^2$$

These F's cannot vanish simultaneously. This implies that $V_{P.E.} > 0$ for all fields and, hence, that supersymmetry is spontaneously broken. An analysis of $V_{P.E.}$ is shown in Figure 1. Note that $V_{P.E.}$ is flat in the Y direction for any values of the parameters. Hence, $\langle Y \rangle$ is undetermined at tree level. We can now discuss recent attempts to build realistic models.

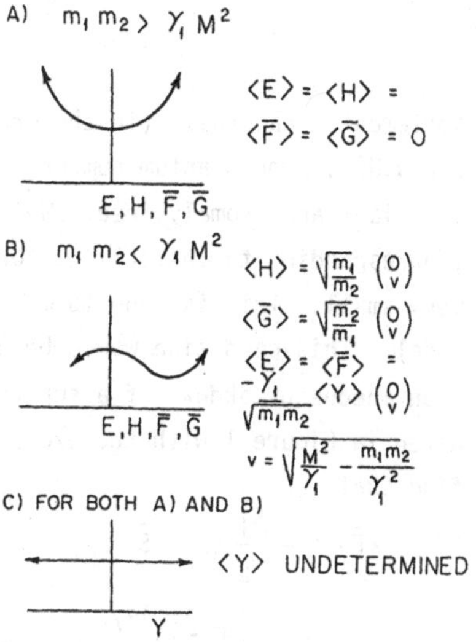

A) $m_1 m_2 > \gamma_1 M^2$

$\langle E \rangle = \langle H \rangle = \langle \bar{F} \rangle = \langle \bar{G} \rangle = 0$

B) $m_1 m_2 < \gamma_1 M^2$

$\langle H \rangle = \sqrt{\frac{m_1}{m_2}} \begin{pmatrix} 0 \\ v \end{pmatrix}$

$\langle \bar{G} \rangle = \sqrt{\frac{m_2}{m_1}} \begin{pmatrix} 0 \\ v \end{pmatrix}$

$\langle E \rangle = \langle \bar{F} \rangle = \frac{-\gamma_1}{\sqrt{m_1 m_2}} \langle Y \rangle \begin{pmatrix} 0 \\ v \end{pmatrix}$

$v = \sqrt{\frac{M^2}{\gamma_1} - \frac{m_1 m_2}{\gamma_1^2}}$

C) FOR BOTH A) AND B)

$\langle Y \rangle$ UNDETERMINED

Fig. 1

Supersymmetric SU(3) × SU(2) × U(1) Models

I) C. Nappi and B. Ovrut [5].

The strategy of this model is outlined in Figure 2A. Sectors 1 and 2 are each O'Raifeartaigh models of the type just discussed. Note that they share Y in common. We hope to choose the parameters such that the sector 1(2) VEVs are as shown in Figure 1A(B). VEV $\langle Y \rangle$ should be undetermined at tree level. To best display the effect of

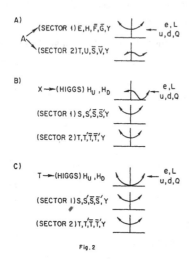

A)

(SECTOR 1) $E, H, \bar{F}, \bar{G}, Y$

A

(SECTOR 2) $T, U, \bar{S}, \bar{V}, Y$

B)

$X \rightarrow$ (HIGGS) H_U, H_D

e, L
u, d, Q

(SECTOR 1) $S, S', \bar{S}, \bar{S}', Y$

(SECTOR 2) $T, T, \bar{T}, \bar{T}', Y$

C)

$T \rightarrow$ (HIGGS) H_U, H_D

e, L
u, d, Q

(SECTOR 1) $S, S', \bar{S}, \bar{S}', Y$

(SECTOR 2) $T, T, \bar{T}, \bar{T}', Y$

Fig. 2

radiative corrections on the masses we couple lepton and quark super-fields e,u,d (right singlets) and L,Q (left doublets) to sector 1 only. It follows that quarks, leptons and their scalar partners will all be massless at tree level. Their masses will arise solely through radiative corrections. Finally, for fermions to get non-vanishing radiative mass, it is necessary to introduce another chiral superfield A which links sectors 1 and 2.

It is one thing to discuss a strategy and another to implement it. The above strategy can be implemented naturally (in the group theoretical sense) if we introduce a new U(1) type quantum number S (sector number). The theory is R-invariant and anomaly free. We choose (unnaturally) the κ coefficient (corresponding to the U(1) factor in the gauge group) to be zero or very small. This is done to make our theory a pure O'Raifeartaigh model. This condition would be natural if our theory arose from the spontaneous breakdown of a supersymmetric GUT. The VEVs are exactly as given in Figure 1 with the exception that $<\bar{F}>$ is no longer zero. We find that

$$<\bar{F}> = -\frac{k_1}{m_2} <A><\bar{S}>$$

$$<A> = \frac{m_2}{k_1} \left(\frac{m_3^2 - m_4^2}{m_4^2} \right)^{1/2} \tag{18}$$

All squark and slepton VEVs vanish. $V_{P.E.}$ evaluated at this minimum is proportional to $(m_3 m_4)$ (electro-weak breaking scale v)^2. Since $V_{P.E.} > 0$, supersymmetry is spontaneously broken. Once the electro-weak scale is chosen (v $\sim$ 240 GeV), the scale of supersymmetry breaking is set by $m_3 m_4$.

The analysis of the vacuum is not complete since $<Y>$ remains un-determined at tree level. This VEV can be evaluated by minimizing the one-loop correction to the effective potential given by the formula

$$\Delta V(<\phi>) = \sum_i \frac{(-1)^F}{64\pi^2} M_i(<\phi>)^4 \ln(M_i(<\phi>)^2/\mu^2) \tag{19}$$

The right side of Eq. (19) is most easily evaluated numerically. The result is shown in Figure 3. From Table 1 we see that Y carries non-vanishing R quantum number. Hence, $<Y> \neq 0$ spontaneously breaks R-invariance. Experimentally $m_{gluino} \gtrsim 1.5$ GeV. Since R-invariance forbids a gluino mass, it follows that we must work in the range of parameters $(m_3 m_4 \sim O(v^2))$ for which $<Y> \neq 0$. This has the consequences that 1) there is a strict upper bound on radiatively induced masses (these are proportional to $m_3 m_4$, and if $m_3 m_4 > O(v^2)$, then $<Y> = 0$) and 2) there is a Goldstone boson (axion) associated with the spontaneous break-

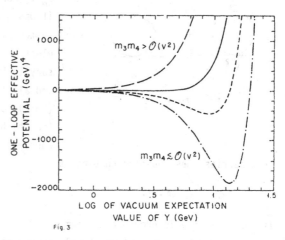

Fig. 3

ing of R-invariance. (This is the price of naturalness.)

We now consider induced masses of sleptons and squarks. These are non-vanishing at the one-loop level. For simplicity consider the right-handed electron scalar e . Graphs contributing to a mass for e are shown in Figure 4. The graphs involving Yukawa couplings only can be evaluated analytically with the result that

$$m_{e,\lambda}^2 = + \frac{\lambda^2}{16\pi^2} \gamma_1^2 \frac{(m_3 m_4)^2}{m_1^2}\left\{1 + 0\left(\frac{m_3 m_4}{m_1^2}\right)\right\} \tag{20}$$

Note that (20) is positive and second order in the supersymmetry breaking parameter $m_3 m_4$. The graphs involving gauge couplings are most easily evaluated numerically. The result is that $m_{e,g}^2 \sim 10^{-2} m_{e,\lambda}^2$. For typical ranges of the parameters $m_e^2 \lesssim + 400(GeV)^2$ which is

YUKAWA COUPLINGS:

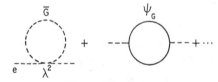

GAUGE COUPLINGS :

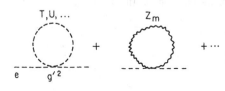

Fig. 4

phenomenologically acceptable. We find similar masses for all other squarks and sleptons.

Now consider the radiatively induced masses of the quarks and leptons. These can be non-vanishing at the one-loop level. For simplicity consider the electron ψ_e. For the scalars to get mass it was necessary to break supersymmetry only. However, for the electron to get mass it is necessary to break both supersymmetry and all the quantum numbers (SU(2), U(1) and S) carried by $\bar{G}$ (to which the electron Yukawa couples). The product of VEVs that carries these quantum numbers is

$$<\bar{G}> \sim <\bar{S}><A><Y>^{\dagger} \tag{21}$$

If supersymmetry is broken but $m_3 m_4 > 0(v^2) (\Rightarrow <Y> = 0)$ and/or $m_3 = m_4 (\Rightarrow <A> = 0)$, then the electron remains massless. If supersymmetry is broken and $<\bar{S}><A><Y>^{\dagger} \neq 0$, then the graph contributing to the electron mass is shown in Figure 5A. We find that

$$m_{elec} \propto \frac{\lambda}{4\pi} \alpha_{weak} k_1 \gamma_1 \frac{<\bar{S}><A><Y>^{\dagger}}{m_1^2} \tag{22}$$

which can be set equal to 1/2 MeV without having to take any parameters very large or very small. Note that, since we must break more symmetries in order for the electron to get mass, there are more heavy mass propagators in Figure 5A than in the corresponding scalar graphs. Hence, m_{elec} is naturally smaller than the masses of its scalar partners. We conclude that <u>the mass sum rule has been avoided!</u> Similar results apply to all leptons and quarks.

Finally, the graph contibuting to the gluino mass is shown in Figure 5B. For typical ranges of parameters $m_{gluino} \lesssim 2$ GeV which is just above the experimental bound. Our model suffers from the fact that various flavor changing neutral current rates are too large.

A)

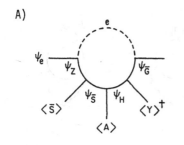

B)

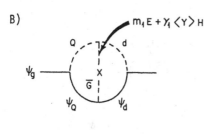

Fig. 5

II) M. Dine and W.Fischler [6].

The strategy of this model is outlined in Figure 2B. Sectors 1 and 2 are similar to those of the preceding model. However, the fields of sector 2 are SU(3)(SU(2)) triplets (singlets). One hopes to choose the parameters such that the VEVs of sectors 1 and 2 are as shown in Figure 1A. The <Y> VEV should be undetermined at tree level. In addition to sectors 1 and 2, one adds two Higgs chiral superfields H_U and H_D which, by coupling them to a singlet superfield X, should acquire non-vanishing VEVs. Quark and lepton superfields couple to H_U and H_D and, hence, their component fields have degenerate, non-zero mass.

This strategy cannot be implemented naturally (in the group theoretical sense). Several terms which break R-invariance are included in the Lagrangian, whereas other allowed terms (X^2, Y^2, etc.) are omitted. Also, some masses are identified by hand. It should be noted, however, that uncancellable divergences are not expected, due to the non-renormalization theorems of supersymmetry. The theory is anomaly free. One chooses κ to be zero or very small. The VEVs are as given in Figure 1. In addition

$$<H_U> = \frac{1}{\sqrt{2}} \begin{pmatrix} v \\ 0 \end{pmatrix} \quad , \quad <H_D> = \frac{1}{\sqrt{2}} \begin{pmatrix} 0 \\ v \end{pmatrix} \tag{23}$$

$$<X> = 0$$

All squark and slepton VEVs vanish. $V_{P.E.}$ evaluated at this minimum is proportional to μ_1^4. Therefore, supersymmetry is broken with the scale set by μ_1^2. It is assumed in all further calculations that $\mu_1^2 \gg v^2$ which will be justified later on.

VEV <Y> can be determined from expression (19) and is found to vanish. Now consider the radiatively induced masses of sleptons and squarks. These are non-vanishing at the two-loop level. Graphs contributing to masses for these fields are shown in Figure 6. These graphs can be evaluated analytically with the result that

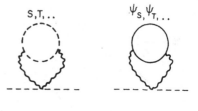

Fig. 6

$$m^2 = \frac{C^2}{2}\left(\frac{\alpha_3}{4\pi}\right)^2 \Lambda_T^2 + \frac{T^2}{2}\left(\frac{\alpha_2}{4\pi}\right)^2 \Lambda_S^2 + \left(\frac{\alpha_1}{4\pi}\right)\left(\frac{Y}{2}\right)^2 \left(\frac{1}{2}\Lambda_S^2 + \Lambda_T^2\right)$$

$$\Lambda_T^2 = +\frac{8\lambda_2^2(\mu_1^2)^2}{m_2^2} \quad , \quad \Lambda_S^2 = +\frac{8\lambda_1^2(\mu_1^2)^2}{m_1^2} \tag{24}$$

where C(T) are SU(3)(SU(2)) Casimir coefficients. Note that m^2 is positive and second order in the supersymmetry breaking parameter. In order for $m_e^2 \gtrsim (16 \text{ GeV})^2$ (the experimental bound) it is necessary to take $\Lambda_S \gtrsim 20$ TeV which justifies the assumption that $u_1^2 \gg v^2$. All scalar masses are phenomenologically acceptable.

Since this theory is not R-invariant at tree level, the gluino mass arises at one-loop. The result is that $m_{gluino} \gtrsim 2$ GeV . Strangeness changing neutral current processes are adequately suppressed in this model.

III) L. Alvarez-Gaumé, M. Claudson, M. Wise [7].

The strategy of this model is outlined in Figure 2C. The superfield content is similar to the Dine et al. model with the exception that the singlet X is replaced by SU(2) triplet T. One hopes to choose the parameters such that all tree level VEVs vanish except for <Y> which is undetermined. Quarks, leptons and their scalar partners will all be massless at tree level.

This strategy can be implemented naturally. The theory is R-invariant and anomaly free. One chooses κ to be zero or very small. All VEVs vanish with the exception of $<Y>$ which is undetermined. It follows that $SU(2) \times U(1)$ is unbroken at tree level. $V_{P.E.}$ evaluated at this minimum is proportional to u^4. Therefore, supersymmetry is broken with the scale set by u^2.

The $(mass)^2$ parameter in the Y direction can be calculated to the one-loop level and is positive. It follows that $<Y> = 0$. Now consider the radiatively induced masses of the sleptons, squarks, Higgs and T scalars. These are non-vanishing at the two-loop level. Graphs contributing to masses for these fields are similar to those in Figure 6. The result is that

$$m^2(R_3, R_2) = \frac{\alpha_3^2}{32\pi^2} M^2 z^2 \left(DC_2C_2\right)_3 + \frac{\alpha_2^2}{12\pi^2} \tilde{M}^2 \tilde{z}^2 \left(DC_2C_2\right)_2$$

$$z = + \frac{a\mu^2}{M^2} \qquad \tilde{z} = + \frac{b\mu^2}{\tilde{M}^2} \tag{25}$$

where $(C_2)_3$, $(C_2)_2$ and D are $SU(3)$, $SU(2)$ Casimir coefficients and the representation dimension respectively. Note that m^2 is positive and second order in the supersymmetry breaking parameter. In deriving (25) it has been assumed that $M, \tilde{M} > \mu >>$ electro-weak scale and that the renormalization point $\Lambda \sim M, \tilde{M}$. It follows that, at this scale, $SU(2) \times U(1)$ is unbroken. To examine physics at electro-weak energies it is necessary to use the renormalization group to scale the set of dimension two operators from $\Lambda \sim M, M'$ down to $\Lambda \sim 240$ GeV. These operators mix in the anomalous dimension matrix. To simplify the calculation one ignores the scaling of the Yukawa couplings. The mass of the u-quarks at scale Λ is found to be

$$m_U^2 = m_{(3,1)}^2 - \frac{|\lambda_U|^2}{4\pi^2} \left[m_{(3,1)}^2 \; \ln\left(\frac{M^2}{\Lambda^2}\right) + m_{(1,2)}^2 \ln\left(\frac{\tilde{M}^2}{\Lambda^2}\right) \right] \tag{26}$$

Taking $M, \tilde{M} \sim 10^2$ TeV ($\mu \sim 10$ TeV, the same as Dine et al.) and $\Lambda \sim 240$ GeV one finds $m_U^2 \sim + 10^2$ GeV. The second term on the right side of (26) is not large enough to change the sign of m_U^2. Similar results hold for all squarks and sleptons. These masses are

phenomenologically acceptable. However, the mass of the Higgs
scalars is found to be

$$m_H^2 = m_{(1,2)}^2 - \frac{3|\lambda_D|^2}{8\pi^2}\left[m_{(3,1)}^2 \ln\left(\frac{M^2}{\Lambda^2}\right) + m_{(1,2)}^2 \ln\left(\frac{M^2}{\Lambda^2}\right)\right]$$

$$-\frac{4|\lambda_e|^2 + \lambda_T^2}{32\pi^2}\ m_{(1,2)}^2 \ln\left(\frac{M^2}{\Lambda^2}\right) \qquad (27)$$

Since $m_{(3,1)}^2 \sim 10\ M_{(1,2)}^2$ one finds, for the above parameters, that
$m_H^2 < 0$. This leads to non-vanishing values for $<H_U>$ and $<H_D>$ and,
hence, to spontaneous breaking of SU(2) × U(1). Unfortunately, these
VEVs do not break the R-invariance of the model. It follows that
$m_{gluino} = 0$ which is unacceptable.

Finally I would like to mention a model constructed by L. Hall
and I. Hinchliffe. These authors extended the gauge group to
SU(3) × SU(2) × U(1) × $\widetilde{U(1)}$, but utilized both the U(1) and the
O'Raifeartaigh modes of supersymmetry breaking.

None of the above models is completely satisfactory from the
point of view of phenomenology. Each has its strengths and weaknesses
as outlined above. However, I think it is fair to say that the nature
of these weaknesses does not appear to be very profound. In view of
its impact on a truly profound problem (the gauge hierarchy),
I believe that supersymmetry remains an important focal point for
future research.

ACKNOWLEDGMENTS

I would like to thank Chiara Nappi for her collaboration on the
first paper reviewed in this talk for for helpful comments during the
preparation of this manuscript. This work was supported by DOE Grant
DE-AC02-76ER0-2220.

55

TABLE 1

	SU(3)	SU(2)	Y	R
E	1	2	1	1
H	1	2	1	0
$\bar{F}$	1	$\bar{2}$	-1	1
$\bar{G}$	1	$\bar{2}$	-1	0
Y	1	1	0	1

REFERENCES

[1] S. Dimopoulos and H. Georgi, HUTP-81/A022.

[2] P. Fayet, Phys. Lett. 69B (1977) 489.

[3] S. Weinberg, HUTP-81/A047 (1981).

[4] L. Girardello and J. Iliopoulos, Phys. Lett. 88B (1979) 85.

[5] C. Nappi and B. Ovrut, Institute for Advanced Study preprint, December 1981.

[6] M. Dine and W. Fischler, Institute for Advanced Study preprint, December 1981.

[7] L. Alvarez-Gaumé, M. Claudson and M. Wise, HUTP-81/A063.

[8] L. Hall and I. Hinchliffe, LBL preprint 14020, February 1982.

THE KAMIOKA PROTON DECAY EXPERIMENT

- PRESENT STATUS -

K. Takahashi (KEK)

for the collaboration

Collaboration

KEK, National Laboratory for High Energy Physics, Oho-machi,
Tsukuba-gun, Ibaraki-ken 305, Japan
H. Ikeda, A. Nishimura, H. Sugawara, A. Suzuki, and K. Takahashi,

Department of Physics, University of Tokyo, Hongo, Tokyo, 113 Japan
K. Arisaka, T. Fujii, T. Kajita, M. Kobayashi, M. Koshiba, M. Nakahata,
and Y. Totsuka,

Institute for Cosmic Ray Research, University of Tokyo, Tanashi, Tokyo,
188 Japan
J. Arafune, T. Kifune and T. Suda

Department of Physics, University of Niigata, Niigata, 950-21 Japan
K. Miyano

Institute for Applied Physics, University of Tsukuba, Sakura-mura,
Ibaraki-ken, 305 Japan
Y. Asano.

Abstract

A report is given of the potential capabilities and the present

status of the Kamioka Proton Decay Experiment. The detector is a 3000

ton water Cerenkov detector which utilizes 1,000 huge, newly developed

20" photomultipliers, and has outstanding energy resolution and good

capability for identifying various decay modes of nucleons. Data

taking is expected to begin in late 1982.

56

§1. Introduction

We are currently building a 3000 ton Water Cerenkov detector

which utilizes 1,000 huge 20" photomultipliers installed uniformly on

the inner walls of a cylindrical water tank. The most characteristic

features of the detector are its outstanding energy resolution and

its high capability to observe and identify most of the expected

decay modes of nucleons. The detector is to be installed in a deep

underground laboratory in the Kamioka Mine in Japan. The detector is

expected to become sensitive in late 1982.

A brief account is given of the potental capabilities and the

present status of the preparation of the experiment, since a report

on the general plans of this project has already been published

elsewhere [1].

§2. Preparation of the Laboratory

The laboratory is currently under excavation at the Kamioka Mine

of Mitsui Mining and Smelting Co. located in central Japan, some 300

Km west of Tokyo. The mine is producing lead and zinc. The budget

for excavation of the laboratory was approved last fall and the

excavation is now being carried out actively. The experimental site

is situated at a depth of $\sim$1000 m directly below the top of a mountain.

Average density of the rock is measured to be $\sim$2.7 g.cm^{-3} so that the

vertical depth is approximately equivalent to 2700 meters of water.

Since the mine is actively working presently and the area of the

laboratory site has been already well investigated for its rock struc-

ture, an excavation of the cavity for the laboratory is progressing

very smoothly. The utility system, electric power and water supply for

the laboratory is very easily accessible and is also under preparation.

(Fig. 1) The laboratory is expected to be finished by late summer of

1982. We, all the collaboration, are grateful for the generous cooperation of the Mitsui Mining and Smelting Co.

§3. Construction of Water Tank and P.M.T. Support System

The 3000 ton pure-water reservoir is a cylindrical steel tank, having a diameter of 15.5 meter and a height of 16 meter. (Fig. 2) Total volume of water contained in the tank is approximately $\sim$2900 m^3. Inner walls of the tank are painted with a specially prepared black-coloured Epoxy with a thickness of more than 250 μm to maintain water purity.

The Cerenkov light is detected by a total of 1,000 large 20" hemispherical photomultiplier tubes installed on the inner walls of the water container. Phototubes are uniformly distributed with a density of 1 PMT/m^2, as shown in Fig. 2. Phototubes are individually supported by a frame in the form of matrix array which is bolted on the wall of the steel tank. With this phototube arrangement, 20% of the inner wall of the tank is covered by photo-sensitive cathode.

§4. Water Supply and Purification System

Very clean natural water is available in the ditch at the tunnel where the tank is located. We have already chemically analysed the water and found that it is pure enough to be the primary water to our Cerenkov detector. Undesirable chemical contents such as Fe^{++}, Ni^{++}, and Co^{++} ions are found to be too little to be of any effect with a level of less than 0.02 ppm. Other contaminating chemicals or organic materials such as bacteria are also much less than that of usual city drinking water. The temperature of the cavity at the mine is about 12°C and is almost constant all the year round. This relatively low temperature is very helpful to suppress undesirable bacteria multiplication in the water inside the tank. Having this very excellent primary water supply and environment, the design of the water purification system is considerably simplified. Fig. 3 shows our water purification system. By this system, water with transparency length of $\gtrsim$ 25 meter (at $\lambda \sim$ 470 nm) or better could be obtained. The maximum processing rate is 10 m^3 hr^{-1}, and it takes a half a month to fill up 2900 m^3 tank.

Water is continuously circulated and purified during normal operation. The purity of the water is continuously monitored by measuring resistivity and density of micro-particles. Transparency of the water is also checked using laser light.

§5. 20" Phototube Developments

The development of 20" photomultiplier tubes has been carried out very successfully by the manufacturer, Hamamatsu T. V. Company, in close cooperation with us [2]. More than 1,000 good quality phototubes are already prepared and have passed the basic specification check. Each phototube is delivered to us with high voltage divider and signal cables out from the anode, molded solidly with a plastic pipe and polyurethane resin on a tube-base, as shown in Fig. 4. In addition to general checks for individual basic characteristics given in Table I., dynamical performance tests, such as long-term gain stability, water proof test, and resolution check with Cerenkov light from cosmic ray muons in water, have been carried out for more than six months. A stability test in air for more than 6-months and more than 4-months in water looks very good. Water proof tests of photo-tubes and their electric connections in a test water tank at KEK has shown that the system works excellently and has no problem so far. Cerenkov data by cosmic ray muons in the test water tank are analysed and found to be consistent with those from Monte Carlo simulation, as shown in Fig. 5 and Table II. These results also give us support for the potential capabilities of the present detector with its high resolution and good identification of various decay modes of nucleons. Some of the Monte Carlo Simulation results are given in Fig. 6 and Table III.

§6. Electronics, Data Acquisition and Calibration System.

Principal parts of our electronics and data taking system are a linear adder, an ADC, transient digitizers, and a PDP-11/60 on-line computer. The signal from each phototube goes first to a linear adder and then split into two. One goes to a 12-channel summing amplifier and the other goes to the ADC system (LeCroy 2285A), which digitizes the total charge from each phototube with 15-bits resolution. After three steps of summing amplifiers, all the currents from 1000 phototubes are summed up. The total sum of phototube currents provides the energy trigger. An event is triggered when the total energy deposition exceeds $\gtrsim$ 100 MeV ($\sim$130 photoelectrons equivalent). The threshold could be lowered down to $\sim$20 MeV.

To identify a muon by observing a delayed electron from $\mu \rightarrow e\nu\bar{\nu}$ decays, the signals are summed up and digitized over a time of $\sim$5 μsec by sampling every 10 nsec with pulse height full width of 9 bits. Fig. 7 shows a block diagram of our electronics and data acquisition system. Considering fundamental requirements in measuring Cerenkov light in the tank, and in identifying necessary decay proc- esses with the capability of 20" phototubes presently available, a 12- channel linear adder circuit with associated buffer amplifiers has been designed and developed at KEK. A basic performance test has been completed. Since a low noise-level of the summing amplifier as

well as buffer amps is very essential, special attention has been
paid to that. The linearity of both summing and buffer amplifiers
has been also well investigated and found to be excellent, as shown
in Fig. 8. An ADC system, LeCroy-2285A, which has become commercially
available recently, has been already tested extensively and preparation of the system is in progress.

The gain of the 1,000 phototubes in the tank is monitored by a
flashing scintillating ball hanging at the center of the tank and at
four different spots when necessary. This plastic scinti-ball contains
50 ppm BBOT, a wave length shifter, and emits isotropic light with a
spectrum of $\lambda=390 \sim 500$ nm when it is excited by a light of $\lambda=300 \sim$
410 nm generated by a Xe-lamp with a proper filter and a fiber light
guide. This scinti-ball monitor system is calibrated independently
by the standard phototube which should already be well checked in its
absolute gain by means of cosmic ray muons and accelerator beam
tests. The scinti-ball system has been tested and it was found that
the isotropy of light can be within several percent with reasonable
care. The scinti-ball system is now under final preparation.

§7. Conclusions

On the basis of these general characteristics of the 20" phototubes, associated electronics and data-acquisition system developed
and tested so far, we have been repeating the Monte Carlo analysis.
The results of these repeated Monte Carlo simulations are essentially
the same as those reported previously [1]. With these results, the
trigger of the events and the expected rate of backgrounds due to
cosmic ray muons and neutrinos were also investigated. Nuclear
effects upon nucleon decays inside the nucleus were also extensively
studied and will be reported elsewhere [3]. The rate of neutrino

background is of order of 0.1 ev/year in the sensitive area of the spectrum of nucleon decay events. Cosmic ray muons can also be rejected and identified by scanning the pattern of each event after filtering background events by means of software cuts.

Our preparation and various developments are progressing well and waiting for completion of the underground laboratory. Our present status can be summarized as follows:

(1) More than 1,000 good quality 20" phototubes are now all prepared and under a final individual check for installation,

(2) Construction of a 3000 ton water tank and a water supply and purification system is at the stage of component assembly.

(3) The electronics and the data taking system are almost ready for installation, and

(4) Excavation of the underground laboratory is in progress and will be finished in late summer of 1982.

If every thing goes well we should be ready for filling the tank in late 1982 and hope to start taking data before the end of this year.

References

[1] T. Suda, a talk given at ν' 81 Conf. at Maui, Hawaii, U.S.A. July 1981. Proc. of ν' 81, P. 224 (1981). and also H. Ikeda et al KEK Preprint 81-23, January 1982.

[2] T. Hayashi et al., Proceedings of the 1981 INS International Symposium on Nuclear Radiation Detectors, March 23-30, 1981 Tokyo, Japan

[3] A. Nishimura and K. Takahashi; to be published.

Table I. General characteristics of Hamamatsu
R1449X 20" photomultiplier

Photo-cathode area	20" in dia.
Shape	Hemispherical
Window material	Pyrex glass (4 ∿ 5 mm)
Photo-cathode material	Bialkali
Focus	188 mm in dia.
Dynodes	13 stages, Venetian blind, 76 mm
Pressure tolerance	6 kg/cm^2, water proof

Quantum efficiency	23% typ. at λ=420 nm
Gain	10^7 at 2,000 Volt
Dark current	150 nA at G-10^7
Dark pulse count	15 kHz at G-10^7
Cathode non-uniformity	less than 20%
Anode non-uniformity	less than 40%
Transit time	90 nsec
Transit time spread	7.7 nsec fwhm
Resolution[*]	1.25
After pulse rate	less than 0.5%/1p.e.[**]

[*] widths of the photoelectron statistics compared with
     Poisson distribution.
[**] after-pulse appears ∿4 μsec after main pulse.

Table II. A comparison of the energy resolution observed
with 20" phototubes for Cerenkov light by cosmic
ray muons in the test water tank, compared with
that expected from Monte Carlo simulation.

20" Phototubes in the water	Peak	Resolution(%)	No. of P.E.
No. 1	911.8±3.7	9.0±0.4	541.7±2.2
No. 2	1027.8±4.7	7.0±0.5	487.8±2.2
No. 3	1108.3±2.9	8.9±0.3	477.4±1.2
No. 4	1115.4±3.7	10.8±0.3	530.1±1.8
Monte Carlo Simulation	————	6.7	772.3

Table III. Average number of photoelectrons and energy resolution for various decay modes of nucleons.

Decay mode	Mean No. of photoelectrons		Energy resolution %	
$p \to e^+\pi^0$,	1264		4.2	
e^+		615		6.7
$\pi^0 \to 2\gamma$		649		6.6
$n \to e^+\pi^-$,	806		17.6	
e^+		623		6.0
π^-		182		70.1
$p \to e^+\eta^0$,	1265		4.0	
e^+		410		6.0
$\pi^0 \to 2\gamma$		855		5.4
$n \to e^+\rho^-$,	870		13.0	
e^+		232		51.1
$\rho^- \to \pi^-\pi^0$		638		21.6
$p \to e^+\rho^0$,	448		30.9	
e^+		230		52.3
$\rho^0 \to \pi^+\pi^-$		218		71.3
$p \to \mu^+\pi^0$,	959		4.6	
μ^+		317		7.5
$\pi^0 \to 2\gamma$		643		6.2
$n \to \mu^+\pi^0$,	523		26.6	
μ^+		316		8.1
π^-		206		67.3
$p \to \mu^+ K_s^0$,	957		4.0	
μ^+		159		8.7
$K_s^0 \to \pi^0\pi^0$		798		4.8
$p \to \mu^+ K_s^0$,	276		37.5	
μ^+		159		8.7
$K_s^0 \to \pi^+\pi^-$		118		86.7

Figure Captions

Fig. 1.   A sketch of the Kamioka Underground Laboratory.

Fig. 2.   A schematic picture of the 3,000 m^3 water tank and phototube

support system.

Fig. 3.   A block diagram of the water purification system.

Fig. 4.   Photographic pictures showing 20" PMT and its water proof

H. V. and signal connector-base.

Fig. 5.   Pulse height distributions of Cerenkov radiation due to

cosmic ray muons detected in the test water tank. Monte

Carlo result is also shown for a comparison.

Fig. 6.   Simulated Cerenkov light rings of $p \rightarrow e^+\pi^o$ (a) and muons (b),

indicated by phototube responses. Tubes with more than 4

photoelectrons are marked by a circle.

Fig. 7.   A block diagram of the electronics and data aquisition system.

Fig. 8.   Results of linearity test of buffer amplifier (a) and

summing amplifier (b) in a newly developed linear adder

circuit.

FIG. 1

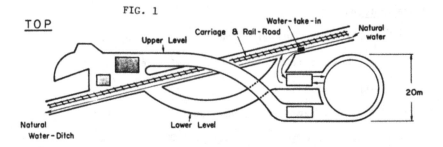

TOP

Upper Level
Carriage & Rail-Road
Water-take-in
Natural water

Natural Water-Ditch
Lower Level

20m

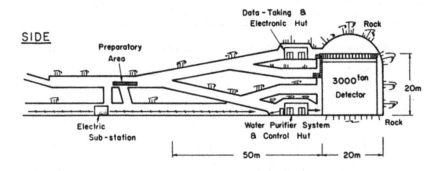

SIDE

Preparatory Area

Data-Taking & Electronic Hut
Rock

3000^{ton} Detector

20m

Electric Sub-station

Water Purifier System & Control Hut

Rock

50m

20m

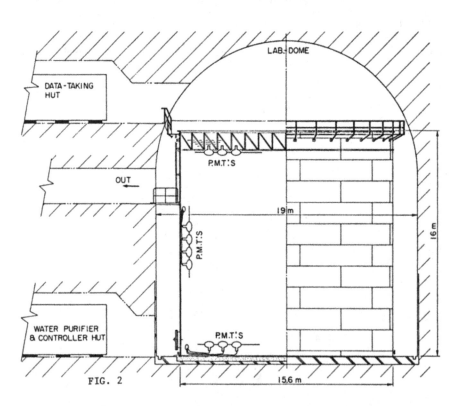

DATA-TAKING HUT

LAB.-DOME

P.M.T.'S

OUT

19 m

16 m

P.M.T.'S

WATER PURIFIER & CONTROLLER HUT

P.M.T.'S

FIG. 2

15.6 m

67

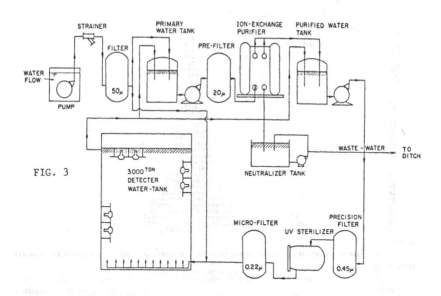

FIG. 3

FIG. 4

68

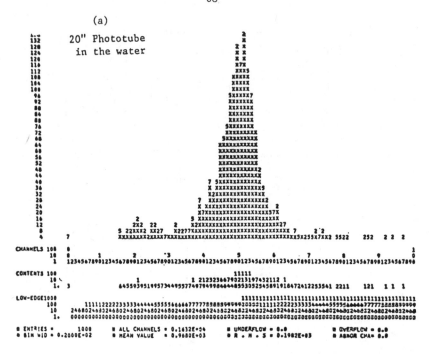

(a)

20" Phototube
in the water

FIG. 5

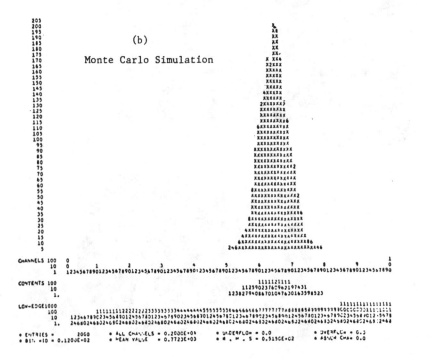

(b)

Monte Carlo Simulation

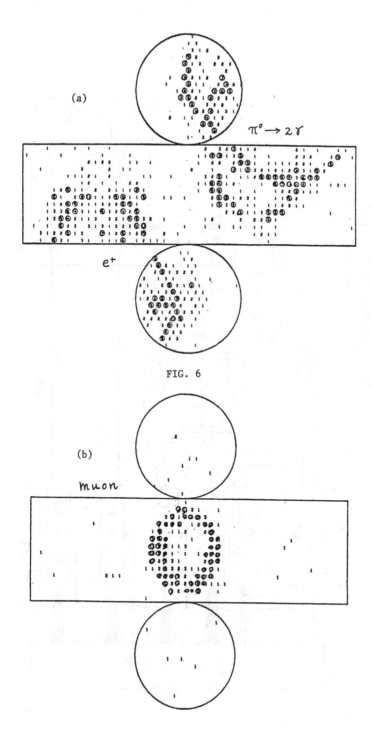

(a)

$\pi^o \rightarrow 2\gamma$

e^+

FIG. 6

(b)

muon

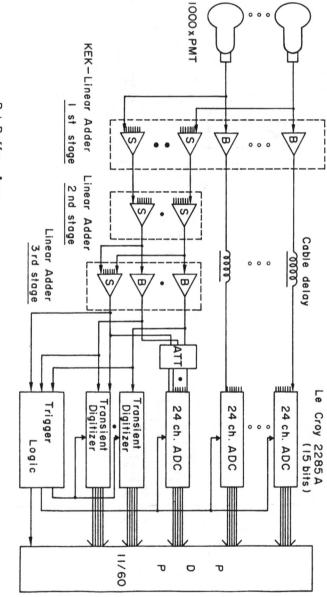

FIG. 7

B : Buffer Amp.

S : 12 channel Summing Amp.

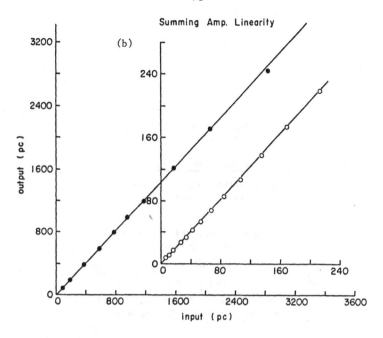

Summing Amp. Linearity

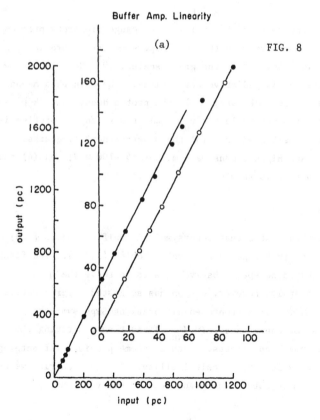

Buffer Amp. Linearity

FIG. 8

GEOMETRIC HIERARCHY

Savas Dimopoulos

Lyman Laboratory of Physics
Harvard University

and

Institute of Theoretical Physics[+]
Stanford University

and

Stuart Raby

Theoretical Division - T-8
Los Alamos National Laboratory

ABSTRACT

We present one approach for solving the gauge hierarchy problem in a grand unified supersymmetric theory. Supersymmetry is broken at a scale of order 10^{12} GeV. Both the grand scale ($\sim 10^{19}$ GeV) and the weak scale are generated via radiative corrections. The main phenomenonological features of the model are: (1) the proton decays into $k^0 \mu^+$ and $k^+ \bar{\nu}_\mu$ and the neutron decays into $k^0 \bar{\nu}_\mu$; (2) the strong CP problem is solved with an invisible axion; (3) the superpartners of quarks, leptons, gauge, and Higgs bosons have masses ~ 50-100 GeV; and (4) the lightest superpartner is stable.

I. INTRODUCTION

It has been suggested that Supersymmetric Unified Models [SUM] can provide a solution to the gauge hierarchy problem [1,2,3]. The first suggestions required non-perturbative effects to solve the problem [1,2,3]. Recently, however, Witten has suggested a perturbative scenario [4]. It requires spontaneously breaking Supersymmetry [SUSY] [5] at a low energy scale ($M \sim M_{weak}$) and then generating the grand scale via one loop effects. Such a scheme provides (if nothing else) the desirable feature of calculability. In our scenario, we use Witten's mechanism to obtain the grand scale. However, the SUSY

72

breaking scale is not M_w but is now an "intermediate" scale $M_I \sim 10^{12}$ GeV. We then generate the weak scale via radiative corrections. We find that the weak scale is of order $f\ M_I^2/M_G$ where f is a function of both gauge and Yukawa couplings. Hence M_I is approximately the geometric mean of M_G and M_{weak}; thus Geometric Hierarchy.

In Section II we discuss the general scenario in the context of the first half of the model. Here we also present the heuristic argument of how it is possible to obtain a weak scale ($M_w \sim 10^2$ GeV) which is naturally much smaller than the scale of SUSY breaking ($M_I \sim 10^{12}$ GeV). This result follows directly from a SUSY decoupling theorem.

In Section III we present the complete model (including all Yukawa couplings). Here we discuss the necessity for having four Higgs multiplets which follows from the requirements: (a) the breaking scheme $SU(5) \rightarrow SU(3) \otimes SU(2) \otimes U(1)$ at M_G, and (b) reasonable amount of proton decay.

II. FIRST HALF

In this section we will introduce the first half of the model as well as the general scenario for solving the gauge hierarchy problem. The gauge group is $SU(5)$ gauge superfield V_G. In the Wess-Zumino gauge we have A_μ (gauge boson), χ (left-handed Weyl gauge fermions or gauginos) and D (auxiliary field); all in the adjoint representation. We also include two matter fields in the adjoint representation, Z_i^j and A_i^j, and a singlet field X. All matter fields are left-handed chiral superfields. For example, the field X contains the components ϕ_X (complex scalar), ψ_X (left-handed Weyl fermion), and F_X (auxiliary field). Finally, in this section we shall also consider the quark, lepton, and Higgs multiplets with all their independent Yukawa couplings set equal to zero. The complete model, with all Yukawa terms, shall be studied in Section III. The quark and lepton superfields are denoted by their $SU(5)$ content. We have 10_i and $\bar{5}_i$ where $i = 1,2,3$ is a generation index. The Higgs superfields are H and

H_1 [5's of SU(5)] and $\bar{H}$ and $\bar{H}_1$ ($\bar{5}$'s).

Consider the superspace potential

$$W = \lambda_1 \text{Tr}(ZA^2) + \lambda_2 X(\text{Tr}\,A^2 - M^2) \tag{2.1}$$

where $M \sim M_I \sim 10^{12}$ GeV. Using W we can evaluate the tree level scalar potential

$$V_{\text{tree}} = |F_X^2| + \sum_{ij} |F_{Z_i}j|^2 + \sum_{ij} |F_{A_i}j|^2 + \frac{1}{2}\text{Tr}\,D^2 \tag{2.2}$$

where $-F_X^* = \partial W/\partial X$, etc; and

$$D = -g([\phi_A^*, \phi_A] + [\phi_Z^*, \phi_Z] + \text{quark, lepton, and Higgs terms}).$$

g is the SU(5) gauge coupling constant. The equations $F_X = 0$, $F_{Z_i}j = 0$ are inconsistent. Thus SUSY is spontaneously broken at the tree level. Minimizing the potential V_{tree} (2.2) we find[F1]

$$\langle\phi_A\rangle = A_0 \begin{pmatrix} 2 & & & & \\ & 2 & & & \\ & & 2 & & \\ & & & -3 & \\ & & & & -3 \end{pmatrix} \tag{2.3}$$

with $A_0 = \lambda_2 M/\lambda_1^2 + 30\lambda_2^2)^{1/2}$. In addition

$$\langle\phi_Z\rangle = Z_0 \begin{pmatrix} 2 & & & & \\ & 2 & & & \\ & & 2 & & \\ & & & -3 & \\ & & & & -3 \end{pmatrix}$$

$$\langle\phi_X\rangle = X_0 \tag{2.4}$$

with $Z_0 = \lambda_2/\lambda_1 X_0$.

The magnitude of Z_0 is however undetermined at the tree level.

Let us now evaluate the one-loop effective potential [6]

$$V_{1\text{ loop}} = V_{\text{tree}} + \frac{1}{64\pi^2} \sum_s (-1)^{2s}(2s+1)M_s^4 \ln M_s^2 \tag{2.5}$$

where s is the spin of the state with mass M_s. M_s are functions of the expectation value Z_0. We find [4]

$$V_{1\text{ loop}} = \frac{\lambda_1^2\lambda_2^2M^4}{(\lambda_1^2+30\lambda_2^2)} \left| 1 + \frac{\lambda_2^2}{\lambda_2^2 + \frac{\lambda_1^2}{30}} \frac{29\lambda_1^2 - 50g^2}{80\pi^2} \ln\frac{Z_0^2}{M^2} \right| \tag{2.6}$$

plus corrections of order M/Z_0 for large Z_0.[f2]

Clearly if $50g^2 > 29\lambda_1^2$, then Z_0 wants to be larger than M. It is quite natural for Z_0 to obtain a value as large as M_{planck} at which point we can no longer trust our perturbation expansion which ignores gravitational effects. Note that Z_0 determines the grand unification scale M_G. We shall see in Section IV that a calculation of the one loop β functions for SU(3), SU(2), and U(1) will in fact determine M_G to be of order M_{planck}. We conclude that the Upside-Down Hierarchy mechanism of Witten's quite naturally allows for a grand scale M_G which is much larger than the fundamental scale $M \sim M_I$ in W [Eq. (2.1)].

Why is it <u>necessary</u>, however, to have the fundamental scale M_I as large as 10^{12} GeV. Moreover, since M_I is also the scale of SUSY breaking, aren't we in danger of dragging all scalar masses up to the scale M_I (the weak scale included). To address these questions let's now consider the tree level mass spectrum and the phenomenon of <u>decoupling</u> of SUSY breaking effects.

We let $<\phi_Z>$ and $<\phi_X>$ take on the value determined by the one-loop potential [Eq. (2.6)]. Inserting these into the superspace potential W [Eq. (2.1)], we obtain

$$W = \lambda_1 Z_0 \, \text{Tr}\left\{ \begin{pmatrix} 3 & & & \\ & 3 & & \\ & & 3 & \\ & & & -2 \\ & & & & -2 \end{pmatrix} A^2 \right\} + \lambda_1 \, \text{Tr}\,(Z'A^2) + \lambda_2 X'(\text{Tr}A^2 - M^2)$$

$$- \lambda_2 X^0 M^2 \tag{2.7}$$

where $\lambda_1 Z_0 = \lambda_2 X_0 \simeq M_G$ and $<\phi_{Z'}> = <\phi_{X'}> \equiv 0$. Thus all the components of A (except for the hypercharge component) obtain a SUSY mass contribution at the scale M_G. The leptoquark components of Z mix with the leptoquark components of V_G (in a SUSY Higgs mechanism) at M_G. The color octet component and SU(2) triplet component of Z obtain at the scale M_I^2/M_G as soon as $<\phi_A> \sim M_I$ is turned on. Finally, one linear combination of X and Z --the hypercharge component of Z (i.e., $\lambda_1 Z_y - \sqrt{30}\,\lambda_2 X/\lambda_1^2 + 30\lambda_2^2)^{1/2}$) mixes with A_y and obtains a mass of order M_I. The other linear combination

$$\tilde{X} \equiv \frac{\lambda_1 X + \sqrt{30}\,\lambda_2 Z_y}{(\lambda_1^2 + 30\lambda_2^2)^{1/2}} \tag{2.8}$$

is massless at the tree level. It includes the pseudo-scalar "invisible" axion, a scalar "scalon" and the goldstino (goldstone fermion). It would be jumping the gun to say anymore about these states now.

We now want to show that the only states which feel SUSY breaking directly in the Lagrangian at the tree level are the scalar fields ϕ_A with mass $\sim M_G$. SUSY breaking effects are induced via the non-zero expectation values of F_X and F_Z. Using the tree level result for $\langle \phi_A \rangle$ [Eq. (2.3)] we obtain

$$\langle F_X \rangle = \frac{\lambda_2 \lambda_1^2 M^2}{(\lambda_1^2 + 30\lambda_2^2)} \sim M_I^2$$

(2.9)

$$\langle F_Z \rangle = \frac{\lambda_1 \lambda_2^2 M^2}{(\lambda_1^2 + 30\lambda_2^2)} \begin{pmatrix} 2 & & & \\ & 2 & & \\ & & -3 & \\ & & & -3 \end{pmatrix} \sim M_I^2.$$

Finally, these terms appear in the Lagrangian, only in the expression

$$\delta = \mathrm{Tr}[(\lambda_1 \langle F_Z \rangle + \lambda_2 \langle F_X \rangle)\phi_A^2]$$

$$= \frac{\lambda_1 \lambda_2^2 M^2}{(\lambda_1^2 + 30\lambda_2^2)} \; \mathrm{Tr} \begin{pmatrix} 3 & & & \\ & 3 & & \\ & & 3 & \\ & & & -2 \\ & & & & -2 \end{pmatrix} \phi_A^2.$$

(2.10)

Comparing with Eq. (2.7) we conclude that the only states which feel SUSY breaking directly in at the tree level are the scalar A states which have mass $\sim M_G$.

This fact leads to the following SUSY decoupling theorem. (Here we present a heuristic argument. It is by now well known that in a SUSY theory, any state, which is massless at the tree level, will remain massless until it knows about SUSY breaking [7]. In our case, SUSY breaking effects due to $\langle F_X \rangle$ and $\langle F_Z \rangle$ [Eq. (2.9)] can only proceed via superheavy states with mass at M_G. As a result, SUSY breaking effects into the light sector of the theory are an expansion in the small parameter M_I^2/M_G^2. For this reason the effective SUSY breaking scale of the light sector of the theory is at most

$$\mu \sim \left(\frac{M_I^2}{M_G^2}\right) M_G \sim 10^5 - 10^3 \; \mathrm{GeV}$$

(2.11)

for $M_G \sim 10^{19}$ GeV and $M_I \sim 10^{12} - 10^{11}$ GeV. It is this scale which will also set the scale of the weak interactions, i.e., $M_W \sim f\mu$ where f is a function of both gauge and Yukawa couplings.

Let's now summarize our approach for solving the gauge hierarchy problem. Given one-dimensional parameter in the Lagrangian of order M_I (and no extremely small dimensionless parameters) we generate both the grand scale M_G and the weak scale M_W via radiative corrections. Since $M_W \sim M_I^2/M_G$ we like to refer to it as an induced geometrical hierarchy.

Before we add all the Yukawa terms into W, let's now consider a few examples of the SUSY decoupling phenomenon.

At this point, quarks, leptons, and Higgs fermions are unable to get mass due to unbroken chiral symmetries. However, scalar quarks, scalar leptons, and Higgs bosons can get mass via gauge exchanges. The dominant contribution to their mass comes from the two loop graphs of Figure 1. As discussed recently by several authors [8], these contributions are positive. They give a scalar mass (μ_S) of order

$$\mu_S \sim \frac{\alpha}{2\pi} \frac{M_I^2}{M_G} \sim 10^2 \text{ GeV} \tag{2.12}$$

where α is the relevant coupling. We note that leptoquark gauge exchanges also contribute without any suppression since the integrals are only cut off at the scale M_G. They are also one loop contributions to scalar masses. These are, however, suppressed by an additional factor of M_I^2/M_G^2 with respect to the two loop result (see Figure 2). Finally, we note that the gauge exchanges of Figure 1 are the dominant contributor to scalar quark and lepton masses for the first two generations. This is because the Yukawa couplings for these states are small. Thus, up to corrections proportional to Yukawa couplings, scalar quarks, and leptons with the same weak hypercharge are degenerate. This is necessary in order to suppress "flavor changing neutral current" type interactions via a SUSY GIM mechanism [9].

In addition to scalars, gauginos can now obtain majorana masses radiatively. Consider the graph of Figure 3. It generates a gaugino mass of order

$$\tilde{m} \sim \frac{\alpha}{2\pi} \frac{M_I^2}{M_G} \tag{2.13}$$

where α is the relevant coupling. This is another example of the SUSY decoupling theorem alluded to above and discussed in Appendix A. Notice from (2.12) and (2.13) that these corrections vanish as $M_G \to \infty$. We also note that gaugino masses violate an R invariance of the model, where $Z \to e^{i\alpha}Z$, $X \to e^{i\alpha}X$, $A \to A$, and W rotates by a phase which is absorbed by a rotation of the anticommuting coordinate θ [10]. This symmetry is, however, spontaneously broken at the scale M_G by $<\phi_X>$ and $<\phi_Z>$, thus allowing gauginos to obtain mass. Finally this R invariance of the model shall be retained in the presence of all Yukawa terms. At that time we shall identify it with the Peccei-Quinn U(1) symmetry which is necessary to avoid a strong CP problem [11].

III. THE COMPLETE MODEL

Before we write down the superspace potential W with all the relevant states and Yukawa couplings, we should discuss some of the new problems which have to be faced at this point. (1) We will introduce the Higgs superfields. In a minimal SUSY model both a 5 and $\bar{5}$ are necessary in order to give mass to both up quarks, down quarks, and leptons. The SU(2) doublet components of these fields must obtain a negative mass-squared via radiative corrections, which will be of order M_I^2/M_G and define the weak scale. (2) The SU(3) triplet components, however, must be heavy in order to avoid rapid proton decay. (3) We also want the breaking pattern SU(5) $\rightarrow$ SU(3) $\otimes$ SU(2) $\otimes$ U(1), as discussed in Section II, to be unaffected by the addition of the Higgs fields. Note that this has been a problem in the past [4,12] (4) We want to solve the strong CP problem, and finally (5) we want to retain the nice feature of standard SU(5) [13], i.e., perturbative unification. We shall then be able to believe our results for the weak angle--$\sin^2\theta_W$. These are the issues.

Let's now discuss the complete model. The superspace potential is given by

$$W = \lambda_1 \, \text{Tr} \, (ZA^2) + \lambda_2 X (\text{Tr} \, A^2 - M^2)$$

$$+ \lambda_3 (\bar{H}_1 AH + \bar{H}AH_1) + \lambda_4 (\bar{H}_1 H + \bar{H}H_1)B$$

$$+ \lambda_5 Y (\bar{H}_1 H_1 + \eta C^2 + M'B - \tilde{M}^2)$$

$$+ g_{ij}^u H \, 10_i 10_j + g_{ij}^d \bar{H} \, 10_i \bar{5}_j. \tag{3.1}$$

This is not the most general potential W consistent with all the global symmetries. For example, the coefficients of the terms $\bar{H}AH_1$ and $\bar{H}_1 AH$ have been set equal for convenience. Changing them as follows

$$(\lambda_3 A + \lambda_4 B)_i^{\ j} (\bar{H}_1^i H_j + k\bar{H}^i H_{1j})$$

would not affect the scenario in an essential way. There are however terms which have been omitted which would affect drastic changes. For example, if we include a term XB^2 then the vacuum would be supersymmetric. If we exclude this term however we can never generate it again perturbatively. It is thus technically "natural" to exclude it.

The new states are:

Y, B, C   singlets

H, H_1 (5)   (3.2)

$\bar{H}$, $\bar{H}_1$ $(\bar{5})$

λ_1, λ_2, λ_3, λ_4, λ_5, η, g^u_{ij}, and g^d_{ij} are dimensionless Yukawa couplings. M, M', and $\tilde{M}$ are all mass parameters of order M_I.

The first two terms were discussed in Section II. Let's now discuss in some detail the new terms in the superspace potential W (3.1). Consider the λ_3 and λ_4 terms. We have four Higgs fields. We want to show that this was necessary in order to satisfy issues (2) and (3) above--no rapid proton decay and $SU(5) \to SU(3) \otimes SU(2) \otimes U(1)$. In order to show this, consider the model with H and $\bar{H}$ only. It is desirable to obtain mass for the color triplet Higgs of order M_G. Thus we take first the superspace potential

$$W_1 = \lambda_1 \text{Tr}(ZA^2) + \lambda_2 X(\text{Tr}A^2 - M^2) + \lambda_3 \bar{H}ZH + \lambda_4 \bar{H}HB. \qquad (3.3)$$

Using the equations of motion we find

$$-F^*_{\bar{H}} = (\lambda_3 \phi_Z + \lambda_4 \phi_B)\phi_H$$

$$-F^*_H = \phi_{\bar{H}}(\lambda_3 \phi_Z + \lambda_4 \phi_B)$$

$$-F^{*\,i}_{Z\,j} = \lambda_1 (\phi^2_A - \frac{1}{5}(\text{Tr}\phi^2_A))_i{}^j + \lambda_3 \phi^i_{\bar{H}}\phi_{Hi} \qquad (3.4)$$

$$-F^*_X = \lambda_2 (\text{Tr}\phi^2_A - M^2)$$

$$-F^*_A = [2(\lambda_1 \phi_Z + \lambda_2 \phi_X)\phi_A] - \frac{1}{5}\text{Tr}[\ ]$$

$$-F^*_B = \lambda_4 \phi_{\bar{H}}\phi_H.$$

In order to minimize the scalar potential we want to make all F components as small as possible. As in Section II, the expectation value of ϕ_Z is undetermined at the tree level since it is always possible to set $F_{\bar{H}} = F_H = F_A = 0$. Moreover, if the expectation value

$$<\lambda_3\phi_Z + \lambda_4\phi_B> = M_G \begin{pmatrix} 1 & & & & \\ & 1 & & & \\ & & 1 & & \\ & & & 0 & \\ & & & & 0 \end{pmatrix}$$

then the color triplet Higgs have mass M_G while the SU(2) doublet Higgs are massless. This is exactly what we want. However, since the equations $F_Z = 0$, $F_X = 0$, $F_B = 0$ are inconsistent, SUSY is spontaneously broken at the scale M. Moreover by minimizing the potential we find $\phi_{\bar{H}}$ and ϕ_H obtain expectation values of order $M \sim M_I$ which is unacceptable. We also find the expectation value for ϕ_A has the general form

$$<\phi_A> \sim M_I \begin{pmatrix} a & & & & \\ & a & & & \\ & & a & & \\ & & & b & \\ & & & & c \end{pmatrix}. \tag{3.5}$$

Hence $<\phi_Z> \sim <\phi_A>$ breaks SU(5) to SU(3) $\otimes$ U(1) $\otimes$ U(1) at M_G.

Thus W_1 (3.3) is unacceptable. Consider now (replacing Z by A in λ_3)

$$W_2 = \lambda_1 Tr(ZA^2) + \lambda_2 X(TRA^2 - M^2) + \lambda_3 \bar{H}AH + \lambda_4 \bar{H}HB. \tag{3.6}$$

$$-F_{Z*i}{}^j = \lambda_1(\phi_A^2 - \frac{1}{5}(Tr\phi_A^2)1)_i{}^j$$

$$-F_{A*i}{}^j = ([2(\lambda_1\phi_Z + \lambda_2\phi_X\phi_A] - \frac{1}{5}Tr[\ ])_i{}^j + \lambda_3\phi_{\bar{H}}^j\phi_{Hi} \tag{3.7}$$

$$-F_{\bar{H}}^* = (\lambda_3\phi_A + \lambda_4\phi_B)\phi_H$$

$$-F_H^* = \phi_{\bar{H}}(\lambda_3\phi_A + \lambda_4\phi_B)$$

with all other F components the same as in (3.4). We have now solved one problem. Since ϕ_H and $\phi_{\bar{H}}$ do not enter the equation for F_Z, only $F_Z = 0$ and $F_X = 0$ are inconsistent. Thus the expectation value for ϕ_A is the same as found previously in Section II [Eq. (2.3)]. We thus preserve the breaking pattern SU(5) $\rightarrow$ SU(3) $\otimes$ SU(2) $\otimes$ U(1). The equations $F_A = 0$, $F_{\bar{H}} = 0$, $F_H = 0$ can always be satisfied. For example, one solution is $<\phi_{\bar{H}}> = 0$, $<\phi_H> \sim \begin{pmatrix} 0 \\ 0 \\ 0 \\ 0 \\ V \end{pmatrix}$ and $<\lambda_3\phi_A + \lambda_4\phi_B> \approx M_I \begin{pmatrix} 1 & & & & \\ & 1 & & & \\ & & 1 & & \\ & & & 0 & \\ & & & & 0 \end{pmatrix}$.

Thus, the color triplet Higgs obtain mass of order m_I. As we shall discuss later, this is perfectly reasonable for the Higgs scalars. However, the color triplet Higgs fermions also obtain mass of order M_I. The mass term $M_I \psi_{\bar{H}} \psi_H$ then generates dimension 5 operators which cause rapid proton decay [14,17]. It is to avoid this catastrophe that we add the new Higgs fields H_1 and $\bar{H}_1$ and finally obtain the superspace potential W (3.1). The color triplet Higgs fields can still obtain mass of order M_I. However, the color triplet Higgs fermion mass is now of the form

$$M_I (\psi_{\bar{H}} \psi_{H_1} + \psi_{H_1} \psi_H).$$ \hfill (3.8)

Such an off-diagonal mass term does not directly contribute to proton decay, since only H and $\bar{H}$ couple to quarks and leptons. Radiative corrections will of course induce a diagonal mass term but by decoupling this can only be of order

$$\frac{M_I^2}{M_G} \psi_{\bar{H}} \psi_H$$ \hfill (3.9)

which, as we shall discuss shortly, is perfectly reasonable.

The λ_5 term with M' = 0 is necessary to avoid a massless goldstone boson which results from the spontaneous breaking of the following global symmetry:

$$H \rightarrow e^{2i\alpha} H$$

$$\bar{H} \rightarrow e^{2i\bar{\alpha}} \bar{H}$$

$$H_1 \rightarrow e^{-2i\alpha} H_1$$ \hfill (3.10)

$$\bar{H}_1 \rightarrow e^{-2i\bar{\alpha}} \bar{H}_1$$

$$10_i \rightarrow e^{-i\alpha} 10_i$$

$$\bar{5}_i \rightarrow e^{-i\bar{\alpha}} \bar{5}_i$$

Finally, the last two terms contain the standard Yukawa couplings of the scalar Higgs with quarks and leptons.

Given the new terms in W (3.1), what can we expect to happen? The discussion of symmetry breaking at the scale M_G in Section II remains essentially unchanged. However, now there are several new directions in field space for which the tree level potential is flat. For example, it is possible to have

$$\langle\phi_H\rangle = \langle\phi_{\bar{H}}\rangle = \begin{pmatrix} 0 & & & & \\ & 0 & & & \\ & & 0 & & \\ & & & 0 & \\ & & & & V \end{pmatrix}$$

$$\langle\phi_{H_1}\rangle = -\langle\phi_{\bar{H}_1}\rangle = \begin{pmatrix} 0 & & & & \\ & 0 & & & \\ & & 0 & & \\ & & & 0 & \\ & & & & V_1 \end{pmatrix}$$

$$\langle\lambda_3\phi_A + \lambda_4\phi_B\rangle \cong \begin{pmatrix} 1 & & & & \\ & 1 & & & \\ & & 1 & & \\ & & & 0 & \\ & & & & 0 \end{pmatrix} M_I \tag{3.11}$$

$$\langle\phi_y\rangle = 0$$

$$\langle\phi_C\rangle = \left(\frac{\tilde{M}^2 + V_1^2 - M'\langle\phi_B\rangle}{\eta} \right)$$

where V and V_1 are undetermined at the tree level. We shall assume for
the rest of this paper that this is the relevant direction. Whether or
not this is actually the case must await analysis of one and two loop
corrections to the effective potential.[F3] Given this ansatz it is
easy to see that there exists a range of parameters for which V and V_1
will be small compared to M_I. Consider the one loop effective
potential for V, $V_1 \gg M_I$. There are terms of order $\ln V^2$ or $\ln V_1^2$
which have negative coefficients proportional to g^2 [SU(5) gauge
coupling] and positive coefficient proportional to λ_3^2. Clearly, if λ_3
is sufficiently larger than g, then V and V_1 do not want to be large.
The scale of V and V_1 will then be determined by the scale of SUSY
breaking effects in the Higgs sector which by decoupling is at most of
order M_I^2/M_G.

Finally, if the negative mass squared corrections to the Higgs
scalars dominate, then the weak scale can be generated. We shall
return to this final point in the next section.

COMMENTS

In a nutshell some of the consequences of the model are the

following:

(1) The strong CP problem is solved via the invisible axion mechanism. R invariance plays the role of the Peccei-Quinn symmetry.

(2) Flavor changing neutral currents are naturally suppressed since scalar quarks with the same $SU(3) \times SU(2) \times U(1)$ quantum numbers are approximately degenerate.

(3) $\sin^2\theta_W \approx .25$ and $M_6 \simeq 10^{20}$ GeV.

(4) Of course, there are several new particles ("inos") with masses ~100 GeV.

(5) The dominant nucleon decay modes are
$$p \to K^0\mu^+, \quad p \to K^+\bar{\nu}_\mu, \quad n \to K^0\bar{\nu}_\mu.$$

This is very different than what happens in the standard model as well as the supersymmetric models in which supersymmetry is broken at ~300 GeV.

ACKNOWLEDGEMENTS

We appreciated illuminating discussions with W.A. Bardeen, J.D. Bjorken, M. Claudson, M. Dine, M. Einhorn, W. Fischler, M. Gell-Mann, H. Georgi, S.L. Glashow, T. Goldman, D. Gross, D. Kessler, E.W. Kolb, J. Preskill, M. Srednicki, L. Susskind, M. Turner, S. Weinberg, G. West, F. Wilczek, M. Wise and E. Witten.

This research is supported in part by the National Science Foundation under Grant No. PHY77-22864.

FOOTNOTES

1.) The auxiliary fields F_X and F_Z are given by the relations:
$$-F_X^* = {}_2(\mathrm{Tr}\phi_A^2 - M^2)$$
$$-F_A^* = \lambda_1(\phi_A^2 - \frac{1}{N}(\mathrm{Tr}\phi_A^2)1)$$

where we have generalized to an arbitrary gauge group SU(N) with Z and A in the adjoint representations. Consider $N = 6$. There is then a solution to the equations $F_X = F_Z = 0$ given by

$$\langle\phi_A\rangle = \frac{M}{\sqrt{6}}\begin{pmatrix} 1 & & & & & \\ & 1 & & & & \\ & & 1 & & & \\ & & & -1 & & \\ & & & & -1 & \\ & & & & & -1 \end{pmatrix}$$

i.e., $\phi_A^2 \equiv \langle \phi_A^2 \rangle = M^2/6$ 1. For N = 5 there is no SUSY solution, however, it is then clear that the minimum will be such that ϕ_A^2 is as close to the identity matrix as possible. This is the intuitive reason why the breaking pattern SU(5) → SU(3) ⊗ SU(2) ⊗ U(1) is preferred. We thank Howard Georgi for this observation.

2.) Since SUSY is spontaneously broken at the scale M, the potential can grow at most logarithmically for $Z^0 \gg M$.

3.) This model can have naturally light Higgs doublets. This is accomplished by the introduction of a "sliding" singlet B whose expectation value affects the Higgs mass matrix. In the tree approximation there are several directions in field space which are degenerate. Our ansatz [Eq. (3.11)] is in one of these directions. Radiative corrections will lift this degeneracy. For example, the graphs of Figure 4 provide the leading contribution to the effective potential in the direction of the sliding singlet B. These corrections are of the form

$$V_{eff} \simeq \pm \lambda_3^2 |F_{\tilde{X}}|^2 \ln \frac{M_G^2}{m_H^2}$$

where $m_H \propto \langle \lambda_3 \phi_A + \lambda_4 \phi_B \rangle$. If the sign is negative then it will be energetically favorable to have light Higgs. We appreciated conversations with M. Dine, W. Fischler, L. Susskind, and E. Witten on this point. Further calculation will thus determine whether our ansatz corresponds to a minimum of V_{eff}.

REFERENCES

[1] S. Dimopoulos and S. Raby, Nucl. Phys. B192, 353 (1981)

[2] E. Witten, Nucl. Phys. B185, 513 (1981).

[3] M. Dine, W. Fischler, and M. Srednicki, Nucl. Phys. B189, 575 (1981).

[4] E. Witten, Phys. Lett. 105B, 267 (1981).

[5] Y.A. Gof'fand and E.P. Likhtam, JEPT Lett. 13, 323 (1971); D.V. Volkov and V.P. Akulov, Phys. Lett. 46B, 109 (1973); J. Wess and B. Zumino, Nucl. Phys. B70, 39 (1974); A. Salam and J. Strathdee, Nucl. Phys. B76, 477 (1974), Phys. Rev. D 11, 1521 (1975), Fortschritte der Physik 26, 57 (1978); P. Fayet and S. Ferrara, Phys. Rep. 32C, 249 (1977); J. Wess and J. Bagger, preprint, Institute for Advanced Study (1981); M.T. Grisaru, W. Siegel, and M. Rocek, Nucl. Phys. B159, 429 (1979); and references therein.

[6] S. Coleman and E. Weinberg, Phys. Rev. D 7, 1888 (1973); S. Weinberg, Phys. Rev. D 7, 2887 (1973).

[7] K. Fujikawa and W. Lang, Nucl. Phys. B88, 61 (1975), S. Weinberg, Phys. Lett. 62B, 111 (1976); M.T. Grisaru, et al., Reference [5]; E. Witten, Reference [2].

[8] M. Dine and W. Fischler, The Institute for Advanced Study preprint (1982); L. Alvarez-Gaumé, M. Claudson, and M. Wise, Harvard preprint HUTP-81/A063 (1982).

[9] S. Dimopoulos and H. Georgi, Harvard preprint HUTP-81/A022 (1981); J. Ellis and D.V. Nanopoulos, CERN preprint Th. 3216 - CERN (1981).

[10] P. Fayet, Nucl. Phys. B90, 104 (1975); A. Salam and J. Strathdee, Reference [5]; J. Wess and B. Zumino, Reference [5]; L. O'Raifeartaigh, Lecture Notes on Supersymmetry, Comm. Dias. Series A, No. 22.

[11] R.D. Peccei and H.R. Quinn, Phys. Rev. Lett. 38, 1440 (1978), Phys. Rev. D 16, 1791 (1977); S. Weinberg, Phys. Rev. Lett. 40, 223 (1978); F. Wilczek, Phys. Rev. Lett. 40, 279 (1978); W.A. Bardeen and S.-H. H. Tye, Phys. Lett. 74B, 229 (1981).

[12] H. Georgi, Harvard preprint HUTP-81/A041 (1981).

[13] H. Georgi and S.L. Glashow, Phys. Rev. Lett. 32, 438 (1974); H. Georgi, H.R. Quinn, and S. Weinberg, Phys. Rev. Lett. 33, 451 (1974); for a review see P. Langacker, Phys. Rep. (1981).

[14] S. Weinberg, Harvard preprint HUTP-81/A047 (1981); N. Sakai and T. Yanagida, Max-Planck-Institut preprint MPI-PAE-PTH 55/81 (1981).

[15] J.E. Kim, Phys. Rev. Lett. 43, 103 (1979); M.A. Shifman, A.I. Vainshtein and V.I. Zakharov, Nucl. Phys. B166, 493 (1980); M. Dine, W. Fischler, and M. Srednicki, Phys. Lett. 104B, 199 (1981); M.B. Wise, H. Georgi, and S.L. Glashow, Phys. Rev. Lett. 47, 402 (1981); and H.P. Nilles and S. Raby, Nucl. Phys. B (to be published).

[16] W.A. Bardeen, unpublished; B. de Wit and D. Freedman, Phys. Rev. Lett. 35, 827 (1975); P. Fayet, Phys. Lett. 70B, 461 (1977); N. Cabibbo, G.R. Farrar, and L. Maini, Phys. Lett. 105B, 154 (1981).

[17] S. Dimopoulos, S. Raby, and F. Wilczek, University of Michigan preprint UM-HE-81-64 (October 1981); J. Ellis, D.V. Nanopoulos, and S. Rudaz, CERN preprint Th. 3199 - CERN (November 1981).

FIGURE CAPTIONS

1. The dominant contribution to scalar quark, scalar lepton, and Higgs boson masses via gauge exchanges.

2. One loop contribution to scalar quark, scalar lepton, and Higgs boson masses via gauge exchanges.

3. One loop contribution to gaugino masses.

4. Two loop corrections to V_{eff} which depend on the Higgs mass m_H.

NOTES ADDED IN PROOF:

We have learned that T. Banks, and J. Polchinski and L. Susskind are working independently on the subject of SUSY decoupling. In addition, we have learned that E. Witten, and T. Banks and V. Kaplunovsky are considering geometrical hierarchy models. Finally we have recently received a preprint by D.V. Nanopoulos and K. Tamvakis who also discuss SU(5) SUMS with color triplet Higgs scalars at the scale $\sim 10^{10}$ GeV.

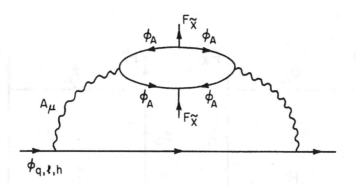

Figure 1

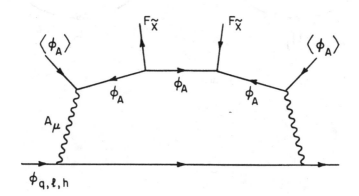

Figure 2

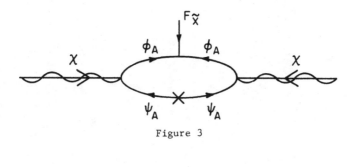

Figure 3

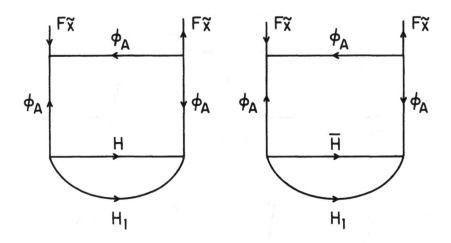

Figure 4

THE HOMESTAKE TRACKING SPECTROMETER[*]

A ONE-MILE DEEP 1400-TON LIQUID SCINTILLATION NUCLEON DECAY DETECTOR

M.L. Cherry, I. Davidson, K. Lande, C.K. Lee, E. Marshall,
and R.I. Steinberg
University of Pennsylvania, Philadelphia, PA 19104

B. Cleveland, R. Davis, Jr., and D. Lowenstein
Brookhaven National Laboratory, Upton, NY 11973

ABSTRACT

We describe a proposed nucleon decay detector able to demonstrate the existence of nucleon decay for lifetimes up to 5×10^{32} yr. The proposed instrument is a self-vetoed completely-active 1400-ton liquid scintillation Tracking Spectrometer to be located in the Homestake Mine at a depth of 4200 mwe, where the cosmic ray muon flux is only $1100/m^2/yr$, more than 10^7 times lower than the flux at the earth's surface. Based on computer simulations and laboratory measurements, the Tracking Spectrometer will have a spatial resolution of $\pm$ 15 cm (0.32 radiation lengths); energy resolution of $\pm$ 4.2%; and time resolution of $\pm$ 1.3 ns. Because liquid scintillator responds to total ionization energy, all neutrinoless nucleon decay modes will produce a sharp ($\pm$ 4.2%) total energy peak at approximately 938 MeV, thereby allowing clear separation of nucleon decay events from atmospheric neutrino and other backgrounds. The instrument will be about equally sensitive to most nucleon decay modes. It will be able to identify most of the likely decay modes (including $n \rightarrow \nu + K^0_s$ as suggested by supersymmetric grand unified theories[1]), as well as determine the charge of lepton secondaries and the polarization of secondary muons.

---

[*] Work supported in part by the U.S. Dept. of Energy under contract DE-AC02-81-ER-40012 and in part by the U.S. National Science Foundation under grant AST-79-08670.

I. GENERAL EXPERIMENTAL ASPECTS OF THE NUCLEON DECAY SEARCH

A. THE SIGNAL

From a model independent point of view, the only reliably known features of the nucleon decay process are (for free protons) a total energy release of 938 MeV and a vanishing vector sum of final state particle momenta. Similar relations exist for bound nucleons, but are somewhat smeared out by the Fermi momentum of the decaying nucleon.

Many theoretical models and calculations for nucleon decay have been proposed.[2] These models provide a wide range of possible nucleon decay final states. For example, on the lepton side, electrons, muons, and neutrinos are expected with varying probabilities, while the decay-product hadrons range from pions and kaons to the vector mesons rho, eta and omega. Further degrees of freedom are proposed in various calculations: predominance of three body decay modes[3], the Goldhaber effect[4] (decay of a bound nucleon to a ~700-MeV electron together with an excited nucleus), three-lepton final states[5], etc.

In this situation, it is evident that the critical requirements for a successful experiment are that it be sensitive to as wide a range of final states as possible and that the model independent features of 938 MeV energy release and final state momentum balance be utilized to the maximum extent possible.

B. THE BACKGROUND

The known backgrounds in nucleon decay experiments are produced by cosmic ray muons (either directly or as a result of cascades produced through their electromagnetic or nuclear interactions) and by cosmic ray neutrinos. Various unknown backgrounds, such as "Kolar events", monopoles, the decay of cosmological relic particles, liberated quarks, etc. are also possible.

The most reliable technique for the suppression of the muon-produced backgrounds is to perform the experiment at as great a depth as possible so as to utilize the rock overburden as a massive muon absorber. Because the remnant cosmic ray muon flux is a rapidly

decreasing function of depth, factors of two or three in depth
typically yield two or more orders of magnitude reduction in the
muon flux and hence in all muon-associated backgrounds.

The neutrino backgrounds are independent of depth and must be
dealt with by detector design alone. In view of the wide variety of
possible nucleon decay final states, the best strategy is to
maximize the information-gathering ability of the detector, i.e.
energy resolution, spatial resolution, time resolution, etc.

C. EXPERIMENTAL GOALS AND METHODS

The prime goal of any nucleon decay experiment at this time
must be the clear demonstration of the existence of the nucleon
decay process. Once the existence of nucleon decay has been firmly
established, it will then be desirable to determine branching ratios,
the nucleon total lifetime, the number, charge state and polarization
of the final state leptons, the nature of the hadronic state, etc.

The design of a nucleon decay experiment involves the optimization of many conflicting requirements. The fundamental requirement
of very large mass imposes financial and physical constraints which
limit the choice of detector. In addition, mine/tunnel safety considerations reduce the desirability of many materials, such as
cryogenic liquids and low flash point solvents. Further compromises
result from the desire to contain nucleon decay events in a finite
sized detector (best with high Z materials) as well as to minimize
the effects of nuclear binding and to provide a significant sample
of free nucleon decays (best with low Z materials).

In view of the complex nature of the nucleon decay problem, we
believe that a variety of experiments, each optimized in particular
ways, will provide the most reliable information on nucleon decay
for lifetimes up to 10^{33} years. If it is necessary to search for
nucleon decay beyond this range, it will probably be appropriate to
concentrate world-wide resources on one or two enormous, highly
sophisticated experiments. However, the results and experience
obtained through performing a reasonable variety of relatively small
experiments will be essential for the proper design of the ultimate
nucleon decay experiment.

II. DESCRIPTION OF THE HOMESTAKE TRACKING SPECTROMETER

A. GENERAL PROPERTIES OF LIQUID SCINTILLATOR

The fundamental property of liquid scintillator that makes it desirable as a nucleon decay detection medium is its ability to convert the kinetic energy of charged particles to visible photons. The scintillation response is linearly related to the energy deposit for all but heavily ionizing particles. Even for such particles, the empirically well-established Birks' law[6] can be used to linearize the scintillator output.

Since the energy deposit needed to produce each optical photon is about 200 eV, a nucleon decay event in liquid scintillator would yield about 5×10^6 photons. With an easily achievable 2% photon collection efficiency and 10% conversion efficiency to photoelectrons, about 10^4 photoelectrons would be detectable, resulting in a theoretical energy resolution of $\pm$ 1%.

Since liquid scintillator responds linearly to energy deposited, a nucleon decay detector utilizing this material will be useful for all possible final state particles (other than neutrinos), thereby permitting a relatively decay-mode independent experiment. The fundamental model independent nucleon decay signature in such a detector would be the release of optical photons equivalent to 938 MeV.

In addition to its ability to determine total energy, liquid scintillator has the property of producing useful optical photons on an extremely rapid time scale. Excellent time resolution (typically $\pm$ 1 ns) is therefore obtained. With this quality of time information, the time-of-flight technique will allow determination of the sense of motion of final state particles and therefore verification of the second model independent nucleon decay feature—vanishing of the sum of the final state momenta.

B. DESIGN FEATURES OF THE HOMESTAKE TRACKING SPECTROMETER

The Homestake Tracking Spectrometer (see Fig. 1) consists of 1406 tons of liquid scintillator contained in 1872 modules each 8 m long, with a 30 cm x 30 cm cross section, and viewed by two 5-inch hemispherical photomultiplier tubes. The PVC counter modules, each of which is lined with an aluminized polyester reflector to provide good light collection, are arranged in stacks of crossed X-Y arrays to provide detailed tracking information for the nucleon decay secondaries. The detector will be housed in a new room on the 4850-foot level of the Homestake mine, adjacent to the existing Penn-Brookhaven water Cerenkov detector-solar neutrino room.

The design parameters of this instrument are summarized in Table I.

III. EXPECTED PERFORMANCE OF THE HOMESTAKE TRACKING SPECTROMETER

The performance of this detector has been carefully evaluated using computer simulation techniques and laboratory tests on a 9-ton prototype stack and on full 8-meter prototype detector elements.

A. ENERGY RESOLUTION

An efficient liquid scintillator having extraordinarily high stability, transparency, and flash point, together with low cost and toxicity has been specially developed for use in the Tracking Spectrometer. Our mineral oil-based scintillator has a light output equal to that of the best commercially available scintillators but provides a superior attenuation length of 7 meters and a transparency lifetime estimated at 30 years.

A typical pulse height spectrum for cosmic ray muons passing through an 8-meter module is shown in Fig. 2. The observed peak corresponds to 54 MeV of ionization energy loss and has a width of $\pm$ 6.8%. Since nucleon decay will release 938 MeV, we expect a 17-fold increase in the total number of detected photoelectrons, or a factor of $\sqrt{17}$ improvement in the overall energy resolution, i.e. a final value of 1.6%. With nonlinearity of the scintillation process, Fermi momentum smearing, and incomplete containment of final state particles properly taken into account in our computer simulations, we expect an energy resolution of $\pm$ 4.2% to be realistically possible for nucleon decay.

Fig. 3 presents the observed pulse height spectrum for a light-emitting diode. The $\pm$ 4.9% width of the peak provides the basis for our estimate of 8800 collected photoelectrons per 938 MeV of energy deposited in the scintillator.

B. TIME AND POSITION RESOLUTION

In Fig. 4 we present results from measurements of cosmic ray muons in a 46-module, 9-ton prototype stack located in a surface laboratory at the University of Pennsylvania. The modules for this stack are 12-foot-long, 1-foot-diameter liquid-scintillator-filled PVC counters, otherwise identical to the proposed Tracking Spectrometer modules. We have triggered one of the twelve-foot modules using pairs of crossed elements located respectively at 45 cm, 180 cm, and 315 cm from one end. For each trigger, we have plotted the pulse height difference vs the time difference for the two phototubes at the ends of the 12-foot module. The X- and Y-projections of the scatter plot are also shown in the figure. From the observed widths of the pulse height and time peaks, we determine the position resolution to be $\pm$ 12 cm and $\pm$ 15 cm, respectively. We therefore have two independent methods for localizing the energy deposited and hence the particle trajectory in each module. Certain events, such as those in which a single module has two independent hits (which may be ambiguous using phototube information from that module alone) will still be interpretable by introducing position information produced by the crossed modules in adjacent layers.

If we use the average time of firing of the two phototubes in each module, we can establish the time of flight of each particle such that localization of the nucleon decay vertex will be possible to an accuracy of $\pm$ 15 cm. Unambiguous determination of the sign of the momentum will likewise be possible for particle trajectories penetrating at least two modules.

C. EVENT SIMULATION USING THE MONTE CARLO METHOD

Computer simulations of the detector response to various nucleon decay final states have been performed. The Monte Carlo program includes a full simulation of the propagation of electromagnetic showers[7], muons, pions, and kaons.

In Fig. 5 we show a typical Monte Carlo simulation of a $p \rightarrow e^+\pi^0$ nucleon decay event as seen in the Tracking Spectrometer. The total visible energy for this event was 913 MeV. In the "energy" view, the location of each number indicates the position of a triggered module, while the number itself indicates the observed energy deposited in that module in units of 10 MeV. The nucleon decay event took place above the point labelled "ND", while the vector from "ND" to "PZ" marks the initial direction of the π^0. Similarly, the vector from "ND" to "E+" indicates the initial direction of the positron. In the "time" view, the time of firing of each module (in ns) is indicated. From the information in this view, the presence of a decay rather than an interaction vertex at the point "ND" is obvious.

The results of 400 simulations for two typical nucleon decay modes are presented in Figs. 6 and 7. Events in which more than 50 MeV appears in the outermost 30 cm of the detector (the "anti" layer) have been excluded. For neutrino emitting modes, neutrino-escape peaks are observed at energies less than 938 MeV. Decay channels with charged lepton secondaries, such as $p \rightarrow e^+\omega$ (Fig. 6) and $n \rightarrow e^+\pi^-$ (Fig. 7), exhibit relatively sharp peaks at 938 MeV, as expected.

The results of the Monte Carlo calculations of the detection efficiencies for various nucleon decay modes are presented in Table II. The table lists average event lengths and detection efficiencies for 80% and for 90% energy containment. To obtain the desired energy resolution of $\pm$ 4.2%, our Monte Carlo calculations indicate that 90% containment is sufficient. The detection efficiencies vary from 35% to 77%, with a mean of 60%. Within a factor of two, therefore, the Tracking Spectrometer constitutes a decay-mode-independent nucleon decay detector. The total fiducial mass is approximately 1400 tons x 60% or 840 tons. We note, parenthetically, that an ideal cubic detector of the same mass would have an efficiency of about 65%.

D. EVALUATION OF BACKGROUNDS

At the 4200 mwe depth of the Homestake laboratory, the

integrated flux of cosmic ray muons over the full spectrometer area is 400,000 muons per year. With an expected 99% rejection of entering muons by the top anticoincidence layer, about 4000 unvetoed muons per year will remain. The tracking (rejection factor >50), timing (>50), and energy resolution (>20) capability of our detector will permit us to recognize and reject these remaining throughgoing muons. Similar arguments convince us that stopping or interacting muons will also be easily rejected. The raw trigger rate for muons of about 1 per minute will allow us to record all muon information on magnetic tape for detailed offline analysis.

The remaining known source of background is cosmic ray neutrino interactions. Neutrinos interacting inside the detector can be shown to be more important than those interacting in the rock surrounding the detector. Because of the excellent energy resolution of our detector, only those neutrino interactions depositing 938 $\pm$ 50 MeV in the detector are potential background problems for nucleon decay. Except for neutral current interactions, the detector will measure the total neutrino energy in the signal generated by the interaction secondaries. The flux of muon neutrinos at 938 MeV is 3.7×10^{-3}/cm^2/sec/sr/100 MeV, while the electron neutrino flux in the same window is 1.7×10^{-3} /cm^2/sec/sr/100 MeV. Using the sum of these fluxes integrated over solid angle and the known neutrino interaction cross section of 0.5×10^{-38} cm^2 leads us to expect a total of 8 neutrino interactions per year under the nucleon mass peak.

We expect that a further factor of 10 suppression of this neutrino background can be made using simple event reconstruction techniques such as momentum-balance determination. The anticipated neutrino background is therefore less than .8 per year. Since atmospheric neutrinos are the dominant source of background in our detector (probably true only for detectors located deeper than 3500 mwe), the background limit of our experiment will be

$$\frac{6 \times 10^{29} \text{ nucleons/ton} \times 1400 \text{ tons}}{.8 \text{ background events/yr}} = 10^{33} \text{ yr.}$$

IV. SUMMARY

The unique features of our Tracking Spectrometer approach to the search for nucleon decay are the extremely good energy and time resolution and reasonable position resolution together with the low muon background environment provided by our mile-deep location.

The value of the $\pm$ 4.2% energy resolution is particularly evident in allowing reduction of the internal cosmic ray neutrino background from a raw rate of 280 events/year in the 0.2-5 GeV region[8] to only 8 per year under the 938 MeV nucleon mass peak. Without good energy resolution, we believe the same factor of 35 reduction could be obtained only through bubble-chamber-quality spatial information.

The ability to distinguish decay from interaction vertices provided by our $\pm$ 1.3 ns time resolution is an additional strong guarantee that true nucleon decay events will be distinguishable from background.

Assuming that one event per year is the lowest event rate at which nucleon decay can be observed, the Tracking Spectrometer will have a sensitivity corresponding to a nucleon total lifetime of

$$\frac{6 \times 10^{29} \times 1400 \text{ tons} \times .6 \text{ (mean detection efficiency)}}{1 \text{ event/yr}}$$

$$= 5 \times 10^{32} \text{ yr.}$$

V. EXPERIMENTAL STATUS AND PLANS

We will be installing a 75-ton, 100-module detector section in our existing laboratory in the Homestake mine during the summer of 1982. Completion of the 1400-ton, 3-stack detector is presently awaiting approval of full funding. With expeditious approval, the complete detector would be operational by the spring of 1984. If the initial physics results with this and other nucleon decay experiments indicate that a larger detector is desirable, the great stability of the rock at Homestake would allow construction of

detectors with cross sections as large as 16 m x 16 m and virtually
unlimited lengths.

REFERENCES

1. S. Rudaz, Proc. 3rd Workshop on Grand Unification, Chapel Hill,
   April 1982.

2. P. Langacker, Physics Reports 72, 4, 185 (1981).

3. M.B. Wise, R. Blankenbecler, and L.F. Abbott, "Three-Body Decays
   of the Proton", preprint SLAC-PUB-2614, Sept. 1980.

4. C.B. Dover, M. Goldhaber, T.L. Trueman, and L.C. Wang,
   preprint BNL 29423 (1980).

5. J.C. Pati and A. Salam, Phys. Rev. Lett. 31, 661 (1973).

6. J.B. Birks, "The Theory and Practice of Scintillation Counting",
   Macmillan, New York (1964), p. 187.

7. T. Stanev and Ch. Vankov, Computer Phys. Comm. 16, 363 (1979).

8. J. Bartelt et al., "Proposal to Build Soudan 2 ... ",
   ANL-HEP-PR-81-12, Sept. 1981, p. 89.

TABLES

I.   Design parameters of the Homestake Tracking Spectrometer.

II. Results of Monte Carlo calculation of detection efficiency
     for various nucleon decay modes. Requiring 90% energy con-
     tainment, a mean detection efficiency of 60% for our 8 m x
     8 m x 24 m detector is obtained.

FIGURE CAPTIONS

1. The proposed 5000 m^3 Homestake experimental room showing the
   Tracking Spectrometer and the Solar Neutrino area. It is
   planned to build the new room adjacent to the existing 2000 m^3
   room at a depth of 4200 mwe.

2. Typical pulse height spectrum for cosmic ray muons passing
   through an 8-meter test module. The muon energy loss is 54
   MeV, while the observed energy resolution is ± 6.8%. The
   observed shape is well represented by the appropriate Landau
   distribution. The resolution observed from a light-emitting
   diode adjusted to produce pulses equal to those of muons at
   this distance is ± 2.9%.

3. Pulse height spectrum for a light-emitting diode located 2.7 m
   from one end of an 8-meter test module. The amplitude of the

light pulses corresponds to 54 MeV. From the observed standard deviation of 4.9%, we calculate that the signal corresponds to 415 photoelectrons. This result implies that, for 938 MeV (the energy deposit expected for most nucleon decay events), we will observe a total of 8800 photoelectrons.

4. Position resolution results obtained from measurements of cosmic ray muons in a 46-module, 9-ton prototype stack. For muons passing at three different locations through a test module, the scatter plot shows the pulse height differences vs the time differences for the two phototubes at the ends of the module. From the observed widths of the projected pulse height and time peaks, the position resolution is determined to be $\pm$ 12 cm and $\pm$ 15 cm, respectively. Two independent methods for localizing the energy deposited and hence the particle trajectory in each module are therefore available.

5. Monte Carlo simulation of a typical $e^+\pi^0$ nucleon decay event as seen in the Tracking Spectrometer. In the "energy" view, the location of each number indicates the position of a triggered module, while the number itself indicates the observed energy deposit in that module in units of 10 MeV. The nucleon decay event took place above the point labelled "ND", while the vector from "ND" to "PZ" marks the initial direction of the π^0. Similarly, the vector from "ND" to "E+" indicates the initial direction of the positron. In the "time" view, the time of firing of each module (in ns) is indicated. From this information, the presence of a decay rather than an interaction vertex at the point "ND" is obvious.

6. Monte Carlo calculation of pulse height spectrum produced by the decay mode $p \to e^+\omega$. Only events in which less than 50 MeV was deposited in the outermost 30 cm of the detector (the "anti" layer) have been plotted. The calculation includes the effects of imperfect energy containment, Fermi momentum smearing, and liquid scintillator nonlinearity and finite resolution. When a muon decay (produced following decay of a π^+) was detected, (detection efficiency = 98%), we corrected for loss of rest mass from the prompt energy release by adding 106 MeV to the observed pulse height.

7. Monte Carlo calculation of pulse height spectrum produced by the decay mode $n \to e^+\pi^-$. The remarks for Fig. 6 apply except that, because the π^- undergoes nuclear capture in liquid scintillator, the full 938 MeV energy release is visible in the prompt pulse.

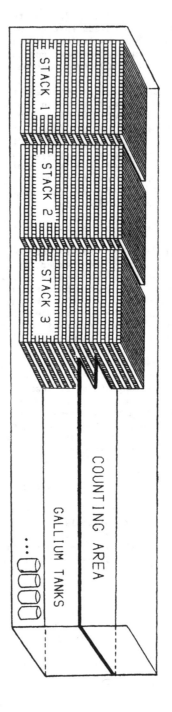

NEW HOMESTAKE EXPERIMENTAL AREA

TRACKING SPECTROMETER

SOLAR NEUTRINO AREA

STACK 1

STACK 2

STACK 3

COUNTING AREA

GALLIUM TANKS

Figure 1

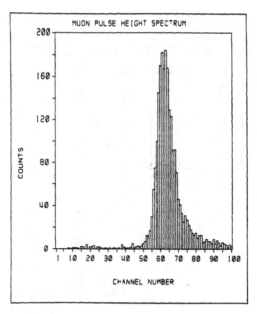

Figure 2

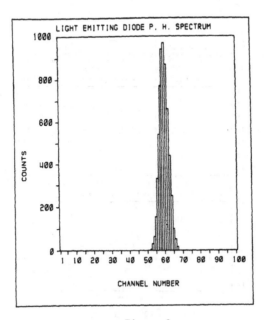

Figure 3

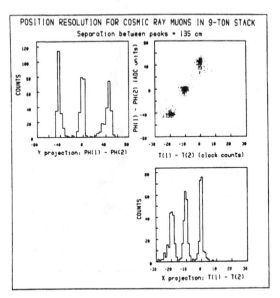

Figure 4

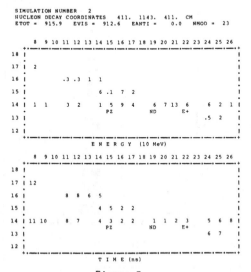

Figure 5

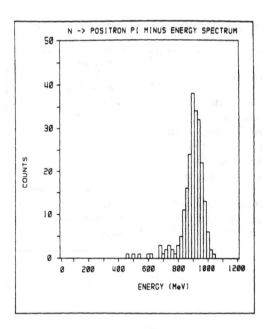

Figure 6

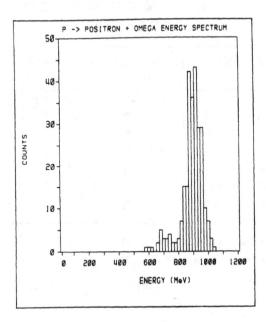

Figure 7

PARAMETERS OF THE HOMESTAKE TRACKING SPECTROMETER

DEPTH:	4200 meters w.e.
RESIDUAL COSMIC RAY MUON FLUX:	1100/m**2/year
OVERALL SIZE:	8 m X 8 m X 24 m
TOTAL MASS:	1406 Tons
PHOTOELECTRONS per 938 MeV:	8800
ENERGY RESOLUTION:	+/- 4.2%
TIME RESOLUTION:	+/- 1.3 ns
SPATIAL RESOLUTION:	+/- 15 cm
	or 0.32 rad. lengths
DECAY MODE SENSITIVITY:	~ ALL
LIFETIME SENSITIVITY:	5 X 10**32 years
COST:	$4.2M

Table I

EVENT LENGTH (L) AND DETECTION EFFICIENCY (e)
FOR 1500 M**3 DETECTOR

DECAY MODE	PERCENTAGE OF ENERGY CONTAINED			
	80%		90%	
	L(m)	e(%)	L(m)	e(%)
P -> E+ PI 0	4.3	48	5.8	35
P -> E+ RHO 0	1.2	82	1.5	77
P -> NU PI +	0.9	88	1.5	77
N -> E+ PI -	3.0	59	4.0	51
N -> E+ RHO -	1.8	74	3.0	59
N -> NU PI 0	2.4	65	3.0	59

e (mean) (8m X 8m X 24m)   69%   60%

e (mean) (11.4m)**3   73%   65%

Table II

BROWN-HET-478

TA-357

COMPLEX ANOMALY-FREE REPRESENTATIONS FOR GRAND UNIFICATION[†]

Kyungsik Kang[*]

Department of Physics

Brown University

Providence, Rhode Island 02912

ABSTRACT

Complete list of the complex representations of Lie groups that satisfy the anomaly freedom and asymptotic freedom has recently been given by E. Eichten, I.-G. Koh and myself.[1] We first give a brief summary of the results of Ref. 1 and then discuss the application of such representations to the composite models based on another recent work by I.-G. Koh, A. N. Schellekens and myself.[2] The role of Fermi-statistics and the method of decompositions of tensor products with definite permutation property are explained in two examples.

[†] Presented at the Third Workshop on Grand Unification, University of North Carolina, Chapel Hill, N.C. 15 - 17 April 1982.

[*] Supported in part through funds provided by the U.S. Department of Energy under Contract DE-AC02-76ERO3130.A009 - Task A.

I. INTRODUCTION

I would like to report on the complete list[1] of representations of all simple Lie groups, which are complex, anomaly-free and asymptotically free. Some of these representations have been applied[2] to the "preon" dynamics in which quarks and leptons are regarded as composites of more fundamental building blocks called preons. I will describe how the wave functions of the composites are constructed out of preons in such a way to satisfy correct Fermi statistics for identical preons. Also the reduction of tensor products with definite permutation property[3] plays a crucial role in determining the symmetry factors of the bound states.

The representations that are used in a grand unified model should satisfy a number of properties. Renormalizability of gauge theories necessitates the use of anomaly-free representations. The anomaly-free condition in SU(5) grand unification theory (GUT)[4] is known to restrict the SU(3) content of fermions to be real. This will make quarks massive in general. The role of anomaly-freedom in preon dynamics[5] is to make it easy to formulate the preon theories through anomaly matching between the confined preons and their composites.

In view of small masses for ordinary quarks and leptons, we expect the usefulness of pseudoreal representations in GUT to be minimal. Existence of complex representations[6] is one criterion that one would require in GUT. In preon dynamics, complex representations are also necessary to get massless bound states. One may picture that quarks and leptons are kept massless by an unbroken chiral symmetry and can get a small mass eventually by an explicit small perturbation. A theory that has no spontaneous chiral symmetry breaking is achieved by excluding any Higgs scalars of elementary type. Indeed, elementary Higgs scalar theories are argued[5] to be "unnatural" because of the lack of asymptotic freedom and other undesirable properties. With complex fermion representations, the gauge invariant bare mass term is not possible to write down.

Within the context of GUT, one usually requires additional constraints such as the representation contains 1, 3 and 3* only of the SU(3) color group;[7] the representation accommodates the desired number of quark-lepton generations; the representation is asymptotically free.[8] In composite models, a new gauge interaction is introduced which becomes strong at a higher energy scale. These kinds of interactions usually have a simple compact group structure. Since quarks and leptons are the singlet states of this new interaction, the preon representations need not be antisymmetric under the new group. On the other hand, it seems reasonable to require that the new interaction is of a confining type just like the color interaction to confine the quarks. Confining property can usually be found in the theory that has asymptotic freedom. The condition that the representation is asymptotically free with respect to the full gauge group was sometimes required[8] in ordinary GUT shceme particularly when the same irreducible representation was allowed to repeat[9] in the anomaly-free combinations. In dynamical models of composite fermions however, preons may combine with themselves so that it is almost necessary to allow repetition of the same irreducible representations. The multiplicity of an irreducible representation will be constrained by the anomaly-free condition and asymptotic freedom.

Thus we may say that the fermion representations either in an ordinary GUT or in preon schemes must satisfy at least the following three conditions: (1) the representation is complex, (2) the representation is free of the triangle anomalies, and (3) the representation is asymptotically free. Complete solutions to these conditions in simple Lie algebra have recently been given in Ref. 1.

2. COMPLEX IRREDUCIBLE REPRESENTATIONS.

Complex representations[10] can be found in gauge groups SU(N) with $N \geq 3$, SO(4N+2), and E_6, of which the latter two groups have anomaly-free representations. Spinor representations of SO(4N+2) = D_{2N+1} with dimension 2^{2N} and the fundamental representation 27 of E_6 are complex. The condition of asymptotic freedom gives the group theoretical constraint[11]

$$\sum_{R_i} T_2(R_i) \leq \frac{11}{2} C_2(G) \tag{1}$$

from the renormalization group equation. In Eq. (1), R_i refers to the irreducible fermion representation, $C_2(G)$ is the quadratic Casimir operator of the adjoint representation G, and $T_2(R_i)$ is defined by

$$T_2(R_i) \dim(G) = C_2(R_i) \dim(R_i) \tag{2}$$

Among simple groups, the only complex irreducible representations that satisfy both anomaly freedom and asymptotic freedom are the following: 16-, 126-, 144-dimensional representations of SO(10); the lowest dimensional spinor representations of SO(14) with dimension 64 and of SO(18) with dimension 256; and 27-dimensional representation of E_6. The representations can be repeated up to 22, 1, 1, 8, 2, and 22 times respectively before exceeding the asymptotic freedom limit. There are no complex irreducible representations in SU(N) which are both anomaly-free and asymptotically free. For SU(N), we may either relax the condition of asymptotic freedom or choose anomaly-free combinations of several complex irreducible representations that satisfy asymptotic freedom.

Following the Cartan labels for the highest weight of an irreducible SU(N) representation, $(a_1, a_2, \cdots a_{N-1})$, where a_i equals the number of columns of the Young tableau with i boxes, a complex SU(N) representation satisfies $(a_1, a_2, \cdots a_{N-1}) \neq (a_{N-1}, \cdots a_2, a_1)$. It is obvious that there are no complex representations in SU(2). It may be interesting to obtain a single irreducible representation of SU(N) that is complex and anomaly-free, even though it may not be asymptotically free. Such a representation would be even more interesting if it can accommodate three generations of quarks and leptons. For this reason, we have made a thorough search[12] for the range of dimensions up to $D = 4 \times 10^9$ for SU(N) with N $\leq$ 16. Altogether we found 28 such representations, of which two lowest dimensional ones occur in SU(6) and SU(5) with $D = 374, 556$ and $D = 1, 357, 827$ respectively. These have the highest weights (0, 5, 0, 0, 4) and (0, 7, 3, 3). We have studied the branching rules of the smallest representation (0, 5, 0, 0, 4) of SU(6) by the method of elementary multiplets[13] for SU(6) $\rightarrow$ SU(5) $\times$ $U^a(1)$ followed by SU(5) $\rightarrow$ SU(3) $\times$ SU(2) $\times$ $U^b(1)$. An exhaustive search was made for possible values of A and B to give the correct charges for the 15 chiral states in one generation from $Q = T_3 + AY^a + BY^b$ where Y^a and Y^b are the hypercharges of $U^a(1)$ and $U^b(1)$ respectively. Though nine different choices of A and B can be made for correct charge assignment, it turned out that the irreducible representation (0, 5, 0, 0, 4) contains only one ordinary generation along with huge number of exotic states.

3. COMPLEX ANOMALY-FREE REDUCIBLE REPRESENTATIONS IN SU(N)

All of the anomaly-free complex irreducible representations have enormous dimensions. In physically interesting theories one can use Eq. (1) to seek out those SU(N) irreducible representations which are asymptotically free and consider the anomaly-free combinations of such representations. There are nine irreducible and complex representations of SU(N) that satisfy the asymptotic freedom. They are given in Table 1.

Table 1.

Young Diagram	Dimension	T_2	C_2	A
$[1]$	N	$\frac{1}{2}$	$\frac{N^2-1}{2N}$	1
$[1^2]$	$\frac{N(N-1)}{2}$	$\frac{N-2}{2}$	$\frac{(N+1)(N-2)}{N}$	$N-4$
$[2]$	$\frac{N(N+1)}{2}$	$\frac{N+2}{2}$	$\frac{(N-1)(N+2)}{N}$	$N+4$
$[2,1]$	$\frac{1}{3}N(N^2-1)$	$\frac{N^2-3}{2}$	$\frac{3}{2}(\frac{N^2-3}{N})$	(N^2-9)
$[2,1^{N-3}]$	$\frac{1}{2}N(N+1)(N-2)$	$\frac{(N-2)(3N+1)}{4}$	$\frac{(3N+1)(N-1)}{2N}$	$\frac{(-N^2+7N+2)}{2}$
$[3,1^{N-2}]$	$\frac{1}{2}N(N-1)(N+2)$	$\frac{(N+2)(3N-1)}{4}$	$\frac{(3N-1)(N+1)}{2N}$	$\frac{(N^2+7N-2)}{2}$
$[1^3]$	$\frac{1}{6}N(N-1)(N-2)$	$\frac{(N-2)(N-3)}{4}$	$\frac{3(N-3)(N+1)}{2N}$	$\frac{(N-3)(N-6)}{2}$
$[2^2]$	$\frac{1}{12}N^2(N^2-1)$	$\frac{N(N^2-4)}{6}$	$\frac{2}{N}(N^2-4)$	$\frac{N(N^2-16)}{3}$
$[1^4]$	$\frac{N(N-1)(N-2)(N-3)}{24}$	$\frac{(N-2)(N-3)(N-4)}{12}$	$\frac{2}{N}(N^2-N-4)$	$\frac{(N-8)(N-3)(N-4)}{6}$

Here we denoted a typical Young diagram having a boxes in each of the first n rows followed by b boxes in each of the next m rows and so on by $[a^n,b^m,\cdots]$. Also given are the dimension D, the second index T_2, the anomaly A and the value of the quadratic Casimir operator C_2. Except for the representations $[1]$, $[1^2]$ and $[2]$, they can be repeated a number of times without losing asymptotic freedom. The maximum allowed values of N from Eq. (1) are 0,0,0,11,9,5,26,6, and 12 respectively.

Relaxing the condition of irreducibility, we can then construct reducible complex representations from anomaly-free combinations of those nine irreducible and asymptotically free representations. Complete list is given in Ref. 1 in two different group of tables.[14] The first group contains the combinations consisting of irreducible representations whose tensor rank is no larger than 2,

$$\bar{n}[1]^* \oplus m[1^2] \oplus \ell[2] \tag{3}$$

where $\bar{n}$, m and ℓ are integers whose magnitudes are bounded by the asymptotic freedom;

$$|\bar{n}| + |m| \ (N-2) + |\ell| \ (N+2) \leq 11N \tag{4}$$

Negative values of the indices $\bar{n}$, m and ℓ are to be interpreted as the appearance of the associated complex conjugate representations with the given multiplicity. The result is further summarized for arbitrarily large value of N and for finite range of N. The second group contains those anomaly-free combinations involving at least one irreducible and asymptotically free representation whose tensor has the rank larger than 2. It turns out however that such solutions exist only for $N \leq 17$. In other words, all reducible complex representations of SU(N) which are anomaly-free and asymptotically free for $N \geq 18$ are only of the type of Eq. (3).

We have given in Ref. 1 complete list of complex reducible as well as irreducible representations of Lie algebras which are anomaly-free and asymptotically free. Complex irreducible representations of SU(N) can indeed be anomaly-free though they are not asymptotically free. In fact, there are no SU(N) complex irreducible representations that satisfy both anomaly- and asymptotic freedom. The solutions given in Ref. 1 are unique up to a pseudoreal term of the form

$$\sum_{i=1}^{9} m_i (R_i \oplus R_i^{*}) \oplus \sum_j n_j r_j \tag{5}$$

where R_i denotes the nine complex representations defined in Table 1 and j runs over all pseudoreal irreducible representations. The multiplicity indices m_i and n_j are non-negative and constrained by the condition for the general solution to be asymptotically free. However, as we have emphasized it before, the role of pseudoreal representations is expected to be minimal for grand unified theories in view of the survival hypothesis.[6]

4. APPLICATION TO PREON DYNAMICS

We have constructed[2] all composite models based on complex, anomaly-free and asymptotically free representations of the gauge groups SU(3) to SU(8), SO(4N+2) and E_6, with no more than two different preons. There are several reasons to postulate that a hidden substructure exists and is responsible for quarks and leptons as the composites. For example, quarks and leptons are appearing in families of several generations; they have a rather remarkable mass-spectrum; there seems to be a natural mass scale of about 100 GeV where spontaneous symmetry breaking occurs and yet elementary Higgs scalars are undesirable as dis-

cussed before. Absence of Higgs scalars will ensure the quarks and leptons to be kept massless because of the unbroken chiral symmetry. It
has been pointed out that[5] such spontaneously unbroken symmetry must
be reflected in the composites and the anomalies of the fundamental
preons have to match with those of the composites. An interesting
aspect of the anomaly matching is that the composites are to appear
with definite multiplicities. If the anomalies can not be matched with
a reasonable set of bound states, one of the two symmetries, either the
confining "metacolor" symmetry or the "metaflavor" chiral symmetry, has
to break. When the confining symmetry breaks one obtains a tumbling
gauge theory.[15] We will assume, supported by arguments given by
others,[16] that this is not what will happen, but that the chiral symmetry will break. In models with a (pseudo) real metacolor representation, the chiral symmetry is likely to break down to a real symmetry.[17]
In such models there are no anomalies left after symmetry breaking and
therefore massless bound states are not required to exist. On the other
hand, if the metacolor representation is complex, the bilinear conden-
sate to break the symmetry does not exist, though it is possible to form
multifermion condensates. These condensates may break the chiral symmetry group to a complex subgroup and their anomalies can be matched by
a set of massless composites. In this case, the original anomaly conditions are met by a combination of composite fermions and composite
Goldstone bosons.[5],[18] The anomaly equations can then be used to determine which complex subgroups are allowed as a result of symmetry
breaking. The advantage of this kind of model is that the symmetry
breaking pattern is not arbitrary but is restricted to a finite and
usually small set of possibilities.

In Ref. 2, we have studied the consequences of the anomaly matching for the unbroken subgroups for the simplest models with complex
metacolor representations of $SU(3)$ - $SU(8)$, $SO(4N+2)$ and E_6, and a few
other models with real representations such as $1 + \bar{1}$, $2 + \bar{2}$ for $SU(3)$,
$SU(5)$ and $SU(7)$. Generally, one has to make further assumptions on the
allowed bound states. The decoupling condition[19],[5] can not be formu-
lated in models with complex metacolor representations. Also one should
note that the bound state dynamics can not be independent of the flavor
group because of asymptotic freedom and because of the correct Fermi
statistics for the composite wave functions. It is needed to assume
that the massless bound state spectrum does not contain radial or
orbital massless excitations, the states containing valence metagluons[20]

and exotic states. Also the number of valence preons in a composite is limited to d if the metacolor group is SU(d).

The massless bound states must have totally antisymmetric wave functions for the identical preons, must be metacolor singlets and must be left-handed spin $\frac{1}{2}$ states. Since the spatial wave functions are symmetric by assumption, the complete symmetry of the preon in a model with p-different preons under the confining metacolor group G_M is given by

$$G_M \times SU(n_1) \times \cdots \times SU(n_p) \times [U(1)]^{p-1} \times SU_L(2) \times SU_R(2) \qquad (6)$$

where n_i is the multiplicity of the ith preon representation and the p-1 U(1) corresponds to the axial preon number left unbroken by G_M-instantons.[21] U(1)-factors are chosen to be metacolor nonanomalous. The SU(2) factors denote the left-and right-handed states of the Lorentz group for particles and antiparticles. The ith preon transforms according to the representations

$$(\underline{d}_i, \ 0, \cdots, 0, \ \underline{n}_i, \ 0, \cdots 0, \ Q_i^1, \cdots Q_i^{p-1}, \ 2, \ 0) \qquad (7)$$

of the symmetry group (6). Here d_i is the dimension of the metacolor representation $\underline{r}_i$ of the preon under G_M and 0 denotes the trivial representation.

In order to construct totally antisymmetric wave functions, it is convenient to use the antisymmetric representations of $SU(N_i)$ for each preon in the bound state where $N_i = d_i \times n_i \times 2$. The use of $SU(N_i)$ maintains the number of components of the preons while it makes the required symmetry of total wave function to be automatic from the permutation properties of the representations. Separate identification of the metacolor, flavor and spin representations of the bound states follows then from the branchings of $SU(N_i)$ representations to those of the subgroup $G_M \times SU(n_i) \times SU_L(2)$. This branching consists of three steps:

$$SU(N_i) \rightarrow SU(d_i n_i) \times SU(2) \qquad (8)$$

$$SU(d_i n_i) \rightarrow SU(d_i) \times SU(n_i) \qquad (9)$$

$$SU(d_i) \rightarrow G_M \qquad (10)$$

Branchings (8) and (9) are special cases of $SU(pq) \rightarrow SU(p) \times SU(q)$ which has been extensibly studied.[22] In particular, completely antisymmetric representation of $SU(N_i)$ corresponding to the m-box Young diagram reduces for the first step of branching (8) to

$$[1^m] = \sum_k Y_{m,k} \times Y^T_{m,k} \tag{11}$$

where $Y_{m,k}$ is any m-box Young diagrams.

The third step of branching (10) is of a different kind.[3] It corresponds to the embedding of a defining representation $\underset{\sim}{r}_i$ with dimension d_i of a group G_M in the fundamental representation of $SU(d_i)$. Also it corresponds to reduction of tensor products of the defining representation $\underset{\sim}{r}_i$ with definite symmetry, which is sometimes called as "plethysm".[23] The branching rules can be obtained from the index sum rules,[24] i.e., the sum rules for the dimension, the second index, the anomaly and the fourth index as well as the congruence numbers,[25] i.e., the generalized n-alities for $SU(n)$ to any Lie algebra. The interpretation of this branching as a symmetrized tensor product allows us to generalize the results to arbitrary rank. For example, all branching rules for Lie groups of the type A_n can be written in terms of Young diagrams without reference to a particular value of n. This implies that the index sum rules can be used for arbitrary n, which makes them much more powerful. Detailed procedure for the third step of branching is described in Ref. 3, along with tables of tensor products with definite permutation properties for seven of the nine irreducible complex and asymptotically free representations of $SU(N)$, for the lowest-dimensional spinor representations of $SO(10)$, $SO(14)$ and $SO(18)$ and for the 27-dimensional fundamental representation of E_6. For all of these representations, the index sum rules can determine the tensor product completely except for one source of ambiguity, the complex anomaly-free representations, since even-order indices are identical for $\underset{\sim}{r}_i$ and $\underset{\sim}{r}_i^*$ and the third order index is zero for anomaly-free representations. In practice, this ambiguity exists only in the group $SO(4n + 2)$ and E_6 since complex anomaly-free irreducible representations of $SU(N)$ occur with extremely large dimensions.[1] This ambiguity is resolved by the use of the congruence class.

Example (A): $(\underset{\sim}{5} + \underset{\sim}{10}^*)_L$ of $SU(5)$

Let us denote the preons $\underset{\sim}{5}_L$ and $\underset{\sim}{10}^*_L$ corresponding to the two complex and asymptotically-free irreducible representations of the metacolor group $SU(5)$ by α and β. They transform according to (7) in which $\underset{\sim}{d}_1 = \underset{\sim}{5}$ and $\underset{\sim}{d}_3 = \underset{\sim}{10}^*$. The anomaly-free combination $\alpha + \beta$ can be repeated up to 13 times without losing asymptotic freedom which must be satisfied by any confining theory. Metaflavor groups of $SU(n_1)$ and $SU(n_2)$ can be

interpreted as a result of such repetition. The bound states correspond
to the metacolor singlets that are made of arbitrary numbers of α, β,
$\bar{\alpha}$ and $\bar{\beta}$. We restrict the total number of these preons in a bound state
to be at most 5 to exclude appearance of exotic states

$$n_\alpha + n_\beta + n_{\bar{\alpha}} + n_{\bar{\beta}} \leq 5 \tag{12}$$

Since metacolor singlets are contained in the pentality-0 states, we require in addition

$$n_\alpha + 3n_\beta + 4n_{\bar{\alpha}} + 2n_{\bar{\beta}} = 0 \quad \text{(modulo 5)} \tag{13}$$

Then there are four candidates for the massless bound states

$$\alpha^5, \ \alpha\bar{\beta}^2, \ \alpha^2\beta, \ \beta^5 \tag{14}$$

The representation for the bound states are to be constructed from these
candidates by identifying left-handed spin $\frac{1}{2}$ states, and by imposing
total antisymmetricity under metacolor-metaflavor-spin transformation.
This is done through the three steps of branchings (8), (9) and (10).
For example, for the state $\alpha\bar{\beta}^2$ we need to identify $SU_R(2)$ singlet of $\bar{\beta}^2$
out of totally antisymmetric $[1^2]$ of $SU(20n)$. The first step (8) gives

$$[1^2] \to [1^2] \times [2] \oplus [2] \times [1^2] \tag{15}$$

where the first factor is the representation of $SU(10n)$ and the second
factor is that of $SU_R(2)$. Clearly the second term of (15) gives right-
handed spin zero so that the $SU(10n)$ representation is $[2]$. Under the
second step of branching $SU(10n) \to SU(10) \times SU(n)$, we get

$$[2] \to [2] \times [2] \oplus [1^2] \times [1^2] \tag{16}$$

The third step is to branch $SU(10) \to SU(5)$ and to see which of the two
$SU(10)$ representations $[2]$ and $[1^2]$ can combine with α to give metacolor
singlet state $\alpha\bar{\beta}^2$:

$$[2] \to [1^4] \oplus [2^2] \tag{17}$$

$$[1^2] \to [2,1^2] \tag{18}$$

This type of branchings can be found easily from the tables of plethysms
given in Ref. 3. It is then clear that only (17) can give $SU(5)$ singlet
for $\alpha\bar{\beta}^2$ so that the $SU(n)$ representation of $\bar{\beta}^2$ is uniquely determined to
be $[2]$. Then the bound state $\alpha\bar{\beta}^2$ transforms according to the representation

$$(0, \ [1], \ [2]^*, \ Q_0 - 2Q_2, \ [1], \ [1^2]) \tag{19}$$

under the symmetry group (6) with p = 2.

For the bound state β^5, there are five identical β's. We take then the totally antisymmetric representation $[1^5]$ of SU(20n) and consider its branching to SU(5) × SU(n) × $SU_L(2)$. The branching of $[1^5]$ under the first step SU(20n) → SU(10n) × SU(2) is

$$[1^5] \rightarrow [3,1^2] \times [3,1^2] \oplus [5] \times [1^5] \oplus [1^5] \times [5] \oplus [4,1] \times [2,1^3]$$
$$\oplus [2,1^3] \times [4,1] \oplus [3,2] \times [2^2,1] \oplus [2^2,1] \times [3,2] \qquad (20)$$

where the first factor in each term is the representation of SU(10n) and the second factor is that of SU(2). Since the bound state β^5 must have spin $\frac{1}{2}$, only the last term is permissible so that $[2^2,1]$ is the SU(10n) representation. Under the second step of branching SU(10n) → SU(10) × SU(n), we get

$$[2^2,1] \rightarrow \cdots \oplus [4,1] \times [3,1^2] \oplus [3,1^2] \times [4,1] \oplus [3,1^2] \times [2,1^3]$$
$$\oplus [3,1^2] \times [3,2] \oplus [3,1^2] \times [2^2,1] \oplus 2[3,1^2] \times [3,1^2] \oplus \cdots \qquad (21)$$

where again the first factor in each term is the SU(10) representation. The third step is to branch SU(10) → SU(5) and see which SU(10) representations contain SU(5) singlet. Of all SU(10) representations in (21), we see from the tables of plethysms given in Ref. 3, only the representation $[3,1^2]$ of SU(10) contains the metacolor singlet,

$$[3,1^2] \rightarrow \cdots \oplus [4,2,1^4] \oplus [3,2,1^5] \oplus [2^5] \oplus [2^3,1^4] \oplus [3,1^7] \qquad (22)$$

where $[2^5]$ is obviously the SU(5) singlet. In this way, we identify the metaflavor representations for the bound state β^5 to be:

$$\underline{R} = 2[3,1^2], [2,1^3], [3,2], [2^2,1], [4,1] \qquad (23)$$

Thus β^5 transforms according to the representation

$$(0, 0, \underline{R}, 5Q_2, [3,2], 0) \qquad (24)$$

under the symmetry group (6) with p = 2. Here $\underline{R}$ denotes the flavor states (23). Similarly, we can find the metaflavor representations of α^5 and $\alpha^2\beta$, which transform respectively under the symmetry group (6) with two preons as

$$(0, [3,2], 0, 5Q_1, [3,2], 0) \qquad (25)$$

and

$$(0, \underline{R}, [1], 2Q_1 + Q_2, [2,1], 0) \qquad (26)$$

where $\underline{R}$ is either $[1^2]$ or $[2]$. The U(1) charges Q_1 and Q_2 are determined from the anomaly-free condition for the $[SU(5)]^2 U(1)$ triangle and they are $Q_1 = 3$ and $Q_2 = -1$. Finally the complete use of anomaly equations determines the indices of the metaflavor representations.

Example (B): 27_L of E_6

Let us take the 27-dimensional $(0,0,0,0,1,0)$ representation of E_6 to be the preon α. The congruence number for the E_6 representation $(a_1, a_2, a_3, a_4, a_5, a_6)$ is defined by

$$C = a_1 - a_2 + a_4 - a_5 \qquad \text{(modulo 3)} \qquad (27)$$

so that 27 has $C = 2$. Then α^3 is an E_6 singlet and is a candidate massless bound state. The α representation can be repeated up to 22 times without losing asymptotic freedom and such repetition introduces a metaflavor group $SU(n)$. The metaflavor representation of the bound state α^3 can be determined uniquely again by imposing Fermi statistics. The first step is to consider the totally antisymmetric $[1^3]$ representation of $SU(54n)$ under the branching $SU(54n) \rightarrow SU(27n) \times SU(2)$:

$$[1^3] \rightarrow [2,1] \times [2,1] \oplus [3] \times [1^3] \oplus [1^3] \times [3] \qquad (28)$$

of which only the first term can give spin $\frac{1}{2}$. The second step is to branch the $SU(27n)$ representation $[2,1]$ into $SU(27) \times SU(n)$:

$$[2,1] \rightarrow [2,1] \times [2,1] \oplus [3] \times [2,1] \oplus [2,1] \times [3] \oplus [1^3] \times [2,1] \oplus [2,1] \times [1^3] \qquad (29)$$

Finally the third step is to see which of the $SU(27)$ representations contain the E_6 singlet under $SU(27) \rightarrow E_6$. Again from the tables of plethysms in Ref. 3, we see that the E_6 singlet is contained in the $SU(27)$ representation $[3]$:

$$[3] \rightarrow (0,0,0,0,3,0) \oplus (1,0,0,0,1,0) \oplus (0,0,0,0,0,0)$$

Thus the bound state α^3 transforms according to the representation

$$((0,0,0,0,0,0), [2,1], [2,1]) \qquad (30)$$

under the symmetry group $E_6 \times SU(n) \times SU_L(2)$.

In Ref. 2, we have given the list of all composite models based on complex, anomaly-free and asymptotic free representations of the confining metacolor groups $SU(N)$ with $3 \leq N \leq 8$, $SO(4N + 2)$ and E_6 as well as a few models with real representations. In all of these cases, the correct imposition of Fermi statistics is very crucial to determine the metaflavor representations of the composites uniquely.

REFERENCES AND FOOTNOTES

[1] E. Eichten, K. Kang, and I.-G. Koh, Brown-HET-464 and Fermilab-Pub-81/83-Thy, J. Math. Phys. (in press).

[2] A. N. Schellekens, K. Kang and I.-G. Koh, Brown-HET-471 and Fermilab-Pub-82/24-Thy, Phys. Rev. D (in press).

[3] A. N. Schellekens, I.-G. Koh, and K. Kang, Brown-HET-469 and Fermilab-Pub-82/23-Thy, J. Math. Phys. (to be published).

[4] H. Georgi and S. L. Glashow, Phys. Rev. Lett. $\underline{32}$, 438 (1974).

[5] G. 't Hooft in "Recent Developments in Gauge Theories," G. 't Hooft et al., eds. (Plenum Press, N.Y. 1980) p. 135.

[6] H. Georgi, Nucl. Phys. B$\underline{156}$, 126 (1979).

[7] M. Gell-Mann, P. Ramond, and R. Slansky, Rev. Mod. Phys. $\underline{50}$, 721 (1978).

[8] P. H. Frampton, Phys. Lett. $\underline{88}$B, 299 (1979).

[9] P. H. Frampton, Phys. Lett. $\underline{89}$B, 352 (1980).

[10] H. Georgi and S. L. Glashow, Phys. Rev. D$\underline{6}$, 429 (1972).

[11] In the special case of equality in Eq. (1), an additional constraint arising from higher orders must be satisfied for the theory to be asymptotically free. The solutions given in Ref. 1 correspond to the inequality in Eq. (1).

[12] Part of the anomaly-free irreducible representations was known to S. Okubo, Phys. Rev. D$\underline{16}$, 3528 (1977) and P. H. Cox.

[13] J. Patera and R. T. Sharp, J. Phys. A; Math. Gen. $\underline{13}$, 397 (1980); R. T. Sharp, J. Math. Phys. $\underline{13}$, 183 (1972); J. Patera and R. T. Sharp, Phys. Rev. D (in press).

[14] Partial lists of such representations have been obtained previously by M. T. Vaughn, J. Phys. G$\underline{5}$, 1317 (1979); M. Popovic, Phys. Rev. D$\underline{23}$, 1872 (1981); C. L. Ong, SLAC-PUB-2778; Y. Tosa and S. Okubo, Phys. Rev. D$\underline{23}$, 3058 (1981).

[15] S. Raby, S. Dimopoulos and L. Susskind, Nucl. Phys. B$\underline{169}$, 373 (1981).

[16] E. Eichten and F. Feinberg, Fermilab-Pub-81/62-Thy.

[17] M. E. Peskin, Nucl. Phys. B$\underline{175}$, 197 (1980). An exception may be models with a real fermion representation that can be screened by metagluons; see for example, R. Babieri, L. Maiani and R. Petronzio, Phys. Lett. $\underline{96}$B, 63 (1980).

[18] Y. Frishman, A. Schwimmer, T. Banks and S. Yankielowicz, Nucl. Phys. B$\underline{177}$, 157 (1981); T. Banks, S. Yankielowicz and A. Schwimmer, Phys. Lett. $\underline{96}$B, 67 (1980); C. H. Albright, Phys.

Rev. D24, 1969 (1981); C. H. Albright, B. Schrempp and F. Schremp, Fermilab-Pub-82/14-Thy. Additional references can be found in Ref. 2 and in I. Bars, Yale Preprint YTP 82-04.

[19] The use of decoupling condition in models with real representations has been criticized by J. Preskill and S. Weinberg, Phys. Rev. D24, 1059 (1981). See also I. Bars and S. Yankielowicz 101B, 159 (1981).

[20] Thus we avoid the exception of the kind discussed by R. Babieri et al., Ref. 17.

[21] G. 't Hooft, Phys. Rev. Lett. 37, 8 (1976); Phys. Rev. D14, 3432 (1976).

[22] C. Itzykson and M. Nauenberg, Rev. Mod. Phys. 38, 951 (1966).

[23] J. McKay, J. Patera and R. T. Sharp, J. Math. Phys. 22, 2770 (1982)

[24] J. Patera, R. T. Sharp, P. Winternitz, J. Math. Phys. 17, 1979 (1976); 18, 1519 (1977); W. McKay, J. Patera, R. T. Sharp, J. Math. Phys. 17, 1371 (1976); W. McKay and J. Patera, Tables of Dimensions, Indices, and Branching Rules for Representations of Simple Lie Algebras (Marcel Dekker, New York, 1981); S. Okubo, J. Math. Phys. 23, 8 (1982).

[25] F. W. Lemire and J. Patera, J. Math. Phys. 21, 2026 (1980); R. Slansky, Phys. Rep. C79, 1 (1981).

NUCLEON DECAY EXPERIMENT AT KOLAR GOLD FIELDS

M.R.Krishnaswamy,M.G.K.Menon,N.K.Mondal,V.S.Narasimham and B.V.Sreekantan
Tata Institute of Fundamental Research,Bombay 400 005 India
N.Ito,S.Kawakami and Y.Hayashi
Osaka City University,Osaka,Japan
S.Miyake
Institute for Cosmic Ray Research,University of Tokyo,Japn

The Kolar Gold Field (KGF) detector has now been in operation for more than a year and some of the preliminary results have already been reported in the previous GUT workshop at Ann Arber and published [1].

The Kolar Gold Mines are situated at a height of about 900 m above sea level at 12.9°N. The surface is reasonably flat with $\sim$ 20 m modulation over a mean horizontal level upto several kilometers. Fig.1 shows a brief sketch of underground locations which had been used in our earlier works; figures are depths in unit of hg/cm^2. The composition of Kolar rock appears to be very special ; the density and other characteristics were given in Fig.1 together with those of the standard rock. The present experiment is at 7000 hg/cm^2 in Kolar rock which corresponds to 7600 m.w.e. of the standard rock.

The results from the earlier observations have provided information about the variation of muon intensity with depth and neutrino induced phenomena deep underground. All the observed events are reasonably understandable except some anomalous events called as "Kolar events" though the rate of the event is very small. The Kolar events are suggestive of the existence of massive ($>$ 3 GeV) and long lived ($\sim 10^{-9}$ sec) particles that are seen to decay into 2 - 3 particles.

1. Detector

A detector specifically designed for the search of "bound" nucleon decay has been set up at a depth of 7000 hg/cm^2 and has been in operation since November 1980. The detector belongs to the class of fine

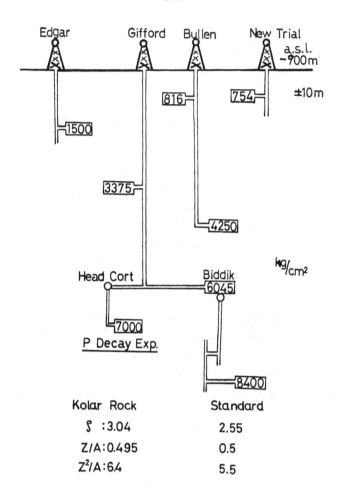

Edgar   Gifford   Bullen   New Trial

a.s.l.
~900 m

±10 m

816   754

1500

3375

4250

kg/cm²

Head Cort   Biddik
6045

7000

P Decay Exp.

8400

Kolar Rock	Standard
ρ : 3.04	2.55
Z/A : 0.495	0.5
Z^2/A : 6.4	5.5

Fig.1   A schematic diagram of Kolar Gold Mines.   The places were
used in our earlier experiments.   Comparison of nature of rocks also
shown.

grain calorimeters with sampling of ionisation under each radiation
length of iron.   As shown in Fig.2, 1600 proportional counters, made
up of iron pipes of square cross-section (10 cm x 10 cm) and wall thickness 2.3 mm are arranged in an orthogonal configuration with alternate
layers made up of 4 m and 6 m long counters to get two views of tracks.
The counters having a central tangusten wire of 100 μ m, are filled with
90 % argon and 10 % methane at about 1 Atm. pressure.   Each layer
of proportional counters has 1.2 cm thick iron absorber below it except

Structure

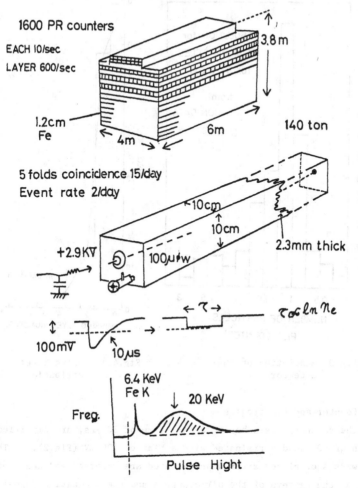

1600 PR counters

EACH 10/sec

LAYER 600/sec

3.8 m

1.2cm
Fe

4m

6m

140 ton

5 folds coincidence 15/day
Event rate 2/day

10cm

10cm

2.3mm thick

+2.9 KV

100μ w

$\tau \propto \ln n_e$

100mV

10μs

6.4 KeV
Fe K

20 KeV

Freg.

Pulse Hight

Fig.2   Schematic diagrams of the detector, proportional counter, pulse height to time width convertor, and pulse height distribution of the counter.

the second and 33rd layers which have 2.4 cm thick iron absorber and the bottom-most layer which has 1" lead below.   The construction of the detector was simplified due to the ruggedness of the counters. The counters and iron plates were piled up layer by layer without any external supports.   The detector measuring 6m long 4 m wide and 3.8 m high has 34 proportional counter layers in all and weighs approximately 140 tons.

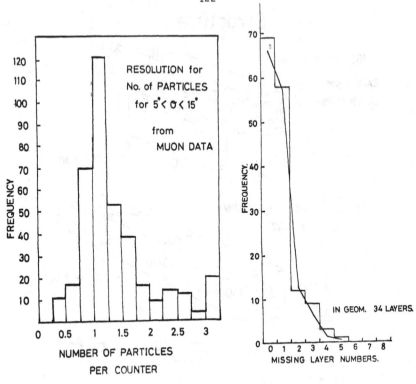

Fig.3   Resolution of the
        detector

Fig.4.   Missing layer
         distribution

2.   Counter and Electronics

The counter pulses having a fall time of 10 μs, are amplified by
a gain of 80 and discriminated with a bias of 100 mV (Fig.2).   The
time widths of pluses after discriminated are recorded and analysed in
terms of the pattern of the hit counters and the ionization deposited
in them in units of equivalent number of minimum ionizing particles.

For the calibration of the counters, it is possible to use a sharp
peak of Fe K-line of 6.4 keV from iron wall of the counter, excited by
the natural radioactive gamma-rays from the surrounding rock.   It is
about one third of single track ($\sim$ 20 keV) as shown in lower part of
Fig.2.   This caribration method is important at deep underground
where the cosmic ray muons are rather rare.

The triggering system is essentialy by 5-fold coincidence of more
than two layers of the counters, however, the top layer, the bottom
layer and the edge counters in the remaining 32 layers are excluded.

In Fig.3, the resolution of the detector for muons incident at small zenith angles is shown by conversion of ionization into equivalent number of minimum ionizing particles.   From the figure, the overall resolution in the energy measurements is estimated as about ∼30 %.

In Fig.4, the missing layer distribution has been shown for small zenith angle muons.   The layer efficiency, which takes into account the individual counter efficiency for different path lengths of tracks as well as dead spaces between counters, is obtained as 0.86 per 34 layers in average.   The detection efficiency of counters is estimated as 97.5 %.

For Proton decay events, the detection efficiency depends strongly on the decay schemes and their branching ratios.   Since it is difficult to estimate exact efficiency, we estimate an efficiency of ∼ 0.5 inclusive of the loss of hadrons due to absorption within the iron nucleus. (see Appendix I).

3.   Results

During the total running time of 387 days, we observed about 700 muons passing through the detector.   A zenith angular distribution of these muons are reasonably understandable as atmospheric origin and neutrino induced muons, similar to the previous report in 2nd WOGU. In this report, we note only the events relevant to nucleon decay phenomena.

The Neutrino Induced Background to Nucleon Decay Events

The neutrino collisions inside the detector can give rise to event configurations and visible energy which are to be expected for the decay of "bound" nucleons.   The basic inputs for these estimates are the cosmic ray neutrino flux and their interaction cross sections. Elastic interaction, N* production and inelastic interaction of both the charged and neutral current type have been considered for the neutrino energy region from 0.3 GeV to 100 GeV.

Table I.   Neutrino Interaction in Detector

Energy(GeV)	Expected No.	Observed No.
0.3 - 1.0	4	3
1.0 - 10	5	2
> 10	1	0

where the energy is a visible energy inside the detector.

Combining these small number of neutrino events with the restriction on the configuration of the secondary tracks (back to back alignment)

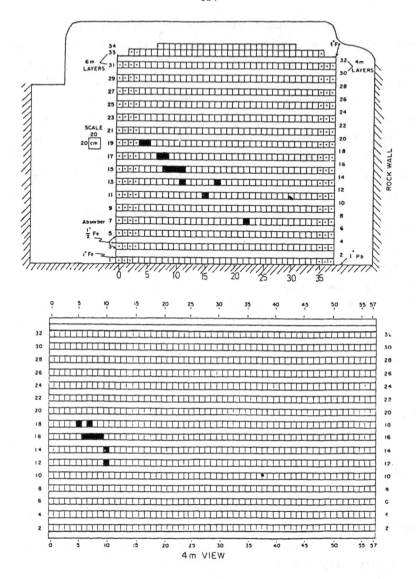

Fig.5   Proton decay candidate, Event No. 587

for nucleon decay, a neutrino induced background seems to be negligible
at present.

A part of elastic interaction and some stopping muons may be conf-
used.   In this experiment, there are 8 events which have one end of
their tracks in the detector.   In principle, there is a equilibrium
state between production and absorption of neutrino induced muons at
deep underground.   Two of them are expected to be stopping muons of

atmospheric origin.   Therefore,3 muons,half of the rest 6,may be due
to the elastic interaction of neutrino which should be added to the
observed No. in Table I.

Candidates of Nucleon Decay Event

There are two candidates ,Event 87 and Event 251,having their vertex well inside the feducial volume of the detector (100 tons).
They have been discussed and published elsewhere [1].   Afterwards,
two more events,Event 587 and Event 722 are obtained as candidates.

Event No.587

As shown in Fig.5, the event is fully confined to the detector
volume.   The axis of the event, zenith angle θ = 58° and azimuthal
angle ϕ = 35° (from North to East crockwise).   The ionization
recorded in each counter is shown in Fig.6, in enlarged scale, in :
terms of the equivalent number of particles.   Since the total number
of the particles is 42.6 , conventional track length method for elect-
ro-magnetic cascade shower gives estimation of energy as 979 MeV.
Also, the pattern of tracks and the penetration indicate that the
event is composed mainly of electrons and photons, i.e.,electromagne-
tic in nature.   These two features are conformity with proton decay
into an electron and $\pi°$ (2 γ).

In principle, this event could be generated by neutrino background
through the following reactions leading to electron or neutral pion,

1. νe ($\tilde{\nu}$e) + Z ----- e⁻ (e⁺) + X   (charged current)

2. ν (ν) + Z ----- $\pi°$ + X   (neutral current)

From the shower pattern seen in this event, it is clear that such an
interaction should occur by a downgoing neutrino with energy transfer
of $\sim$ 1 GeV.   The rate of such cases are estimated as less than 0.5
per year.   In order to make such pattern of cascade shower, the
Monte Carlo simulations show that the large fluctuation is needed.
In view of these points, we consider this event also a candidate of
nucleon decay.

Event No. 722

The event is shown in Fig.7 with ionization for individual counters.
Total number of the tracks of the event is 39.8 and the energy is
estimated as 915 MeV if one uses similar method to the previous event.
Another way of energy estimation is,taking into account energy loss
in the argon gas inside the counters (39.8 x 20 keV = 796 keV),and
multiply by the mass ratio of the gas to iron absorber in the detector
(796 keV x 1000 = 796 MeV) assuming energy loss in both materials.

126

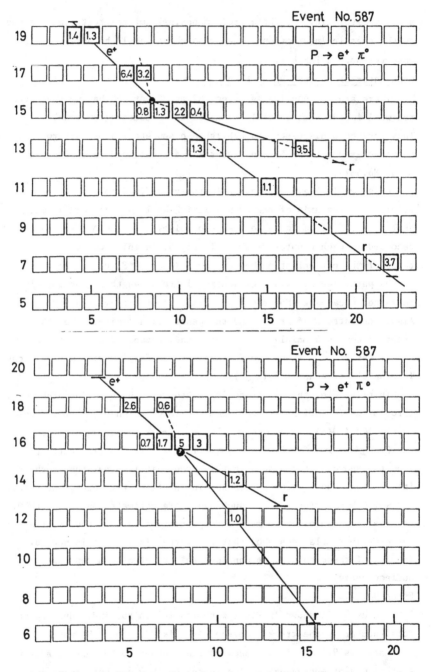

Fig.6 Event 587 in enlarged scale, with the ionization recorded
in each counter, in terms of the equivalent number of particles.
Lines are drawn for the case of proton decay into e^+ and π^0.

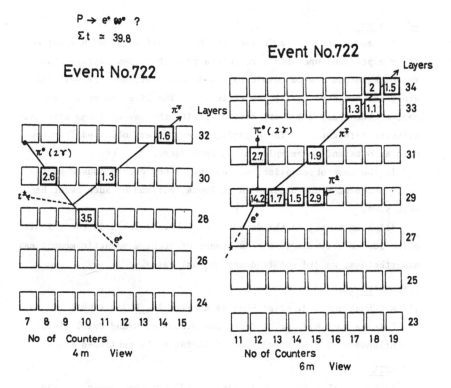

Fig. 7 Candidate event 722

In this event,it is necessary to use four particles to explain
the pattern of the tracks,because of somewhat complicated configurations.   Out of these 4 tracks,2 tracks are one layer tracks respectively.   Therefore,there are large ambiguities in the other side
of the projected pattern, for example, pion in 29th layer has no
information in the 4 m view.   Similar condition is also for the
downwards-going electron in 27th layer.   The lines drawn for these
cases in Fig.7 have large ambiguities.   The details of the event
are still discussed in our group, therefore,the explanation of the
event written in Fig.7 is not final.

Lifetime Estimates

We consider the fiducial volume of 100 tons inside the detector
and assume a value of 0.5 for the detection efficiency(inclusive of
meson absorption in the host nucleus) in estimating the lifetime of
bound nucleons in standard models like SU(5), SO(10).   Taking into
account all 4 candidates observed in 387 days as nucleon decay,we get
8×10^{30} years as estimated nucleon lifetime.

CONCLUSION

The accumulated data in the past KGF experiments and new nucleon decay experiment operated more than a year, have demonstrated that the backgrounds due to cosmic ray muons and neutrinos are well understood and could be easily identified.   The lifetime of nucleon and part of the decay modes are estimated with the use of large size calorimeter detector.   The 4 candidate events are obtained from the data as outstanding examples of nucleon decay.

In the same observation, we searched 2 GeV events caused by the oscillation of bound neutron in nucleous, but non of such event has been seen.

All long tracks in the detector have been checked in their ionization losses relevant to the existence of running magnetic monopoles, nevertheless, we did not find any unusual cases.

REFERENCES

[1] Krishnaswamy M.R. et al Physics Letters 106 B 339 (1981)
    Krishnaswamy M.R. et al Proc. 2nd W.Shop on Grand Unif. (1981)
    Krishnaswamy M.R. et al Proc.of ICOBAN to be published

APPENDIX

Auther of this paper would like to add two more pages for the general explanation of proton decay phenomena which is normaly not easy to give a clear and definit explanation for each event.
One problem is on the bending angles of decay event due to the Fermi momentum of about 250 MeV/c in the nucleous.   The other problem is the nuclear effect to the bound nucleon decay as to the probability of interactions of secondary mesons.   These are shown in Fig.8 and 9 respectively.[2]

Another point of discussion may be focused onto the appearance of low energy electromagnetic cascade showers.   The variation of the pattern of the showers by the fluctuation of the cascade development reflects largely on the accuracy of the estimation of the energy. These problems are shown in Fig.10 and Table II.[3]

REFERENCES IN APPENDIX

[2] J.Arafune   Private communication
[3] K.Kasahara Private communication

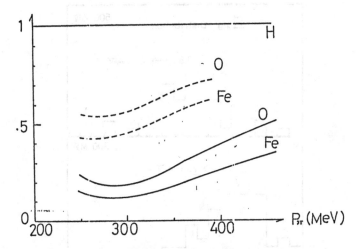

Fig. 8. Going out probability of pions from host nucleus in which bound nucleon has decayed. Full line shows without any interaction and dotted line is after some interaction with other nucleons.

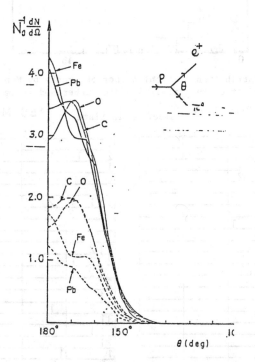

Fig.9. Distribution of deflection angle due to Fermi momentum of nucleon in the various nuclei, calculated for the assumption of proton decay into positron and pion.(- - ,with absorption)

130

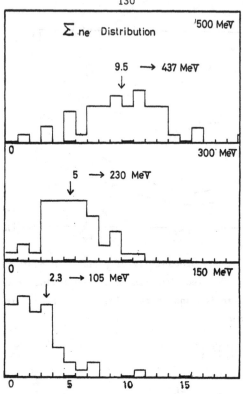

$\sum_{ne}$ Distribution

'500 MeV

9.5 → 437 MeV

0   300 MeV

5 → 230 MeV

0   150 MeV

2.3 → 105 MeV

0   5   10   15

Fig.10 Distribution of total number of tracks by Monte Calro simulation, sampled at every 2 r.l. Average number and estimated energy also given

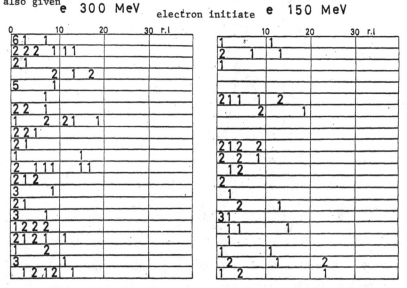

e 300 MeV   electron initiate   e 150 MeV

Table II   20 examples of Monte Carlo simulation at every 2 r.l.
number of particles are given (no number = no particle)

FROM FLUX QUANTIZATION TO MAGNETIC MONOPOLES

Blas Cabrera

Physics Department, Stanford University, Stanford, California 94305

Abstract

Superconductive technologies developed over the last decade now allow detailed observation of flux quantization in five centimeter diameter rings. The theoretical similarities between flux quantization and Dirac magnetic charges make superconductive systems natural detectors for these elusive particles. Recent work on Grand Unification theories, the topic of this workshop, strongly suggests the existence of stable supermassive magnetically charged particles. These particles would be nonrelativistic, weakly ionizing, and very penetrating; and thus may have eluded previous searches. A new superconductive detector designed to look for an extraterrestrial flux of such particles has been operating for more than six months and the first results include a single event. It is consistent in magnitude with the passage of a particle possessing a single Dirac unit of magnetic charge.

Superconductive technologies primarily developed at Stanford
University over the last decade have led naturally to very sensitive
detectors for magnetically charged particles. Much of the work is
based on ultra-low magnetic field shielding achieved using expandable
superconducting cylinders. Application of ultra-low field technology
allows full utilization of modern SQUID (Superconducting QUantum Interference Device) sensitivities. After briefly describing the ultra-low
field technology, I will present data clearly demonstrating flux
quantization in a five centimeter diameter superconducting ring and
show why this data necessarily indicates sufficient resolution for
detecting the passage of Dirac magnetic charges. Several detectors
based on this idea will be described. The first was designed to
measure any nonzero magnetic charge associated with the niobium spheres
from the Fairbank-Phillips-LaRue-Hebard fractional charge search. More
recently a new detector has been used to search for massive magnetically
charged particles suggested by Grand Unification theories, the topic of
this conference. It has yielded a single candidate event. I will end
by briefly describing work on a new, larger three axis detector and a
design for the next generation of yet larger superconducting detectors.

2.   Ultra-Low Magnetic Fields

To begin, I will briefly describe how ultra-low magnetic field
regions are made.[1] Cylinders made of pure 0.007 cm thick lead foil are
tightly folded in accordion fashion, as shown in Fig. 1. These are
slowly cooled through their superconducting transition temperature
and trap the ambient magnetic field. In an ideal superconductor one
would expect the Meissner effect to spontaneously exclude all magnetic
fields; however, in practice one finds that for thin foils pinning
forces prevent the trapped flux vortices from moving. A more correct
picture is thus of magnetic flux lines pinned strongly to their original
locations at the moment when the shield was cooled through its transition temperature. Once superconducting any subsequent changes in the
external magnetic field will be exactly cancelled by supercurrents
on the outer surface of the shields. Therefore, by mechanically
expanding these tightly folded cylinders while maintaining them
superconducting, one obtains a region of lower magnetic field,

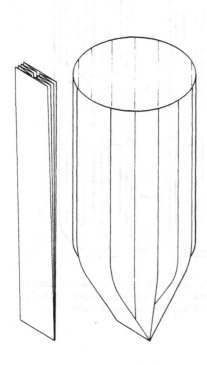

Figure 1. Shields are cooled as tightly folded accordions and expanded with a spherical plunger.

as shown schematically in Fig. 2. To obtain the lowest magnetic fields a bootstrap technique is used, as outlined in Fig. 3. New tightly folded shields are carefully cooled inside of previously expanded ones, and after three or four cycles field levels below 10^{-7} gauss are obtained routinely.

An experimentally measured magnetic field profile is shown in Fig. 4 for a completed shield measuring 20 cm in diameter and 100 cm in length. The three field components along the cylinder axis are scanned using a 5 cm diameter sensing loop coupled to a SQUID. In addition to the magnetic charge detectors described here, these shields have been applied to three other experiments at Stanford: the relativity gyroscope experiment[2] – superconducting gyroscopes placed in earth orbit to measure small inertial frame precessions predicted by general relativity;

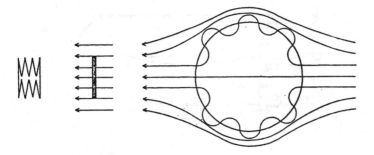

Figure 2. Schematic representation of magnetic field before and after the expansion of a lead shield.

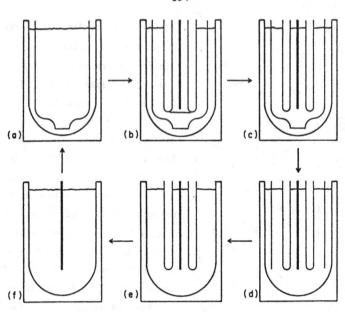

Figure 3. Shield cooling cycle: (a) open shield, (b) insert inner shield, (c) cool inner shield, (d) tear out bottom of outer shield, (e) remove outer shield and (f) remove vacuum-jacketed tube.

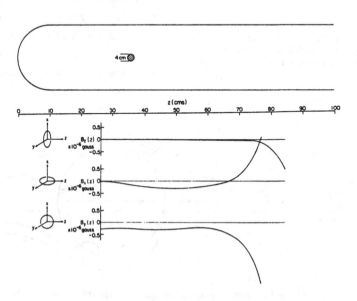

Figure 4. Three mutually perpendicular magnetic field component profiles along the axis of 20 cm diameter superconducting shield.

the He^3 electric dipole moment experiment[3] - an alternate approach
from the neutron measurements to look for a direct violation of time
reversal invariance; and the h/m$_e$ experiment[4] - a determination of
Planck's constant over the electron mass at the several ppm level
using a rotating superconducting ring.

Several interesting features are seen in Fig. 4. First, all field
components are exponentially attenuated upon entering the shield from
the open end, and second, the residual trapped fields seen after the
leakage fields no longer dominate are entirely horizontal. The
exponential attenuation is understood by expanding the magnetic scalar
potential Φ_m, which satisfies LaPlace's equation inside a semi-
infinite cylinder, in terms of Bessel functions radially, sines and
cosines azimuthally, and exponentials along the axis. Thus, Φ_m is
expressed as a sum of terms

$$J_m(k\rho/a) \begin{Bmatrix} \sin m\theta \\ \cos m\theta \end{Bmatrix} e^{-kz/a} \tag{1}$$

Using the boundary condition that the radial component of the field
must vanish at the cylinder radius a, one obtains two dimensional
series expansions for the field components where each coefficient in
the exponential, k'_{mn}, is the nth zero for the derivative of the mth
Bessel function. The smallest of these is k'_{11} = 1.82, and thus
any external field must fall off at least as fast as $e^{-1.82}$ =0.162
per radius into the shield. This exponent agrees with the transverse
measurements in Fig. 4 within a geometric uncertainty of 0.5%. For
a purely vertical field the leading coefficient for the k'_{11} component
is exactly zero and slowest attenuation is now given by k'_{01} = 3.83.
Again the measurements for the vertical component are in general
agreement; however, the scan is not kept exactly along the cylinder
axis so the k'_{11} term is still present and soon dominates.

The second feature, i.e. entirely horizontal residual trapped
fields, is experimental evidence for strong pinning forces. Imagine
an arbitrarily directed magnetic field penetrating the accordion folds
in the tightly folded shield as it is cooled through its transition.
The folds are very close together. Thus, if the trapped flux vortices
do not move during the expansion procedure, one would expect to find
the magnetic field lines going horizontally across inside the open
shield, exactly as is observed experimentally.

Constructing magnetometers capable of measuring ultra-low
magnetic fields has been the most difficult part of the work. It has
been relatively straightforward to obtain lower magnetic fields by
simple modifications of the expansion technique; however, it has proven
much more difficult to make instruments with sufficiently low background magnetic contamination levels to observe the residual fields in
the shields rather than the field associated with the instrument itself.
The instrument used for measuring the shield magnetic field properties
in Fig. 4 is capable of measuring absolute fields at the 10^{-9} gauss
level. It is a flip coil assembly made almost entirely out of fused
quartz and housed in an aluminum vacuum can. Rotation is achieved
using a dacron string wrapped around a quartz pulley, and the tension
in the string is maintained with a fused quartz spring. Four turns
of 0.005 cm diameter niobium wire are wound inside of the 2 mm
diameter quartz tubing formed into a 5 cm diameter circular coil
form. The twisted pair niobium leads come out through the axis and
go up to a SQUID sensor some 35 cm above the coil. We shall return
to this instrument later, where I describe its use as a magnetic
charge detector.

3.   Flux Quantization

In this section I will summarize a simple theoretical derivation
for flux quantization. It will include all that is needed in the
next section for understanding the coupling of a hypothetical Dirac
magnetic charge to a superconducting ring. In 1957 Gorkov[5] demonstrated theoretically that the Ginzberg-Landau (GL) equations for
superconductivity are an exact consequence of the Bardeen-Cooper-
Schreiffer (BCS) theory[6] in the limit of the temperature T approaching
the transition temperature T_c of the superconductor. Global properties
such as flux quantization are straightforward consequences of the GL
equations but often become impossibly difficult to derive from the
full microscopic BCS theory. I shall assume that the global properties
derived from GL are valid even well below T_c, as observed experimentally.

The GL supercurrent density equation

$$\vec{j} = \frac{\hbar e^*}{2im^*} \left(\psi^* \vec{\nabla}\psi - \psi\vec{\nabla}\psi^*\right) - \frac{e^{*2}}{m^* c}\psi^*\psi\,\vec{A} \qquad (2)$$

looks identical with the nonrelativistic single particle Schrodinger equation; however, the interpretation is different. The particles participating in the supercurrent are the Cooper pairs and have twice the mass $m^* = 2m$ and charge $e^* = 2e$ of the electron. In addition, ψ, called the order parameter, now represents a coherent many-body state of Cooper pairs with $\psi^*\psi = n_s/2$ representing the local pair density, thus half the superelectron density n_s. Writing $\psi = |\psi|e^{i\phi}$, we find

$$\psi^* \vec{\nabla}\psi - \psi\vec{\nabla}\psi^* = 2i\psi^*\psi\,\vec{\nabla}\phi \; . \qquad (3)$$

A line integral of $\vec{\nabla}\phi$ will depend only on the end points. Thus, upon integrating around a closed path, the phase change $\Delta\phi$ must equal $2\pi n$, where n is an integer, in order to maintain a single-valued function ψ along any path as required physically. This requirement is exactly the origin of flux quantization and is experimentally observable whenever the coherence of the order parameter extends around a closed path entirely within the superconductor.

Making these substitutions into Equ. 2 we obtain

$$\vec{j} = \frac{\hbar e n_s}{2m}\vec{\nabla}\phi - \frac{e^2 n_s}{mc}\vec{A} \; . \qquad (4)$$

Before taking the line integral around a closed path it can easily be shown that the current density $\vec{j}$ vanishes inside of a thick superconductor by taking the curl of both sides twice (note that n_s remains constant to a high degree throughout the superconductor)

$$\text{curl curl }\vec{j} = -\frac{e^2 n_s}{mc}\text{ curl }\vec{B} \; . \qquad (5)$$

Using Maxwell's equation curl $\vec{B} = 4\pi/c\,\vec{j}$ and taking div $\vec{j}$ zero for a time independent current

$$\nabla^2\vec{j} - \frac{1}{\lambda^2}\vec{j} = 0 \; , \qquad (6)$$

where

$$\lambda = \sqrt{\frac{mc^2}{4\pi e^2 n_s}} \tag{7}$$

is the London penetration depth, typically 300-500 Å. Solutions for
the supercurrent density $\vec{j}$ in Equ. 6 fall exponentially to zero over
a characteristic length λ into the superconductor.

Now integrating Equ. 4 around a closed path we obtain

$$\frac{4\pi\lambda^2}{c}\oint_\Gamma \vec{j} \cdot d\vec{\ell} = n \frac{hc}{2e} - \int_{S_\Gamma} \vec{B} \cdot d\vec{s} \tag{8}$$

where S_Γ is the area bounded by the path Γ. Then for a superconducting
ring thick compared to λ we can always find a path Γ along which $\vec{j}$
vanishes. Thus, the magnetic flux through the ring must be an integer
number of

$$\phi_o = \frac{hc}{2e} = 2.07 \times 10^{-7} \text{ gauss}\cdot\text{cm}^2 , \tag{9}$$

the flux quantum of superconductivity.

Flux quantization was first observed in 1961, at Stanford by
Fairbank and Deaver and independently in Germany by Doll and Nabauer.[7]
Using the ultra-low field technology and modern SQUID sensitivities
S. Felch, J. T. Anderson, and I have recently demonstrated flux
quantization in a single turn 5 cm diameter ring made of 0.005 cm
diameter niobium wire. The data shown in Fig. 5 were taken with the
h/m_e apparatus and show the ring biased within several millikelvin of
its superconducting transition temperature T_c. They were taken with
a second superconducting loop of higher transition temperature closely
coupled to the ring and connected to a SQUID sensor. The potential
barriers separating the various flux quantum states are reduced in
height as the temperature approaches T_c until finally the kT thermal
energy associated with the normal electrons is sufficient to occasionally kick the ring from one quantum state to the next. If the temperature is too low, e.g. far left of the figure, no transitions are seen;
whereas, if the temperature is too close to the critical temperature
the transitions between states occur too rapidly to follow with the
1 Hz bandwidth used. There is a regime, 2 to 3 millikelvin below T_c,
where the transition rate is slow enough so that each of the quantum

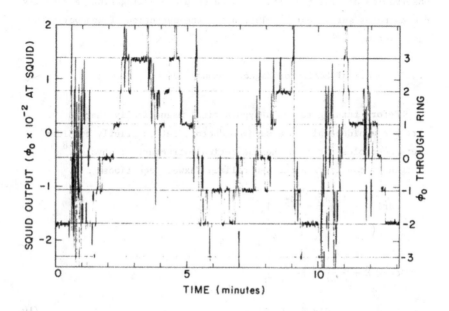

Figure 5. Thermally induced transitions between quantum states, clearly demonstrating flux quantization. The passage of a single Dirac charge through this 5 cm superconducting ring would induce a current corresponding to two flux quanta change.

states are clearly shown. The kT thermal excitations carry the ring through no more than 6 or 7 quantum states because the energy associated with the flux quantum $\phi_o^2/2L$ is smaller than kT by a factor of nearly 10. This 5 cm diameter ring is the first measured of large enough area and thus large enough self-inductance L to satisfy this condition.

As discussed in the next section the coupling of a Dirac magnetic charge to this ring would change the quantum state by 2. Thus, direct observation of flux quantization in a superconducting ring necessarily demonstrates sufficient resolution for the detection of monopoles in that system.

4.   Theoretical Basis for Magnetic Monopoles

Before turning to how a hypothetical magnetically charged particle would couple to a superconductor, let us briefly consider where monopoles fit into the theoretical structure of physics.[8] I begin by considering the generalized Maxwell equations

$$\text{curl } \vec{B} - \frac{1}{c} \frac{\partial \vec{E}}{\partial t} = \frac{4\pi}{c} \vec{j}_e$$

$$\text{div } \vec{E} = 4\pi\rho_e$$

$$\text{curl } \vec{E} + \frac{1}{c} \frac{\partial \vec{B}}{\partial t} = -\frac{4\pi}{c} \vec{j}_m$$

$$\text{div } \vec{B} = 4\pi\rho_m \quad . \tag{10}$$

The magnetic charge density ρ_m and the magnetic current density $\vec{j}_m$ are normally set to zero from lack of experimental evidence for magnetically charged particles. The minus sign in the $\vec{j}_m$ term insures that the continuity equation for magnetic charges and currents has the same form as for electric charges and currents. The symmetrization of Maxwell's equations in this way is not a strong reason for believing in the possible existence of magnetic charges. One can define new electric and magnetic fields as linear combinations of the present electric and magnetic fields

$$\vec{E}' = \vec{E} \cos \theta + \vec{B} \sin \theta$$

$$\vec{B}' = -\vec{E} \sin \theta + \vec{B} \cos \theta , \tag{11}$$

where θ is an arbitrary angle, and then

$$\rho'_e = \rho_e \cos \theta$$

$$\rho'_m = -\rho_e \sin \theta . \tag{12}$$

Thus every charged particle would carry both electric and magnetic charge in a universal ratio $\rho'_m/\rho'_e = -\tan \theta$. If we pick $\theta = \pi/4$, Maxwell's equations become symmetric, but no new physics has been added. We must thus find a more fundamental reason for suggesting the existence of magnetic charges.

In 1931, Dirac[9] found such a reason by asking whether the existence of magnetically charged particles could be made consistent with quantum mechanics. Dirac considered a single electron in the field of a magnetic charge and found that for the electron wavefunction to remain single-valued a quantization condition must exist between the elementary electric and magnetic charges

$$eg = \frac{1}{2} \hbar c . \tag{13}$$

Thus, if magnetic charges existed they would explain the experimentally observed quantization of all electric charges. At that time no other theoretical explanation for this observed quantization existed.

We can understand the Dirac quantization using a similar argument to the flux quantization derivation by considering the Schrodinger current density for a single electron in the field of a magnetic charge

$$\vec{j} = \frac{\hbar e}{2im} (\psi^* \vec{\nabla} \psi - \psi \vec{\nabla} \psi^*) - \frac{e^2}{mc} \psi^* \psi \vec{A} . \tag{14}$$

Then for any closed path Γ not through the monopole

$$\frac{mc}{e^2} \oint_\Gamma \frac{\vec{j} \cdot d\vec{\ell}}{\psi^* \psi} = n \frac{hc}{e} - \int_{S_\Gamma} \vec{B} \cdot d\vec{s} . \tag{15}$$

Flux quantization is not found here because for atomic distances $\vec{j}$ cannot be set to zero. However, if we take two surfaces S_Γ and S'_Γ each bounded by the path Γ and with the two together enclosing the monopole, then the current density term of Equ. 15 is identical for both. Subtracting the two equations we obtain

$$k \frac{hc}{e} = \oint_{S_\Gamma - S'_\Gamma} \vec{B} \cdot d\vec{s} = 4\pi g \ , \tag{16}$$

where k is an integer and the closed surface integral equals the total flux emanating from the pole. Thus, Eq. 13 is obtained for k = 1.

This Dirac condition does not totally symmetrize electric and magnetic charges because the elementary magnetic charge is predicted to be much stronger than the elementary electric charge. From Equ. 13

$$g = \frac{1}{2\alpha} e \ , \tag{17}$$

137/2 times greater than the electric charge e. Therefore, two magnetic charges a certain distance apart feel a force which is $(137/2)^2$ greater than that between two electric charges the same distance apart. The coupling constant, $g^2/\hbar c = \alpha \left(\frac{g}{e}\right)^2 \approx 34$, would thus be stronger than any known force.

Note also from Equ. 16 that the flux emanating from a Dirac charge is

$$4\pi g = \frac{hc}{e} = 2\phi_o \ , \tag{18}$$

exactly twice the flux quantum of superconductivity! This should not be surprising, for Dirac imposed the same single-valuedness condition on a single electron wavefunction in the magnetic field of a pole as we used above in deriving flux quantization from the coherent many-body Cooper pair order parameter. The Cooper pairs possess twice the electron charge, accounting for the factor of two in Equ. 18. Thus the observation of flux quantization in a superconductive system immediately demonstrates sufficient resolution for Dirac magnetic charge detection.

The Dirac suggestion motivated many experimental searches for particles possessing magnetic charge, but no convincing candidates were found.[10]

Over the last decade work on unification theories has unexpectedly yielded strong renewed interest in monopoles. In 1974 't Hooft and independently Polyakov[11] showed that in a large class of unification theories magnetically charged particles are necessarily present. These include the standard SU(5) Grand Unification model. The modern theory predicts the same long range field as the Dirac solution, however, now the near field is also specified leading to a calculable mass. The Dirac theory was not able to predict a mass. The standard SU(5) model predicts a monopole mass of 10^{16} GeV/c^2, horrendously heavier than had been considered in previous searches. I leave a detailed discussion of these theories, the topic of this volume, to the many excellent papers by expert authors.

These supermassive magnetically charged particles would possess qualitatively different properties[12]. These include necessarily non-relativistic velocities from which follow weak ionization and extreme penetration through matter. Thus such particles would have escaped detection in previous searches. To understand why a superconducting ring makes a natural detector for such slow monopoles we next compute the coupling of a magnetic charge to a superconducting ring.[13]

5.   Monopole Coupling to Superconducting Ring

Consider a single charge g moving at velocity v along the symmetry axis of a superconducting wire with radius b located in the x-y plane. Using Maxwell's generalized equation for the monopole current

$$\text{curl } \vec{E} + \frac{1}{c} \frac{\partial \vec{B}}{\partial t} = -\frac{4\pi}{c} \vec{j}_m \tag{19}$$

and taking a surface integral over the area S_Γ bounded by a path Γ everywhere in the interior of the wire, we have

$$\oint_\Gamma \vec{E} \cdot d\vec{\ell} + \frac{1}{c} \frac{d\phi}{dt} = -\frac{4\pi g}{c} \delta(t). \tag{20}$$

Neglecting the finite response time of the superelectrons and assuming the wire diameter to be many λ, we set $\vec{E} = 0$ along Γ and obtain

$$\phi(t) = -4\pi g \, \theta(t) , \tag{21}$$

where we have taken $\phi = 0$ for $t = -\infty$. ϕ is the sum of the flux from the monopole, ϕ_g, and that from the induced supercurrent, $\phi_s = -I(t)L$, where L is the self-inductance of the ring. ϕ_g is given by

$$\phi_g(t) = 2\pi g \left[1 - 2\theta(t) + \frac{\gamma vt}{\sqrt{(\gamma vt)^2 + b^2}} \right], \qquad (22)$$

and we can immediately find ϕ_s. Substituting $4\pi g = 2\phi_o$ we find the induced current $I(t)$ in the ring to be given by

$$I(t) = \frac{\phi_o}{L} \left[1 + \frac{\gamma vt}{\sqrt{(\gamma vt)^2 + b^2}} \right]. \qquad (23)$$

Fig. 6 shows the continuous change of the current in the ring from zero to a value corresponding to two flux quanta. The change occurs over a characteristic time given by $b/\gamma v$. Thus, for particle velocities of several hundred kilometers per second and a loop radius of several centimeters transition times would be several hundred nanoseconds.

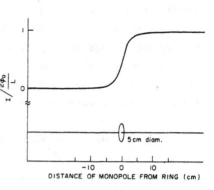

Figure 6. The induced current in a superconducting ring for an axial monopole trajectory.

For an arbitrary nonrelativistic particle trajectory, we may formulate the problem geometrically in the following way. Note that the flux from the monopole coupling to the ring is given by the solid angle subtended by the ring as seen from the monopole. As the magnetic charge approaches the ring, the induced current will be proportional to this flux. There are two distinct cases: the particle trajectory passes through the ring or it does not. In the former case the induced current begins at zero, reaches ϕ_o/L when the particle is in the plane of the ring and continues asymptotically to $2\phi_o/L$. Thus, any such plot of $I(t)$ looks similar to Fig. 6 and has the same limits. In the latter case, the induced current will always remain less than ϕ_o/L as the trajectory approaches the ring, become zero for the point intersecting the plane of the ring, and turn negative on the other side asymptotically returning to zero. Thus, if a magnetic charge g passes through the ring, the number of flux quanta threading the ring will change by two; whereas, if the particle does not pass through, the flux will remain unchanged. Finally if the magnetic charge passes through the bulk superconductor, such as the wire of the ring, it would leave a trapped doubly quantized vortex, and some intermediate total current would persist.

A superconductive system based on these properties is sensitive only to magnetic charges and thus makes a natural detector. The passage of any known particle possessing electric charge or a magnetic dipole would cause very small transient signals but no dc shifts. Moving particles such as the proposed supermassive monopoles can be detected by monitoring the current in a superconducting loop. Since experimental resolutions better than 10^{-4} ϕ_o are standard using commercially available SQUIDs (Superconducting QUantum Interference Devices)[14] it is clear that unambiguous detectors for all magnetically charged particles can be made using existing techniques. We next describe several such devices.

6. Magnetic Charge Measurements on Niobium Spheres

I first became interested in magnetic monopoles[15] about ten years ago when Fairbank and Hebard, and later LaRue and Phillips,[16] began to suspect that fractional electric charges might exist in matter. I remembered a paper written by Schwinger[17] in which he suggested that quarks might be magnetically charged, accounting for their large binding energies without introducing a new force. Thus, it seemed interesting to directly measure the magnetic charge of the niobium spheres by passing them through a superconducting loop.

The detector is shown schematically in Fig. 7 and consists of a magnetically clean quartz fiber with a sample holder on the end which is passed through a small superconducting coil. In turn the coil is coupled to a SQUID which measures any induced current changes. As samples are passed through the ring, one looks for a dc shift in the output indicating a net magnetic monopole charge. Fig. 8 shows the output from the passage of a single niobium sphere which had previously been measured by Fairbank, et al, to be consistent with a +1/3 e fractional electric charge. As the sphere passes through the detector, the dipole field associated with its trapped magnetic flux, a few milli-

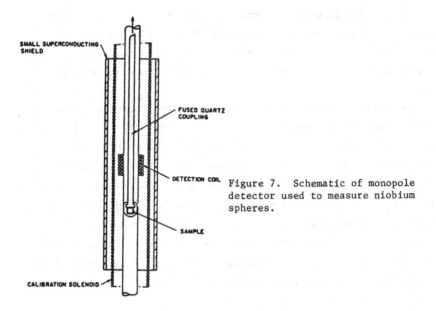

Figure 7. Schematic of monopole detector used to measure niobium spheres.

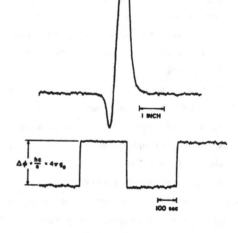

Figure 8. Passage of niobium sphere through monopole detector.

gauss in magnitude, is clearly seen. However, after it passes on
through to the other side, the dc level returns to its original value,
indicating zero magnetic charge associated with the sphere. A calibration signal of magnitude corresponding to the passage of a single
Dirac charge g is also shown in the figure and is 50 times larger than
the noise level of the device. Over the last decade I have measured
twelve spheres, two of them consistent with fractional electric
charge just prior to the magnetic charge measurement and the others
from the same batches. All of the measurements have shown unambiguous
zeros. These measurements have continued from time to time, with the
aim of bracketing two unchanged fractional electric charge measurements with a magnetic charge measurement.

After completing the initial measurements I made a literature
search to determine the possibility of setting lower monopole density
limits in the matter. I found that Alvarez, et al,[13] had just completed an exhaustive search through some 30 kilograms of matter.
These samples included exotic substances such as moon rocks, sediment
from the bottom of the ocean, and snow from the magnetic poles. It
seemed very difficult to compete!

It was interesting that the Alvarez detector, also using a superconducting loop but no SQUID, was about 1000 times less sensitive than
my own to a single pass. I could easily detect a Dirac charge on a
single pass, whereas the Alvarez apparatus used a multiple pass technique to obtain the necessary resolution. At that time I could think
of no advantage to the single pass detector. An advantage suddenly
became evident several years ago when the possible existence of supermassive magnetically charged particles arose from Grand Unification
theories. If such massive monopoles do not come to rest at the
surface of the earth, one would have to look for a particle flux and
a single pass detector is essential. I next describe the first such
detector.

7. First Superconductive Detector for Moving Monopoles

This first detector was the existing magnetometer for the ultra-
low field work described in section 2. It consists of a four turn
5 cm diameter loop made of 0.005 cm diameter niobium wire. The flip
coil is positioned with its axis vertical and is connected to the
superconducting input coil of the SQUID. The passage of a single

Dirac charge through the loop would result in an 8 ϕ_o change in the flux through the superconducting circuit, comprised of the detection loop and the SQUID input coil (a factor of 2 from 4 πg = 2 ϕ_o and of 4 from the turns in the pickup loop). Shown in Figure 9, the SQUID and the loop are mounted inside an ultra-low field shield, and in turn these are mounted inside a single mumetal cylinder. The combined shielding provides 180 db isolation from external magnetic field changes and an ambient field of 5 x 10^{-8} gauss.

The voltage output of the SQUID electronics, which is directly proportional to the supercurrent in the detection loop, is processed through a two pole 0.1 Hz low-pass filter onto a strip chart recorder. In addition, several times per day digital voltmeter readings are taken to guard against recorder failures.

The detector sensitivity has been calibrated in three independent ways: a) by measuring the SQUID response to a known current in calibration Helmholtz coils and calculating their mutual inductance to the superconducting loop ($\pm$ 4%); b) by estimating the self-inductance of the supercon-ducting circuit ($\pm$ 30%); and c) by directly observing flux quan-tization within the superconducting circuit ($\pm$ 10%). All three methods agree within their independent uncertainties.

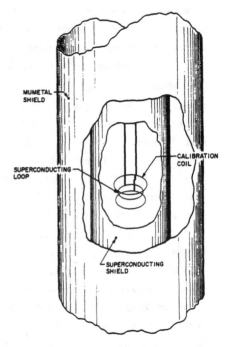

Figure 9. Schematic of flip coil magnetometer in the moving monopole detector configuration.

Two additional effects influence the detector response. First, a magnetic monopole whose trajectory intersects the transformer loop in the SQUID, the twisted leads from the SQUID to the loop, or the loop wire itself, would produce shifts of arbitrary magnitude. Computing the average area ratio of the loop to the remainder of the transformer circuit, indicates such events will be suppressed by a factor of 25 compared to loop events. Second, a particle traversing the supercon-ducting shield will leave doubly quantized vortices wherever the trajectory intersects a wall. The effect is a magnetic field change inside the shield and an applied flux change across the loop. The total induced current change in the loop is

$$\Delta I = \frac{8\phi_o}{L}\left(\delta - f \; \frac{A_\ell}{A_s}\right), \tag{24}$$

where $A_\ell/A_s = 0.06$ is the ratio of the loop to the shield areas, $\delta = 1$ for a trajectory intersecting the loop and 0 for one that misses, and f is a geometric factor which depends on the trajectory impact para-meter and inclination angles and has maximum value 1 for axial trajec-tories through the shield and a minimum value 0 for transverse ones. Current changes of $(0.06)8\phi_o/L$ or less will be observed for trajectories that pass through the shield but not the loop. The probability for such events with $\Delta I \geq (0.02)8\phi_o/L$ is about 10 times larger than for the loop.

A total of 151 days of data were recorded as of March 11, 1982, and the device has remained in continuous operation. Several intervals throughout a continuous one month time period are shown in Figure 10(a), where no adjustment of the dc level has been made. Typical disturbances caused by daily liquid nitrogen and weekly liquid helium transfers are evident. A single large event was recorded (Fig. 10(b)). It is consis-tent with the passage of a single Dirac charge within a combined uncer-tainty of $\pm$ 5% (resulting from the calibration uncertainty and the distribution of geometric factor f). It is the largest event of any kind in the record. In Figure 11 are plotted the 27 spontaneous events exceeding a threshold of 0.2 ϕ_o, which remain after excluding known disturbances (Fig. 12), such as transfers of liquid helium and nitrogen. An event is defined as an offset with stable levels for at least one hour before and after. Only six events were recorded during the 70% of the running time when the laboratory was unoccupied.

The following sources for spurious events have been considered:
(a) Line Voltage Fluctuations caused by two power outages and their accompanying transients failed to cause detectable offsets.
(b) RF Interference from the motor brushes of a heat gun failed to produce any offsets when operated in close proximity to the detector.

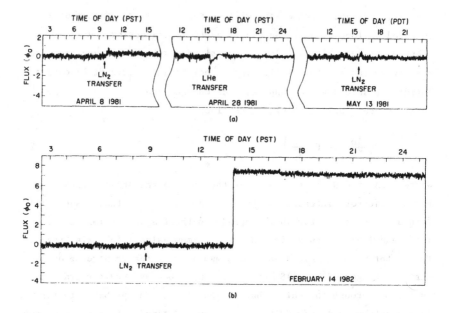

(a)

(b)

Figure 10. Data records showing (a) typical stability and (b) the candidate monopole event.

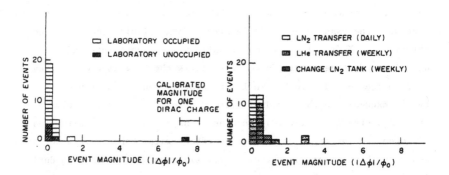

Figure 11. Histogram of all spontaneous event magnitudes.

Figure 12. Histogram of magnitudes for all events directly correlated with dewar maintenance.

(c) <u>External Magnetic Field Changes</u> are exponentially attenuated by 180 db, primarily from an exponential factor of $e^{-1.83\ z/a}$, where z = 72 cm is the distance in from the open top of the shield and a = 10 cm is the shield radius.

(d) <u>Ferromagnetic Contamination</u> is minimized using clean bench assembly techniques and checked with magnetometer measurements within the shield.

(e) <u>The Critical Current</u> of the loop has been measured to be at least a thousand times greater than $8\phi_o/L$ and is typically 10^8 times greater.

(f) <u>Mechanically Induced Offsets</u> have been intentionally generated and are probably caused by shifts of the four turn loop wire geometry, which produce inductance changes and inversely proportional current changes. Sharp raps with a screwdriver handle against the detector assembly cause offsets. The magnitude distribution of 31 raps is shown in Figure 13. The induced event most similar to the record of February 14 is shown in Figure 14. There is a transient spike and a clear drift over the ensuing hour.

(g) <u>No Seismic Disturbance</u> occurred on February 14, 1982.

(h) <u>Energetic Cosmic Rays</u> depositing $\leq$ 1 GeV/cm in traversing the wire would raise the local wire temperature by only $\lesssim 0.01$ K, but a 5 K change is needed to reach the critical temperature.

(i) <u>A Practical Joke</u> would be very difficult, since the SQUID offset potentiometer dial setting had been recorded prior to the February 14 event and was unchanged after the event.

It seems the easiest way to attribute the February 14 event to a spurious cause is to find a way to move it from Figure 11 to Figure 13, i.e., find a possible mechanical impulse source. A spontaneous and large external mechanical impulse does not seem a possible cause for the event; however, a more subtle internal release of stress in the apparatus cannot be ruled out. Regardless, as of June 30, 1982, the experiment has set an upper limit of 3.5×10^{-10} cm^{-2} sec^{-1} sr^{-1} for the isotropic distribution of any moving particles with magnetic charge greater than 0.06 g.

An observational upper bound on the density of supermassive monopoles is given by limits on the local "missing mass"[18]. Visible matter has a measured local density of 0.09 Ms/pc^3 (solar mass per cubic parsec), whereas the calculated mass density given by the velocity distribution out of the galactic plane is 0.14 Ms/pc^3. This

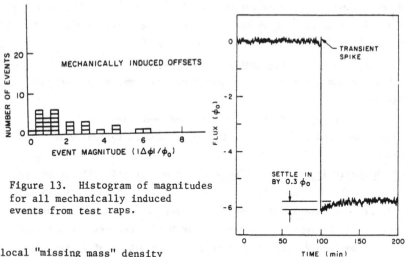

Figure 13. Histogram of magnitudes for all mechanically induced events from test raps.

Figure 14. Closest mechanically induced event to the candidate event (Fig. 10(b)).

local "missing mass" density estimate of ~ 0.05 Ms/pc^3 is in good agreement with the halo mass estimates extrapolated back to our local galactic radius, which give 0.04 Ms/pc^3. If we assume this entire "hidden mass" to be made up of monopoles of mass 10^{16} GeV with isotropic velocities of order 300 km/sec, as suggested by Grand Unification theories,[13] the number passing through the earth's surface would be ~ 4 x 10^{-10} cm^{-2} sec^{-1} sr^{-1}.

Although the particle flux suggested by the single event is barely within the galactic mass bound, there remain several important problems. Turner, Parker, and Bogdan[19] argue that the existence of the 3 microgauss galactic magnetic field limits the maximum allowable isotropic monopole particle flux to value 10^5 smaller. However, Dimopoulos, Glashow, Purcell, and Wilczek[20] suggest an enhanced local source gravitationally bound to the solar system and slowly fed from the sun.

Perhaps the most important question is, can conventional ionization devices with their much larger sensing areas detect the passage of single Dirac charges with velocities of order 10^{-4} to 10^{-3} c?[21]

8. Larger Sensing Area Superconductive Detectors

In closing I will briefly describe the most recent work at
Stanford. A group including M. Taber, S. Felch, R. Gardner, J. Bourg,
and me has continued to monitor the 5 cm diameter detector. A sensitive accelerometer[22] mounted directly on the instrument has been
added and is also continuously recorded. We are also developing two
new systems of larger cross section. The first, along the lines of
the present system but with improved mechanical stability, contains
three mutually orthogonal loops, each twice the diameter of the
present one. In addition to a tenfold increase in cross section,
most particle trajectories would intersect at least two loops,
providing valuable coincidence information. Figure 15 is a top view,
along the axis of the cylindrical superconducting shield. It shows
the three loops and the circular calibration coil. The three dots
represent the twisted pair leads which run parallel to the shield
axis. We expect this three axis detector to be operational in July
or August.

On a longer time scale, we are also designing a new system, based
on existing techniques, which will use several square meters of a

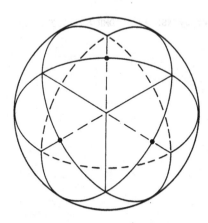

Figure 15. Schematic top view of
three axis coil detector.

thin superconducting sheet in the
form of a cylinder for recording
magnetically charged particle
tracks. It is analogous to a
photographic emulsion which records
electrically charged particle
tracks. A magnetic charge, traversing the cylinder twice in most
cases, would record signatures
consisting of doubly quantized
trapped flux vortices on the walls
(Fig. 16). As long as the sheet
remains superconducting, the strong
flux pinning due to lattice and
surface defects will prevent any
motion. The trapped flux pattern,
about one per cm^2 in a 10^{-7} gauss
field, will be periodically
recorded using a small scanning

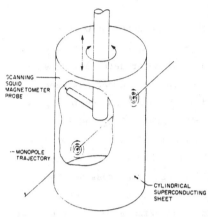

coil coupled to a SQUID. Two
simulated scans, over a 10 cm by
10 cm area, with appropriate SQUID
noise included are shown in Figure
17 (a) and (b). In (b) a new
doubly quantized vortex has appeared
and is most clearly represented
when (a) is subtracted from (b) as
shown in (c).

With these new detectors,
either we will soon begin to see
more events which can now be made
very convincing, or we will set
particle flux limits 100 times
smaller than suggested by the
earlier single event. Either way,
this is a very exciting time for
our group.

Figure 16. Schematic of large
area scanning detector.

9.   Acknowledgments

This work has been funded in part by NSF grant DMR 80-26007,
NASA and the DOE.

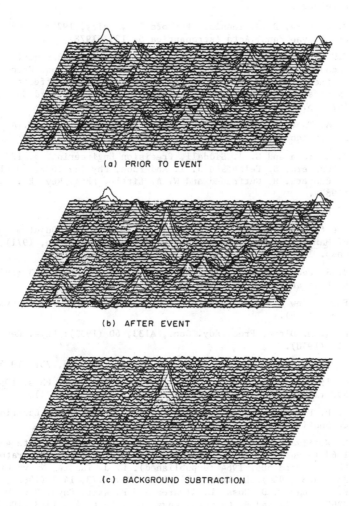

(a) PRIOR TO EVENT

(b) AFTER EVENT

(c) BACKGROUND SUBTRACTION

Figure 17. Simulated data from scanning detector.

10. References

[1] B. Cabrera, Ph.D. Thesis, Stanford University, 1975; B. Cabrera and F. van Kann, Acta Astronautica $\underline{5}$, 125 (1978).

[2] C.W.F. Everitt, Experimental Gravitation, ed. B. Bertotti (Academic Press, New York, 1973), p. 331; J. A. Lipa, Proceedings of the International School of General Relativity Effects in a Physics and Astrophysics: Experiments and Theory (3rd Course), 1977, p. 129.

[3] M. A. Taber, Ph.D. Thesis, Stanford University, 1978; M. A. Taber, J. Physique $\underline{C6}$, 192 (1978).

[4] B. Cabrera and G. J. Siddall, Precision Engineering, $\underline{3}$, 125 (1981); B. Cabrera, S. Felch and J. T. Anderson, Physica $\underline{107B}$, 19 (1981); B. Cabrera, H. Gutfreund and W. A. Little, Phys. Rev. B1, June (1982).

[5] L. P. Gorkov, Sov. Phys. -JETP $\underline{9}$, 1364 (1959).

[6] See for example: A. L. Fetter and J. D. Walecka, Quantum Theory of Many-Particle Systems (McGraw-Hill, San Francisco, 1971), Chap. 13.

[7] B. S. Deaver and W. M. Fairbank, Phys. Rev. Lett. $\underline{7}$, 43 (1961); R. Doll and M. Nabauer, ibid. $\underline{7}$, 51 (1961).

[8] For review see: P. Goddard and D. I. Olive, Rep. Prog. Phys. $\underline{41}$, 1357 (1978).

[9] P. A. M. Dirac, Proc. Roy. Soc., A133, 60 (1931); Phys. Rev., $\underline{74}$, 817 (1948).

[10] A. S. Goldhaber and J. Smith, Rep. Prog. Phys. $\underline{38}$, 731 (1975).

[11] G. 't Hooft, Nucl. Phys. $\underline{79B}$, 276 (1974) and Nucl. Phys. $\underline{105B}$, 538 (1976); A. M. Polyakov, JETP Letts. $\underline{20}$, 194 (1974).

[12] J. P. Preskill, Phys. Rev. Lett. $\underline{19}$, 1365 (1979); G. Lazarides, Q. Shafi and T. F. Walsh, Phys. Lett. $\underline{100B}$, 21 (1981).

[13] L. W. Alvarez, Lawrence Radiation Laboratory Physics Note 470, 1963 (unpublished); P. Eberhard, Lawrence Radiation Laboratory Physics Note 506, 1964 (unpublished); L. J. Tassie, Nuovo Cimento $\underline{38}$, 1935 (1965); L. Vant-Hull, Phys. Rev. $\underline{173}$, 1412 (1968); P. Eberhard , D. Ross, L. Alvarez and R. Watt, Phys. Rev. $\underline{D4}$, 3260 (1971); and B. Cabrera, Phys. Rev. Lett. $\underline{48}$, 1378 (1982).

[14] Model BMS SQUID system, manufactured by S.H.E. Corp., San Diego, California; for theory see: J. Clarke, Proc. of the IEEE, $\underline{61}$, 8 (1973).

[15] B. Cabrera, Proceedings LT14, $\underline{4}$, 270 (1975); B. Cabrera, AIP Conference Series, $\underline{44}$, 73 (1978).

[16] G. S. LaRue, W. M. Fairbank, and A. F. Hebard, Phys. Rev. Lett. $\underline{38}$, 1011 (1977); G. S. LaRue, W. M. Fairbank, and J. D. Phillips, Phys. Rev. Lett. $\underline{42}$, 142, 1019 (E) (1979); G. S. LaRue, J. D. Phillips, and W. M. Fairbank, Phys. Rev. Lett. $\underline{46}$, 967 (1981).

[17] Schwinger, Phys. Rev. $\underline{173}$, 1536 (1968).

[18] For review see: S. M. Faber and J. S. Gallagher, Ann. Rev. Astron. Astrophys. $\underline{17}$, 135 (1979).

[19] M. S. Turner, E. N. Parker and T. J. Boydan, Enrico Fermi Institute Preprint No. 82-18.

[20] S. Dimopoulos, S. L. Glashow, E. M. Purcell and F. Wilczek, Harvard University Theoretical Physics, Preprint No. A 016.

[21] S. Geer and W. G. Scott, Cern $\bar{p}p$ note 69; K. Hayashi, to be published; S. P. Ahlen and K. Kinoshita, to be published.

[22] MODEL 7701-100 Accelerometer, Endevco Corp., San Juan Capistrano, CA.

COMPOSITE/FUNDAMENTAL HIGGS MESONS[*1]

Howard Georgi

Lyman Laboratory of Physics
Harvard University

Why is the $SU(2) \times U(1)$ breaking scale so small? Why are there several generations? The hierarchy puzzle and the flavor puzzle have mystified unifiers for decades, but they seem particularly acute in unified theories based on $SU(5)$ (and its simple relatives $SO(10)$ and $E(6)$).

Supersymmetry, (see Dimopoulos) may perhaps address the first puzzle (though the present indications are inconclusive at best) but it does not seem to have anything with the second. More ambitious attempts to address both puzzles go under the name of "extended technicolor" or ETC. Technicolor (TC) is a generic term for the strong gauge interactions which are (allegedly) responsible for the dynamical breakdown of the $SU(2) \times U(1)$ symmetry [2]. The mechanism is the same as the mechanism of chiral symmetry breakdown in the color $SU(3)$ interactions of QCD. Technifermions, which feel the TC force, transform according to a chiral representation of $SU(2) \times U(1)$, the left-handed (LH) and right-handed (RH) components transforming differently. When these technifermions are bound by the TC interactions, they develop dynamical masses which break $SU(2) \times U(1)$. The Higgs mesons are, in a sense, composites.

Technicolor is a very attractive idea. It directly addresses the gauge hierarchy puzzle. If the TC gauge coupling is small at the unification scale, the TC confinement scale Λ_{TC} (which determines the $SU(2) \times U(1)$ breaking) is exponentially smaller than the unification scale, because the coupling constant changes slowly [3]. The trouble with technicolor is that it is hard, without introducing light

158

fundamental scalars, to get the SU(2) xU(1) breaking into the normal
quark and lepton mass matrix. It is hard to break all the global
chiral symmetries which keep the quarks and leptons massless.

In ETC models, the TC gauge group is extended so that quarks and
leptons are in ETC multiplets with technifermions [4,5]. If the ETC
symmetry breaks down to TC at an ETC scale larger than Λ_{TC}, the ETC
gauge interactions can break all the global chiral symmetries and
generate quark and lepton masses without any fundamental scalar mesons.

The ETC idea is extremely ambitious. A successful ETC model would
describe physics in terms of a small number of gauge couplings and some
strong interaction parameters which are calculable in principle, if not
with present technology. Unfortunately, there are difficulties. It is
not clear what breaks ETC. Does this require yet another TC mechanism?
In many toy models, there are unacceptably light Goldstone or pseudo-
Goldstone bosons associated with TC chiral symmetry breaking [4,5].
Most toy models produce too large a θ parameter in QCD or else yield an
axion [6]. Most toy models produce unacceptably large flavor changing
neutral current effects [5,7]. Finally, the most serious difficulty is
that there are no models which are not toys. No one has been able to
build a model with anywhere near enough structure to describe the
complicated quark and lepton mass matrix, at least not without putting
the complication in by hand in the form of unreasonable dynamical
assumptions [8].

Holdom had an idea which solves some of these problems by raising
the ETC scale [9], M_{ETC}. The standard estimate of M_{ETC} is based on the
fact that the quark and lepton masses are proportional to the vacuum
expectation value (VEV) of the technifermion mass operator, $\bar{\Psi}\Psi$, which,
if TC is asymptotically free, scales with dimension three. Thus on
dimensional grounds, we expect the VEV to be

$$<\bar{\Psi}\Psi> \simeq \Lambda_{TC}^3. \tag{1}$$

Then quark and lepton masses induced by the ETC interactions are of
order

$$m \simeq \Lambda_{TC}^3/M_{ETC}^2. \tag{2}$$

Holdom noted that if TC is not asymptotically free, but instead has a
nontrivial ultraviolet (UV) fixed point, then $\bar{\Psi}\Psi$ scales with some
anomalous dimension D at short distances and (1.2) is modified to

$$m \simeq \Lambda_{TC}^D/M_{ETC}^{D-1}. \tag{3}$$

If D is less than three, M_{ETC} can be large, which makes the ETC interactions less phenomenologically troublesome. But the difficulties of producing a realistic mass matrix remain.

Shelly Glashow and I suggested a radical extension of Holdom's idea [10]. If D is precisely equal to one, M_{ETC} can be arbitrarily large. In particular, it can be identified with the grand unification scale M_G. This identification requires a complete reassessment of the ETC idea. It is quite natural to have fundamental scalar mesons and additional fermions with a mass of order M_G. The global chiral symmetries of the light quark and lepton fields can all be broken through their Yukawa couplings to the fields associated with heavy (mass ~ M_G) particles. Thus there is no need for ETC as an additional gauge interaction at all. The role of ETC can be played by the interactions of light quarks and leptons with heavy particles at M_G.

Suppose that the TC gauge coupling constant is G and that the β function for the TC theories has the form discussed by Holdom [9], IR free at the origin with a simple zero at $G = G^*$. The point $G = G^*$ is a nontrivial UV fixed point. We are eventually going to be interested in the strong coupling phase $G > G^*$, but before discussing this phase, we consider what happens if $G = G^*$. Since $\beta(G^*) = 0$, this is true at all scales and the theory is scale invariant. In general, we might expect the $G = G^*$ theory to be a complicated strongly interacting theory which does not admit a simple particle interpretation because all the fields scale with anomalous dimensions. There is some evidence that such behavior is possible in theories with nontrivial infrared fixed points [11].

There is, however, a different possibility which in some ways is much simpler. The $G = G^*$ theory may be a free theory (certainly a massless free theory is scale invariant). But it need not be the same free theory as the G = 0 TC theory. Indeed, we would not expect it to be. Instead, the theory may be described by some effective fields.

If this is the nature of the TC theory at $G = G^*$, we might expect that for $G > G^*$, we can rewrite the TC theory in terms of the effective fields which become free for $G = G^*$. Then, in terms of these effective fields, the theory should be described by some effective coupling constant g(G), such that $g(G^*) = 0$. At short distances, this effective theory should be much easier to deal with than the original theory because the coupling is small. The effective theory is asymptotically free.

What we would like to do, then, is to find the rules for going from the original theory with a nontrivial UV fixed point at $G = G^*$ to the effective asymptotically free theory with a UV fixed point at $g = 0$. Before discussing this in detail, we will make one general comment. In the effective theory, since the coupling is small near the fixed point, we can calculate the β function in perturbation theory, and we will find, in any theory, that it vanishes like some power of g. $\beta(g)$ has a multiple at $g = 0$. Then unless the function $g(G)$ has very special properties, $\beta(G)$ must also have a multiple zero. As discussed in reference 10, this behavior is important to our understanding of the gauge hierarchy puzzle.

What is the effective theory? We cannot answer the question with certainty, or even prove that an effective theory exists. But if we assume the existence of an asymptotically free effective theory, we can formulate a set of consistency conditions which the effective theory must satisfy. For definiteness, let us assume that the technifermions of the TC theory are N Dirac fermions transforming according to a complex representation of the TC gauge group. The generalization to Majorana fermions transforming according to an arbitrary anomaly free representation of TC should be obvious.

The theory with N massless Dirac fermions has an $SU(N)_L \times SU(N)_R \times U(1)$ symmetry. If we write all the fermions as LH fields, we can characterize their transformation properties under $SU(N)_L \times SU(N)_R \times TC$ as follows: the LH technifermions transform like an N of $SU(N)_L$ and like some representation R of TC; the antiparticles of the RH technifermions transform like an $\bar{N}$ of $SU(N)_R$ of TC. In an obvious notation, the LH fermions are

$$(N,1,R) + (1,\bar{N},\bar{R}). \qquad (4)$$

The effective theory should satisfy the following consistency conditions:

(1) It is renormalizable, since we have not introduced any scale.

(2) It is a non-Abelian gauge theory, since it is asymptotically free [12]. We will call the non-Abelian gauge symmetry of the effective theory "effective technicolor" or eTC.

(3) It should have the same $SU(N) \times SU(N) \times U(1)$ global symmetry as the TC theory.

(4) It must contain fermions with the same anomalies with respect to the $SU(N) \times SU(N) \times U(1)$ generators as the anomalies of (4).

The fourth condition may require some explanation. We can imagine, following 't Hooft [13], gauging the global symmetries of the original TC theory and canceling the anomalies of the fermions of (4) with spectator fermions which transform trivially under TC. The effective theory must also be anomaly free, thus the eTC fermions must have the same anomalies as the TC fermions. Note that we do not have to worry about the fact that the global symmetries may (and indeed will) be spontaneously broken, because the effective theory should be equivalent to the original at all moments, including momenta large compared to the confinement scale, Λ_{TC}, where dynamical symmetry breaking takes place.

Unfortunately, conditions 1-4 are not sufficient to completely determine the effective theory. But in a class of TC theories, there is a particularly simple way of satisfying these conditions. Suppose that the representation R of the TC group has dimension r, but the TC group is some subgroup of SU(r) such that the TC theory is not asymptotically free. For example, we might have r = 3 with TC the U(1) x U(1) subgroup of SU(3) generated by T_3 and T_8, with discrete symmetries which enforce the equality of the two U(1) coupling constants. If the eTC gauge group is swollen to SU(r), and if this swelling is enough to restore asymptotic freedom, then the conditions 1-4 are satisfied if the LH eTC fermions transform as

$$(N,1,r) + (1,\bar{N},\bar{r})$$

under SU(N) x SU(N) x SU(r). It is not clear, by any means, that this kind of swelling is the only way to satisfy the conditions. But it is the simplest. We will assume that this swelling mechanism takes place. We note in passing, that the swelling solution satisfies elementary decoupling requirements obtained by allowing some of the technifermions to have masses which can be varied from zero to values large compared to Λ_{TC} [13,14].

One might object to the SU(3) swelling solution for the eTC gauge group of a U(1) x U(1) TC theory because the 3 dimensional representation of the U(1) x U(1) gauge group is reducible. The TC theory with N massless Dirac 3's thus has a larger symmetry than SU(N) x SU(N) x U(1). There is a separate SU(N) x SU(N) for each component of the 3. What happens to the extra symmetries?

The SU(N) x SU(N) symmetry which we have discussed and imposed on the eTC theory is the one which acts equally on all the components of the 3 of TC. In other words, it commutes with the discrete symmetries which mix up the various components of the three. The generators of

the extra symmetries transform nontrivially under the discrete
symmetries. This is a clue to the answer.

In the eTC theory, the discrete symmetries are promoted to become
part of the eTC gauge symmetry. And the eTC gauge theory is confining.
Thus the physical Hilbert space contains only states which are
singlets under the eTC SU(3) and thus under any discrete subgroups.
Thus the extra symmetry generators cannot possibly be relevant. They
take physical states to unphysical states.

This is a particularly clear case of a very general feature of
swelling solutions. Whenever the TC gauge group can be regarded as a
subgroup of the eTC group, there are TC singlet states which are not
eTC singlets. We must assume that the dynamics pushes these states
out of the physical Hilbert space (like colored states in confining
QCD). In the $U(1) \times U(1)$ example, these dynamics spontaneously break
the extra $SU(N)$ symmetries (which are not present in the eTC theory).
The Goldstone bosons decouple from the physical states for the same
reasons.

The extra $SU(N)$ symmetries are analogous to the chiral $U(1)$
symmetry in QCD. The extra $SU(N)$ currents are not invariant under the
discrete symmetries of the TC theory just as the conserved chiral $U(1)$
is not color gauge invariant. The Goldstone bosons associated with
the spontaneous breaking of the extra $SU(N)$ symmetries decouple from
the physical states just as the Goldstone boson of the axial $U(1)$
decouples.

So far we have discussed the gauge symmetry and the fermions of
the effective theory. But there may also be scalar mesons. These may
be of three kinds. There may be scalars which transform nontrivially
under the eTC gauge group but trivially under the global symmetries.
There may be scalars which transform nontrivially under both the eTC
gauge group and the global symmetries. Finally, there may be scalars
which transform trivially under the eTC gauge group but nontrivially
under the global symmetry group. It is in this last kind of effective
scalar multiplet that we will find the physical Higgs scalar.

The representations of effective scalars which can appear are
constrained by conditions 1-4. The strongest constraint is that the
scalar self couplings not spoil asymptotic freedom. This constraint is
particularly strong if there are scalars of the last and most
interesting type, which transform trivially under the eTC gauge group.
We must assume that this last type of scalar exists, because scalars

transforming nontrivially under eTC are of no use as Higgs mesons.
They cannot couple to quarks and leptons which are eTC singlets. We
call the eTC singlet scalars, the "effective Higgs mesons".

The self couplings of the scalar mesons could not be
asymptotically free were it not for their eTC gauge couplings and
their Yukawa couplings to the effective eTC fermions. For the
effective Higgs mesons, which have no eTC gauge couplings, the Yukawa
couplings are crucial. The effective Higgs mesons must, therefore,
have Yukawa couplings to eTC singlet mass operators of the fermions.
In the swelling scheme, this means that the effective Higgs scalars,
if there are any, must transform like $(N,\bar{N},1)$ under $SU(N) \times SU(N) \times eTC$.
This finally, is the connection to the speculation of reference 10.
The technifermion mass operators of the TC theory scale with dimension
$D = 1$ at short distances because in the effective eTC theory, they are
associated with canonically scaling effective Higgs scalars.

Even when the effective Higgs scalars have the appropriate Yukawa
couplings, the constraint of asymptotic freedom is not trivial.
Indeed, the effective theory is not asymptotically free in the strong
sense that the origin of the multidimensional coupling constant space
is a stable fixed point, because for generic values of the eTC gauge
coupling g, the Yukawa coupling f and the Higgs scalar coupling λ, the
Yukawa coupling goes to zero at short distances too fast, faster than
the gauge coupling, leaving the scalar couplings to fend for
themselves [15]. The couplings can go to zero at short distances only
if the ratios f/g and λ/g^2 are fixed at special values. This is called
an eigenvalue condition [16]. It should, perhaps, not surprise us
that these couplings are related by an eigenvalue condition. After
all, in the original TC theory, there is only one coupling, G. All
the couplings in the effective theory must be functions of G and
therefore be related to one another. An eigenvalue condition is just
a particularly simple example of such a relation.

In the applications of effective technicolor to grand unification,
the eTC theory (and the TC theory to which it is equivalent) are low
energy approximations, valid for scales small compared to the
unification scale M_G. The full complexity of the theory shows up only
at the scale M_G. But of course, even in the low energy theory, there
are fields which do not take part directly in the eTC interactions,
the quark and lepton fields and the $SU(3) \times SU(2) \times U(1)$ gauge fields.
If we are to justify our name for the effective Higgs scalars, we must

argue that they can couple appropriately to the low energy fields. To the extent that these couplings are weak, we can ignore the more complicated question of how they modify the eTC theory itself.

The gauge couplings are simple. They are determined by the gauge symmetry, which just incorporates some subgroup of the $SU(N) \times SU(N) \times U(1)$ global symmetry of the eTC theory.

More interesting are the renormalizable couplings of the effective Higgs to quarks and leptons and the $SU(N) \times SU(N) \times U(1)$ breaking self couplings of the effective Higgs. These are related to $SU(N) \times SU(N) \times U(1)$ breaking terms induced in the TC theory through the couplings of the light technifermions and the light quarks and leptons to heavy fundamental bosons and fermions with mass of order M_F: These couplings also produce a variety of nonrenormalizable interactions, but these we ignore because they are unimportant at low energy.

In the TC theory the technicolor scale Λ_{TC} arises dynamically through dimensional transmutation [17]. There are no explicit mass terms even though scale invariance is broken, because none are allowed by the symmetries. In the eTC theory, scalar meson mass-squared terms of order Λ_{TC}^2 are allowed. Whether they are positive or negative, the effective Higgs meson develop nonzero VEVs. This happens because of their Yukawa couplings to the effective technifermions mass operators, which develop VEVs of order Λ_{TC}^3, like $\bar{\Psi}\Psi$ in the color $SU(3)$ theory. The Yukawa couplings then induce a VEV for the effective Higgs scalars.

We expect this mechanism to break the $SU(N) \times SU(N) \times U(1)$ global symmetry down to $SU(N) \times U(1)$. Thus there are N^2-1 Goldstone or pseudo-Goldstone bosons. In the simplest possible case in which $N = 2$ and the LH (or RH) technifermions transform like a doublet under the weak $SU(2)$, there are no pseudo-Goldstone bosons at all. All the surviving Higgs scalars (the true Higgs, two more neutral scalars and a charged scalar) have mass of order Λ_{TC}.

We have argued that a suitable technicolor model with a nontrivial UV fixed point can be reinterpreted in terms of an asymptotically free effective technicolor model with light effective Higgs scalars whose couplings are related to the effective technicolor gauge coupling by an eigenvalue condition. Yukawa couplings of these effective Higgs mesons to quarks and leptons can be induced by interactions at the unification scale, M_G.

Thus the effective Higgs meson is both composite and fundamental. In the TC theory, it must be interpreted as a bound state of technifermion and antitechnifermion. But in the equivalent eTC theory, the effective Higgs scalar is a "fundamental" field, scaling canonically at short distances.

Clearly, much work remains to be done to validate and quantify these dynamical speculations and to pin down the connection between the TC theory and the eTC theory. But the eTC langauge allows us to organize our ignorance into a few strong interaction parameters. This is enough to allow us to use these ideas in model building.

A unified model based on these ideas must have the following unusual properties:

1. The TC theory must be in a strong coupling phase. Remember
    that it is the coupling constant of the original TC gauge
    group which is related to the other couplings at unification.
    But that means that the TC gauge coupling must be large at
    M_G, while unless there are many families, the SU(3), SU(2) and
    U(1) couplings are small. We need a nontrivial relationship
    between the gauge couplings at unification.

2. The effective Higgs mesons must not mediate baryon number
    violating interactions. This is a very strong requirement.
    If, for example, the technifermions form complete SU(5)
    families, the effective Higgs mesons will surely mediate
    proton decay, because the Higgs doublet must appear with a
    color triplet with baryon number violating couplings in a
    complete SU(5) representation (such as 5 or 45). The only way
    we know to ensure that there are no dangerous effective
    Higgs mesons is to require that all the technifermions have
    zero color triality, or simpler still, no color at all.

3. But that means treating color and weak SU(2) very unsym-
    metrically! This seems to strike at the very heart of
    unification. One might expect that it will get us into
    trouble with the low energy coupling renormalizations which
    are known to work pretty well in the simplest scheme in which
    SU(2) and SU(3) are treated symmetrically [18].

In addressing these problems in the subsequent sections, we will adopt a rather extreme philosophy about the nature of the theory at the unification scale: Namely that we do not know how Nature judges simplicity at 10^{-29} cm, and therefore that we can assume that the theory is as complicated as may be necessary, so long as it produces a simple and realistic structure at larger distances. In particular, that means that we will not ask a particular type of question (i.e. Are the vacuum expectation values (VEVs) which break the unifying symmetry natural, in any sense?) because the answer is always the same (i.e. Yes! We will complicate the theory until they are natural, without changing the low energy theory.) "Sufficient unto the day is the evil thereof". How can we build such a model? One solution to the first problem suggests itself immediately. If the generators of the low energy SU(3), SU(2) and U(1) are sums of N properly normalized generators of the unifying group, while the TC generators consist of only a single similarly normalized generator, then the technicolor coupling α_{TC} is larger than the other α's at M_G by a factor of N. This is familiar from the SU(2) x U(1) model. When several gauge subgroups are broken down to a diagonal subgroup (generated by the sum of the original generators), the inverse α's add,

$$\frac{1}{\alpha_{diagonal}} = \sum_i \frac{1}{\alpha_i} . \tag{6}$$

This in turn suggests that we take the unifying gauge group to be a product of simple factors connected by a discrete symmetry. Consder, for example, the product of some number of SU(5) groups, with a discrete symmetry which cyclicly permutes the factors. We can take the low energy SU(3), SU(2) and U(1) generators to be the sum of the SU(3), SU(2) and U(1) generators from several SU(5)'s while the TC generators might be, for example the T_3 and T_8 generators of an SU(3) subgroup of a single SU(5).

So far so good, but why should there be several SU(5)'s? Again, an obvious answer suggests itself. There are several generations of quarks and leptons. Perhaps we should have one SU(5) for each generation and another for technicolor. Then the observed multiplicity of families would be related to the large ratio of TC to color coupling constant.

Taking this view seriously, we get a clue to the fermion content of the theory. For each SU(3), we need a 10 and $\bar{5}$ of left-handed (LH) fermions, plus possibly some representation which is real with respect to all the SU(5)'s. Note that the discrete symmetry requires the fermions to be the same for each SU(5) group.

This fermion structure, in turn, suggests a solution to the second problem. We have (so far) at least four SU(5)'s. One each for the e, μ and τ generations and another for TC. We will label them

$SU(5)_e$, $SU(5)_\mu$, $SU(5)_\tau$ and $SU(5)_T$. The low energy $SU(3) \times SU(2) \times U(1)$ gauge symmetry must include the $SU(3) \times SU(2) \times U(1)$ subgroups of the $SU(5)_e$, $SU(5)_\mu$ and $SU(5)_\tau$. Technicolor, we are assuming, is the $U(1) \times U(1)$ subgroup of the $SU(3)$ subgroup of $SU(5)_T$. But if we are to build effective Higgs doublets, the technifermions must transform under $SU(2) \times U(1)$. Thus, we must assume that the low energy $SU(2) \times U(1)$ also involves the $SU(2) \times U(1)$ subgroup of $SU(5)_T$. This is exactly what we want to avoid the embarrassment of instanton proton decay. There are, in the 10 and $\bar{5}$ of $SU(5)_T$, technifermions which are $SU(2)$ doublets, but no technifermions which carry color.

As we anticipated, however, we cannot stop here. We have mucked up the analysis of reference 18 in two ways. The $SU(2)$ and $U(1)$ coupling constants are smaller than the color coupling at M_G because they are diagonal sums of four factors while $SU(3)$ is a diagonal sum of only three. Furthermore, there are fewer color triplets in the low energy theory than $SU(2)$ doublets, so the $SU(3)$ coupling is even more asymptotically free, relative to the $SU(2)$ and $U(1)$ couplings, than in the standard theory. Both these effects tend to increase $\sin^2\theta$ and decrease the unification scale. The second is a disaster. Proton decay is much too fast.

There are several possible options for completing the model fragment. They all involve adding at least one extra component of the semisimple group (so that it contains five simple factors, if there are three families) to reimpose the symmetry between the $SU(3)_{color}$ and $SU(2) \times U(1)$ subgroups. Thus we add to the fragment at least one additional $SU(5)$, which we will label $SU(5)_A$ (A for abnormal). This $SU(5)_A$ contributes to color $SU(3)$ but not to $SU(2) \times U(1)$. The low energy $SU(2) \times U(1)$ is the diagonal sum of four $SU(2) \times U(1)$ subgroups, e, μ, τ and T, and the low energy $SU(3)$ is likewise a diagonal sum of four $SU(3)$ subgroups, e, μ, τ and A. Thus the coupling constants are equal at M_G.

In the $10 + \bar{5}$ of LH fermions associated with $SU(5)_A$, there is a pair of abnormal quarks. They are abnormal because they do not carry the low energy $SU(2) \times U(1)$ quantum numbers. Thus, in particular, they are electrically neutral. If this pair of quarks is light, the renormalization of the low energy couplings is essentially the same as in the standard $SU(5)$ model.

The trouble with a model of this kind is that the abnormal quarks must be hidden somehow. If they are light, we would see their bound

states with normal quarks as fractionally charged particles. Either
the abnormal quarks must be very heavy or better still confined into
systems with zero color triality. For example, one might imagine that
the $SU(2) \times U(1)$ subgroup of $SU(5)_A$ is broken down to a $U(1)$ analogous
to the ordinary $U(1)$, but with a larger coupling like TC, so that it is
confining. There is nothing really wrong with this option, but it
seems unattractive because there is no gauge symmetry which keeps the
abnormal quarks light. It is possible to use this idea in $SU(5)$ with
yet another $SU(5)$ (and an associated family of abnormal techniquarks),
but it is much simpler to extend all the $SU(5)$'s to $SO(10)$'s.

Suppose that the $SO(10)_A$ is broken down to $SO(4)_A \times SO(6)$ where the
$SO(6)$ is further broken down to $SU(3)$ (which is combined with the
other $SU(3)$ subgroups to form $SU(3)_{color}$) and a $U(1)$ associated with a
charge K. Under these subgroups the abnormal 16_A of LH fermions
consists of a color singlet with $K = 1$ and a color triplet with
$K = -1/3$, both of which are doublets under one of the $SU(2)$ subgroups
of $SO(4)$, and a color singlet with $K = -1$ and color antitriplet with
$K = -1/3$ which are doublets under the other $SU(2)$ subgroup of $SO(4)_A$.
Thus, these are eight four component fields, a pair of color singlets
with $K = 1$ and a pair of color triplets with $K = -1/3$ all massless
because they transform like chiral doublets under $SO(4)_A$. If the $U(1)$
is strong and confining, we would expect that at some mass Λ_A, the
chiral symmetries will be broken spontaneously and the abnormal quarks
will develop dynamical mass. They will be bound into systems with
$K = 0$ (and thus color triality zero) and mass of order Λ_A.

In this theory, the abnormal quarks cause no phenomenological
embarrassment. Their $K = 0$ bound states are electrically neutral and
have color triality zero, so they do not produce fractionally charged
particles or baryon number violating interactions.

We find the idea of the $SO(10)^5$ model attractive not only because
it shows that our ideas can be implemented without phenomenological
disasters, but also because it establishes a connection between the
gauge hierarchy puzzle and the multiplicity of flavors. There must be
multiple flavors in order that the TC theory be in the strong coupling
phase that produces the effective Higgs mesons. In principle, the
number required to give the right TC gauge coupling is even calculable,
although in practice, this is a strong coupling calculation far beyond
our present abilities. We may hope, however, to use the $SO(10)^5$ theory
to attain some understanding of the quark and lepton mass matrix. To

do so, we must flesh out the model.

The cyclic symmetry constrains the form of the fermions and scalar mesons in the model. The fermion representations we need are each associated with a single SO(10) subgroup. The light LH quarks and leptons are in 16's, one for each SO(10). We will label these

$$16_{fL} \tag{7}$$

where f = e, μ, τ, T or A, labels the SO(10) group under which the fields transform. In addition, we will need fermions transforming like SO(10) 45's,

$$45_{fL}, \tag{8}$$

in the same notation. These are real representations, so there are SO(10) invariant bare mass terms for these fields, of order M_G.

We also need scalar fields transforming like 16's under the separate SO(10)'s, denoted by

$$16_{fS}. \tag{9}$$

In addition, we need scalar fields which transform nontrivially under adjacent pairs of the SO(10)'s. In an obvious notation, these are

$$(16_f, \overline{16}_{f'})_S, (45_f, 45_{f'})_S \tag{10}$$

where f,f' run over the adjacent pairs e,μ; μ,τ; τ,T; T,A; A,e. All these scalars have mass or VEV of the order of M_F. These fields are all we need to build a model which is complete, except perhaps in the sense that the VEV's to be discussed below may not be obtainable by minimizing a potential involving only these. For the reasons discussed earlier, we do not much care. We can always complicate (?) the theory to produce the desires VEV's, but (7-10) incorporate all the essential physics.

We will assume that the VEV's of the scalar fields are the most general VEVs consistent with the pattern of symmetry breaking discussed above and are all of order M_G. Thus, for example, the 16_{eS} has a nonzero VEV for its neutral SU(2) x SU(3) singlet component (the RH neutrino component). A VEV for a $(16_f, 16_{f'})_S$ produces a mass mixing between some irreducible components of 16_f and $16_{f'}$. There is one independent VEV for each irreducible component that the 16_f and $16_{f'}$ have in common. Thus the $(16_e, \overline{16}_\mu)_S$ has six independent VEVs, one for each of the irreducible SU(3) x SU(2) x U(1) components of the 16. The $(16_\tau, \overline{16}_T)_S$ has only three independent VEVs, one for each of the irreducible SU(3) singlet components. The SU(3) triplet components of

16_τ and 16_T cannot mix since one carries color while the other carries technicolor. The $(45_A, 45_e)$ has only a VEV which mixes the SU(3) octets and VEVs which mix singlets.

We are now in a position to see how the structure outlined in the previous section gives rise to the light particle masses. The basic mechanism is that the heavy particle interactions generate couplings between light fermions and light technifermions which in turn produce couplings of the light fermions to the effective Higgs mesons of the eTC theory. In the SO(10)[5] model, the most general couplings of the fields (7-10) to each other are adequate to break all the global chiral symmetries of the fermions fields. Thus, when the SU(2) xU(1) symmetry breaks spontaneously at Λ_{TC}, we expect all the fermions to get mass. To see what kind of mass matrix we expect, we must look at the diagrams in detail.

Before we do this, let us identify the fermions which get mass of order M_G. The 45_{fL}, as noted above, gets a bare mass term of order M_G. Also, the VEVs of the 16_{fS} ($f = e, \mu, \tau$ or T) and the $\overline{16}_{fS} 16_{fL} 45_{fL}$ Yukawa couplings induce mass mixings of order M_G between the SU(3) x SU(2) singlet components of the 16_{fL} and the 45_{fL}. All the neutral SU(3) x SU(2) singlets get mass of order M_G in this way.

The fermions in the 16_{AL} get dynamical mass of order Λ_A. The mechanism has nothing to do with the eTC theory, so we need not discuss them further. This leaves, as light fermions, only the three normal families, the 16_{EL}, $16_{\mu L}$ and $16_{\tau L}$ minus their respective RH neutrinos, and the lepton doublet and charged singlet in the 16_{TL}. We will call the charged particle in this doublet the T lepton, or $T^\pm$.

The $T^\pm$ is in the same SO(10) multiplet as the technifermions. Thus it can get mass through the $SO(10)_T$ gauge interactions. A LH lepton can emit a gauge boson and become a LH technifermion. This can emit an effective Higgs boson and become a RH technifermion which in turn can absorb the gauge boson and become a RH lepton.

All the quarks and leptons can get mass through their Yukawa interactions. The exchanged scalar mesons are the SU(3) singlet parts of the 16_{fS} multiplets. These can (and do) mix with the corresponding components of the 16_{TS}, because they transform the same way under the unbroken symmetries. Then these scalars can communicate between ordinary quarks and leptons and technifermions.

But what is most exciting about this is that these diagrams can produce a quark mass matrix with the right properties to explain the observed quark masses and mixing angles, without unnaturally small parameters. A mass hierarchy for the quarks results because the mixing of the scalar fields in 16_{fS} with those in 16_{TS} decreases as f gets farther from T on the circle of flavor. The mixing angles are small if the mixing between the 45_{fL} of superheavy fermions is small compared to their bare mass. Yet a nonzero Cabibbo-KM mixing does result, in general, because the charge 2/3 quark mass matrix is related to the mixing of the charged components of the scalars, while the charge -1/3 mass matrix is related to the different mixing of the neutral components.

It is probably premature to discuss the mass matrix in any detail without a more fundamental understanding of the theory at M_G and above. Nevertheless, we hope to return to these questions in future work.

Is this model, or one like it, the answer to the gauge hierarchy puzzle? We find the idea of effective Higgs mesons very attractive, but clearly, such theoretical work will be needed to test our dynamical speculations and tie down the parameters in the effective theory. Further, while the $SO(10)^5$ model is a tantalizing step towards a solution of the flavor puzzle, the unified theory still looks very complicated. Ultimately we will want some more fundamental answers.

But in the meantime, can we find experimental tests of the idea which do not depend on the details of the eTC theory? One striking prediction is the existence of a normal lepton doublet, the T^- and its neutrino, with no corresponding normal quarks. The T^- is probably not too heavy. We might expect a contribution to the T mass of order

$$m_\tau^2/m_\mu \simeq 30 \text{ GeV}. \tag{11}$$

Note also that there are semistable technibaryons, decaying only by the analogs of baryon number changing interactions. Further, while there are technifermions which should slow up at very high energies, there are no pseudogoldstone bosons.

*This research is supported in part by the National Science Foundation under Grant No. PHY77-22864.

REFERENCES

[1] This talk is based on two papers, H. Georgi and I.N. McArthur, HUTP-81/A054; and H. Georgi, HUTP-81/A057; both to be published in Nucl. Phys. B.

[2] L. Susskind, Phys. Rev. D $\underline{20}$, 2619 (1979). See also the review, E. Farhi and L. Susskind, Physics Reports $\underline{74}$, 277 (1981). S. Weinberg, Phys. Rev. D $\underline{13}$, 974 (1976) and D $\underline{19}$, 1277 (1979).

[3] H. Georgi, H.R. Quinn and S. Weinberg, Phys. Rev. Lett. $\underline{33}$, 451 (1974).

[4] S. Dimopoulos and L. Susskind, Nucl. Phys. B155, 237 (1979).

[5] E. Eichten and K. Lane, Phys. Lett. 90B, 125 (1980).

[6] E. Eichten, K. Lane and J. Preskill, Phys. Rev. Lett. $\underline{45}$, 225 (1980).

[7] S. Dimopoulos and J. Ellis, Nucl. Phys. B182, 505 (1981).

[8] B. Holdom, Phys. Rev. D $\underline{23}$, 1637 (1981).

[9] B. Holdom, Phys. Rev. D $\underline{24}$, 1441 (1981).

[10] H. Georgi and S.L. Glashow, Phys. Rev. Lett. $\underline{47}$, 1511 (1981).

[11] W. Caswell, Phys. Rev. Lett. $\underline{33}$, 244 (1974).

[12] A. Zee, Phys. Rev. D $\underline{7}$, 3630 (1973); S. Coleman and D.J. Gross, Phys. Rev. Lett. $\underline{31}$, 851 (1973). See also the Postscript.

[13] G. 't Hooft, lecture given at the Cargèse Summer Institute, 1979. See also, S. Coleman and B. Grossman, Harvard preprint HUTP-82/A009.

[14] J. Preskill and S. Weinberg, Phys. Rev. D $\underline{24}$, 1059 (1981).

[15] T.P. Cheng, E. Eichten and L.F. Li, Phys. Rev. D $\underline{9}$, 2259 (1974).

[16] N.P. Chang, Phys. Rev. D $\underline{10}$, 2706 (1974).

[17] S. Coleman and E. Weinberg, Phys. Rev. D $\underline{7}$, 1888 (1973).

[18] H. Georgi, H.R. Quinn and S. Weinberg, Phys. Rev. Lett. $\underline{33}$, 451 (1974).

Search for Proton Decay - The HPW* Deep Underground Water Ĉerenkov Detector[+]

R. Morse

Physics Department

University of Wisconsin-Madison

Abstract

The data gathering and background characteristics of a totally sensitive 850 ton Water Cerenkov Detector containing some 5×10^{32} nuclei and located at a depth of 1700 mwe in the Silver King Mine in Park City, Utah is described. The device contains 704 5" PMT's located throughout its volume with a density of ~ 1 PMT/m^3, and is surrounded with an active shield made from 6 m proportional wire tubes. The detector is optimized to be sensitive to the principle decay modes predicted by Grand Unified Theories, and to suppress the backgrounds that are associated with cosmic ray muon and neutrino interactions occurring both in the detector and in the surrounding rock. Using the Cerenkov light patterns from the proton decay products, the timing sequence and pulse height of the hit PMT's allow nucleon decays to be separated from backgrounds to a lifetime limit of 4×10^{32} years in about 3 years of running. Data taking will start about June 1982.

*J. Blandino[1], U. Camerini[3], D. Cline[3], W.F. Fry[3], J.A. Gaidos[2], W.A. Huffman[1], G. Kullerud[2], R.J. Loveless[3], A.M. Lutz[1], R. March[3], J. Matthews[3], R. McHenry[2], A. More[3], R. Morse[3], J. Negret[2], T.R. Palfrey[2], T. Phillips[2], D.D. Reeder[3], C. Rubbia[1], A.H. Szentgyorgyi[3], R.B. Willmann[2], C.L. Wilson[2], D.R. Winn[1], W. Worstel[1].

[1]Harvard, [2]Purdue, [3]Wisconsin

[+]Invited Talk; Third Workshop on Grand Unification - University of North Carolina, April 15-17, 1982.

I.   Introduction

The Harvard-Purdue-Wisconsin Collaboration (HPW) is currently
(June 1982) starting operation of a deep underground water Cerenkov
detector which will measure nucleon decays to a lifetime of 4×10^{32}
years. If the nucleon decay rates are close to present limits, and
decay via $N \to \ell\pi$ modes, then as many as ~ 50 events/year may be
recorded and various branching ratios established.

The detector is located in a deep mine under a mountain in
Park City, Utah, and sits at an elevation of 2,135 meters and has an
overburden of 1700 mwe. The research chamber or "stope" is
12 m x 15 m in area and 10 m high, and has a thick concrete pad upon
which the detector is built. Access is via a narrow-gauge battery
powered railroad cart that provides access to the surface thru a
5 km horizontal tunnel or "drift" with a one-way trip taking about
20 minutes. A "not-to-scale" sketch of the elevation view of the
detectors relationship to the mountain is shown in Figure 1. The
facility has "unlimited" amounts of very clean water and "limited"
amounts of 440 3ϕ power about 140 meters away at the Thaynes Shaft.
The chamber enjoys a constant year round temperature of 8.3°C /
and 98% humidity.

The University of Utah group (Keuffel, et. al.) excavated this
site in the Silver King Mine in 1964, and the site was operated for
10 years as a neutrino laboratory. The stope is fully roof bolted
and the lab has a 2 ton crane available.

The computer control and data logging portion of the detector
uses both a PDP-11/45 and LSI-11/23 which sits in a trailer at the
portal of the 5 km access tunnel and talks to a second computer,
also an LSI-11/23 inside the stope via a 2 Megaband CATV co-axial

data link. Addition communications is provided by a 13 pair "twisted-pair" cable. This feature allows much of the routine monitoring and data logging aspects of the experiment to be performed above ground. Items such as fire, smoke, gas, heat, ozone and "all-OK" signals are routinely sent out via the link and any emergency situation shuts down the detector, and activates a "halon" fire system if necessary. In addition, there will be sufficient computing power at the surface location to do off-line analysis, and scanning of the reconstructed data.

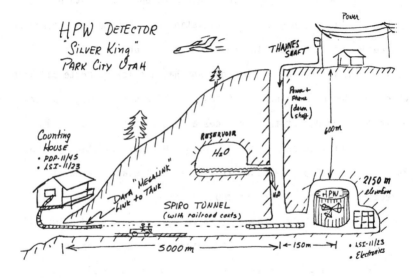

Fig. 1   Cross-Section View of Experimental Site.

II. Detector

The detector consists of a cylindrical tank 12 m in diameter and 7 m high filled with 850 tons of very pure water - some 4.6×10^{32} nuclei. Charged particles and $\pi^\circ \to 2\gamma$ are detected via their Cerenkov light and their Cerenkov cones are used to reconstruct the particles direction, energy loss, and particle type. This information is then used to reconstruct the event type and identify nucleon decay and reject particle backgrounds. The device is particularly sensitive to $p \rightarrow e^+\pi^\circ$ and $n \rightarrow e^+\pi^-$ decays, where the decay signature is "back to back" mono-energetic tracks of about 0.5 Mp each - a signature which is distinctive to any background except for the effects of Fermi momentum. See Figure 2 for a "typical" $p \rightarrow e^+\pi^\circ$ decay pattern.

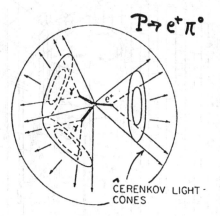

Fig. 2   Čerenkov Light Cones for a $P \rightarrow e^+\pi^\circ$ Decay in H_2O.

To detect this Čerenkov light in an efficient uniform manner, the tank volume is filled with 704 5" EMI PMT's with an acceptance of 3π, the tubes being arranged in a cubical lattice with 1 meter spacing which fills the entire tank volume.

A cut away view of the detector is shown in Figure 3. The cylindrical redwood tank is lined with a "Hypalon" liner to maintain the highest water purity. All internal surfaces are covered with aluminized teflon mirrors with a 90% reflection coefficient. This feature coupled with the volume distribution of PMT's within this mirror lined surface guarantees that the efficiency for detection of proton decays is nearly independent of the decays spatial location within the tank. This "infinite" lattice guarantees that all light is collected - a feature that is crucial if the calorimetry aspects of the detector are to be fully exploited.

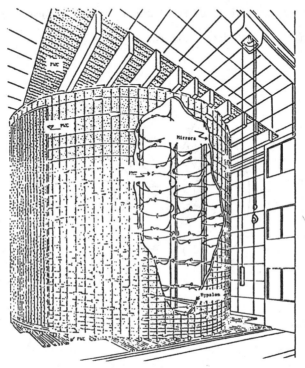

Fig. 3   Cut-away View of the Cerenkov Water Detector.

The PMT's are potted into water tight aluminum cans which are attached to 1 m diameter stainless steel hoops - 4 PMT's per hoop, and the hoops are suspended 1 m apart vertically on plastic covered stainless steel cables from the top of the tank. Each array contains 8 hoops and 32 PMT's. The HV and signal cables run thru water-tight polyethylene tubes and the base of each tube is maintained at a positive pressure by passing air thru a small inner tube into the

base and returning the air thru a second small tube into a bubbler
monitor system.

The water is filtered thru dual-bed de-ionization tanks both
during the filling and recirculation phase to achieve highest purity.
The recirculation rate is 160 ℓ/min thereby treating the entire
volume of 8 x 10^5 ℓ every three days. This continuous circulation
together with a tight roof and hypalon tank liner will provide
$10^7 \Omega$-cm water and a mean alternuation length of >25 m at λ = 420 mm.
A 3 ton mini-version of the HPW detector (8 PMT's) has been in
operation now for over a year during which period both water quality
and maintenance requirements have been continuously studied.

III. Active Shield

The entire active volume of the detector is to be surrounded
with an active shield consisting of gas filled aluminum proportional
tubes and concrete. This shield serves the following purposes:
i) the proportional tubes tag all charged particles including muons
entering the tank, ii) the concrete converts neutral hadrons, and
produces shower form all hadrons making their detection easier,
and finally, iii) the shield will screen out any small levels
of radioactivity in the mine, although from the Utah group's
experience this should not be a problem. The active shield consists
of an inner wall of wire proportional tubes next to the volume of
water, followed by a concrete shield of up to 1 m or 3.3 interaction
lengths in thickness, followed by a second layer of tubes. The
floor of the detector will be covered with a single layer of tubes
which because of tank support members covers about 70% of the area.

The top of the tank will be covered 100% by the proportional tubes, but instead of the concrete shielding and second layer of tubes the top 60 cm of the water in the tank will function as the shield providing about 0.77 interaction lengths. This region is optically separate from the rest of the tank, and has 4π mirrors to trap the Čerenkov light inside the region and is viewed by the top 4 PMT's of every string of hoops. For the walls we expect a 98.6% hadron rejection and a 10^5 photon-rejection while for the roof we expect 84% hadron rejection and 82% photon-rejection.

The bottom of the tank has no active shield, so while muons and charged tracks are no problem the neutrals produced by ν-interaction in the concrete pad can feed undetected neutral hadrons into the active volume so that a larger fiducial cut will have to be made on data accepted near the bottom of the tank.

The tubes themselves are 6 m long with a cross-section of 15 cm x 5 cm and have two 50μ Au-W sense wires. A cross-sectional view of the tubes is shown in Fig. 4. The large gap guarantees at least 300 ion pairs per track crossing so that even when operated at HV = 2500 with an 80/20 A/CH_4 gas mixture their efficiency is in excess of 99.9% for min-ionizing muons. This gas mix is also very fast so that all signals are collected within 400 ns of the particles transit of the tube. At this relatively low gas gain of 2×10^4 the tubes are extremely quiet and have singles rates comparable to the expected cosmic ray muon fluxes.

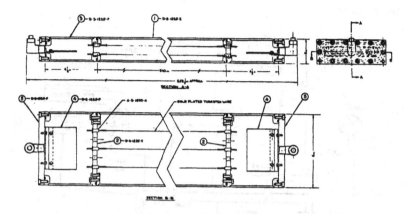

Fig. 4   The 6 m Wire Proportional Tubes.

IV. Electronics/Triggers

Dynode pulses from the PMT's are discriminated at the 1-2
photoelectron level and sent to the trigger summing circuits. A
delayed signal is also sent to the multiple time digitizers (MTD's).
At the same time delayed and integrated anode signals from the
LH-0033 "op-amps" in the PMT bases are sent directly to the ADC's.
The discriminators have computer controlled thresholds and each
channel can have its "singles" rates monitored by the computer to
help in checking on the PMT's proper operation. See Fig. 5 for a
block diagram of the electronics.

Timing is provided by the MTD's which records in very fast
RAMS (256 x 4) the arrival time of discriminator signals from each
PMT at a 125 MH_z rate, giving time bins of 8 ns and a time range of
$2\mu s$ (256 x 8 ns). Each MTD has 32 individual inputs and simultaneous
hits on all channels are allowed, and the unit can record up to
60 hits on any input. The system provides an accuracy of ± 4 ns
on the timing signals, and interpolating techniques are currently

being developed which can increase this accuracy by a factor of
two or more.

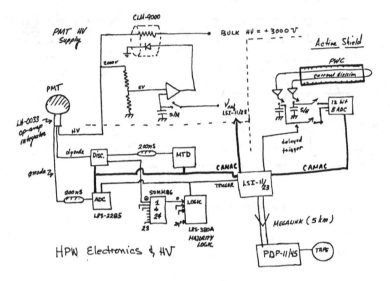

Fig. 5   Block Schematic of the Data/Trigger Electronics.

Trigger information is provided by 20 summing circuits which
adds the outputs from 24 discriminators and produces an output
current of 1 mA per hit PMT into a 50Ω load. These 30 signals are
then fed into a standard LRS 380-A majority logic system so that an
event trigger may be initiated in approximately 200 ns. At this
point "start" signals are sent to the MTD's and LRS 2280 ADC system.

PWC information from the active shield is also read in at this
time after suitable decay ($\sim$ 2μs) to insure that all of the charge
due to particles passing thru the chambers has been collected.
The system uses some 2000 tubes, and each tube has 12 m of wire.
Charge division with about 1% accuracy is used for the longitudinal
co-ordinate and each wire end connects to a charge amplifier with
a long time constant. Later when the delayed strobe arrives, analog
switches are closed and a voltage proportional to the charge is

stored on sample-and-hold capacitors. Later the data is digitized using a 4096 channel sampling ADC for every 256 channels. The data handling has "zero-supression" so that only hit wires, their pulse heights, and address or locations are sent onto CAMAC.

The status of the number of neighboring hit PMT's supplied by the summing circuits, plus the status of the number of neighborhoods hit determined by the majority logic, together with the status of the hit pattern seen in the active shield now forms the basis of a "event-trigger". The system as envisioned offers much flexibility and only after data taking begins can the optimum strategies be decided.

All of the above described data channels are finally formatted onto standard CAMAC channels and read into an LSI 11/23 computer for further processing. This LSI also monitors the PMT's HV system which sends HV to the 704 tubes, the state of the detector, and the fire, smoke, gas, line voltage conditions in the experimental area. A 2 megaband link connects the detectors LSI to a PDP-11/45 at the mine's portal where data are logged and some offline scanning and reconstruction of events is done. After installation and shakedown periods, the experiment will be conducted from the control room above ground at the mine portal.

V.   Calibration

Primary calibration will be provided by a pulsed 336 nm N_2 laser which is capable of 200 ps pulses. The light signal is transmitted via fibre optics to a "Ludox-Ball" inside the tank which fans the light outward isotropically or in any desired pattern. Once the ludox ball is located inside the tank, all of the PMT's can be "timed-in". Constrained fits can be done by using many different location for the ball. Further, by operating the laser with various aperture stops any number of photons can be directed to any PMT so that the 1-2 photo-electron discrimination levels can be set quite routinely. In addition to thresholds, the ADC counts vs light levels can be established for each tube and this information can be compared with muon tracks to provide an absolute light vs energy calibration for the system.

The laser calibration system also serves to monitor the water's transparency and purity level. Further plans call for using a tunable dye-laser which would permit measurement of the attenuation length at different frequencies, or a measurement of $\lambda_{abs}(\omega)$.

A final calibration of the PMT system will be provided by the cosmic ray muons traversing the tank. Their trajectories thru the tank will be uniquely set by the proportional chambers in the active shield, and there will be many thousand such tracks to "tune on" so that in principle any systematic effects in the reconstruction routines will be understood. These tracks will also provide an energy calibration of the system.

VI. Reconstruction

The task of the on-line/off-line reconstruction program is to: i) reject a large portion of the 1 Hz cosmic ray muon flux thru the detector, ii) collect and recognize as many nucleon decays as possible, iii) reject the fake nucleon decays by careful pattern analysis and energy resolution, and iv) to cope at some level with "unknown" events.

a) Signals

The proton decay signature occurring in the sensitive volume is:
1)   The total energy released is approximately 938 MeV,
2)   The decay products have no net momentum - except from Fermi - momentum contributions,
3)   The event is entirely contained with the volume,
4)   The decay axis is randomly distributed.

Recognition of these signatures in water is possible via the Cerenkov light the decay products emit. Charged particles with a

$\beta \geq 0.75$ will emit Cerenkov light in cones of half angle $0 \leq 42°$ with respect to their direction. This light travels thru the detector in a cone shaped pattern where it triggers the array of PMT's as shown in Figure 6. The energy (range) of a stopping particle is determined by the total number of photoelectrons collected. Monte Carlo results indicate that we can obtain a total energy resolution of $\Delta E/E = 0.2$ at $E = 1$ GeV by collecting 5-10% of the available Cerenkov light. The direction of this light permits the use of criteria 2, and 3 to identify the proton decays from the more obvious backgrounds while the $M_{\pi^0 e}{}^+ \sim M_p$ cut will help eliminate the more subtle ν-interaction backgrounds. Reconstruction of the event from the Cerenkov light will permit a vertex determination to about 0.2-0.4 m. Finally reconstruction permits a determination of the decay axis to $\theta < 14°$.

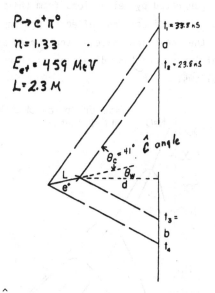

Fig. 6   Ĉerenkov Light Timing Pattern.

We list some expected numbers of photoelectrons collected for proton decays (similar results obtain for neutron decays):

Table 1

Mode	Number of p.e.	<ASYM>	<SPH>
$p \rightarrow e^+\pi^\circ$	600 ± 96	.16	.52
$p \rightarrow e^+\eta^\circ$	593 ± 90	.17	.75
$p \rightarrow e^+\rho^\circ$	243 ± 54	.24	.67
$p \rightarrow \mu^+\pi^\circ$	410 ± 91	.32	.78

and we expect $\mu^+ \rightarrow e^+$ from the π-μ-e sequence to yield 18 ± 10 p.e.

About 40% of the collected photons from proton decays
($p \rightarrow e^+\pi^\circ$) are generated by reflections from the mirrors. The time
distribution of these direct and reflected photons is shown in
Figure 7 where the scale and shape of the curves are set by the PMT
grid spacing of 4.5 ηs and tank dimension of 50 ηs. Reconstruction
proceeds as follows:

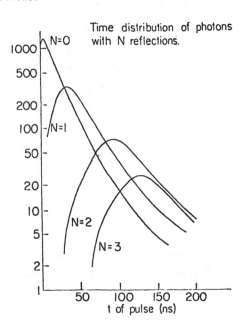

Fig. 7   Time Distribution of Direct and Reflected Photons.

i)   <u>short times - direct and 1st reflections</u> - are utilized to
     initially determine the approximate vertex and decay axis.
     These PMT hits are predominately (80%) from direct photons,
     and once - reflected photons, almost all of which are
     unambiguous.

ii) <u>all times - direct and many reflections</u> - are used thru
     iterative processing, the vertex position and alignment of
     decay axis and particle directions are more accurately
     defined and consistency-checked as the time scale is
     expanded.

iii) Finally the direct and multiply reflected photons are
     incorporated into the energy resolution.

Following reconstruction we can define the following proton
decay tests:

i) $<ASYM> = (N^+ - N^-)/(N^+ + N^-)$ where $N^+(N^-)$ are the number
     of photoelectron collected forward (backward) of the
     assigned vertex,

ii) <SPH> or the "sphericity" defined in the same way as in
     e^+e^- events.

These parameters show values of 0.7 and 0.1 respectively for
ν-induced background such as $\nu_e p \rightarrow ne^+\pi°$, while they have
considerably different values for the proton decay modes given in
Table 1, showing in fact average values of $<ASYM> < 0.3$ and
$<SPH> > 0.5$ respectively.

Conservative estimates suggest that we can expect to resolve
cleanly 1 event per month for nucleon lifetimes of the order of
10^{31} years. If the proton lifetime is more than and order of
magnitude longer the backgrounds from ν-interactions will ultimately
be the limiting factor. In this region where on an "event-by-event"
basis one cannot tell directly the proton decay from ν-interactions,
then one gains statistics only as $\sqrt{t}$ or as $\sqrt{V}$ where V is the volume
since the number of ν-interaction like the proton decays scale
directly with both V and with observing time t.

b)   <u>Background/Active Shield</u>

The expected background in the detector from various cosmic ray
sources is shown in Figure 8, the most important source at 1700 mwe

being atmospheric muons at a rate of ~ 1 H$_z$ so that a fast on-line rejection scheme good to the 1 to 1/10 percent level must be employed to get the number of triggers down from the basic 10^5/day to 10^2 - 10^3/day. Here a "typical" muon that reaches the detector has an energy of 0.8 TeV at the earth's surface with µ-bremsstrahlung causing a spread in primary energies. This flux of vertical muons although very high at this depth is sharply peaked forward toward the vertical and falls sharply with a $\cos^3\theta$ zenith dependence so that most muons will pass thru the tank

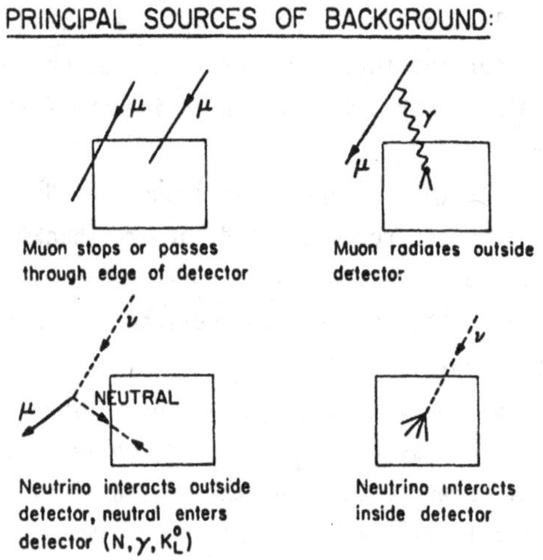

PRINCIPAL SOURCES OF BACKGROUND:

Muon stops or passes
through edge of detector

Muon radiates outside
detector

Neutrino interacts outside
detector, neutral enters
detector (N,γ,K$_L^0$)

Neutrino interacts
inside detector

Fig. 8   Background Sources - µ and ν Interactions.

vertically. However, most muons including the corner cutters can be rejected by the PWC's in the active shield, and by pulse height patterns and the total number of tubes hit. Muons tend to hit few PMT's but also tend to have higher numbers of photoelectrons per PMT, while the light from proton decays is much more diffuse. See Figure 9.

Other less numerous but potentially much more troublesome backgrounds are:

    i)   stopping muons,
    ii)   µ-bremsstrahlung and µN interactions producing both
          charged and neutral hadrons in the tank or surrounding rock,
    iii) νN interaction both in the tank and surrounding rock,
    iv)   natural radioactivity.

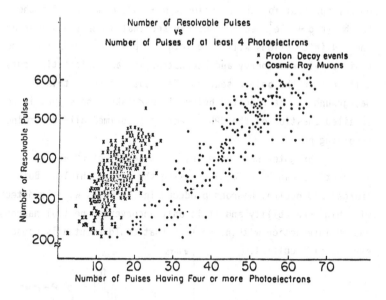

Fig. 9   Pulse Heights for μ and p-decay Events.

The stopping muons will be tagged in the active shield and should cause reconstruction problems only if their tracks are short, but then they should be cut on visible energy. Natural radioactivity in the rock though a problem for calorimeter-type detectors should cause no problem for the water Cerenkov detector since the energy depositions are too small, and since the water detector samples all of the energy. Radioactivity should really only effect the PWC's on the outside of the active shield, though the previous tenants, Keuffel et. al., had no problem with natural radioactivity.

Atmospheric neutrinos comprise the other known source of background which unlike accelerator beams contain appreciable ν_e and $\bar{\nu}_e$ components as well as ν_μ components. Neutrino interactions in the detector in the 0.5 - 2.0 GeV energy region will produce final states that mimic proton decays. For example, $\nu_e p \rightarrow \eta e^+ \pi^\circ$ which occur in the tank where $M_{e^+ \pi^\circ} \sim M_p$ look much like a $p \rightarrow e^+ \pi^\circ$ event with Fermi-momentum smearing, and it is backgrounds like this, that while they can be reduced with good measurement, ultimately limit the detecting power of the apparatus. For water Cerenkov devices with the standard quoted resolutions this limit sets in about

7-8 x 10^{32} years. As an example Fig. 10 shown the $\mu\pi^\circ$ visible
energy spectrum for the reaction $\nu_\mu N \to \mu\pi^\circ p$ compared with the
benchmark $p \to \pi^\circ e^+$ decay. One can see that a certain number of
the $\nu_\mu N$ interactions will feed into the $p \to \pi^\circ e^+$ channel on the
basis of visible energy and reconstruction and orientation cuts
will then be the only resource. The level at which this
background can be rejected before it overtakes the signal is now a
detailed question of the HPW detectors experimentally determined
resolving power.

On a brighter note however, the majority of μN and νN
interactions can be eliminated using the active shield. Both
charged and neutral hadrons produced in the rocks will be detected
with high probability and it is only unconverted neutral hadrons,
and νN interaction within the tank that could eventually cause
serious difficulties.

Fig. 10 The "Insidious"
ν-Backgrounds - an Example.

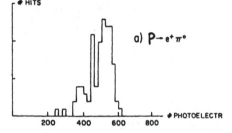

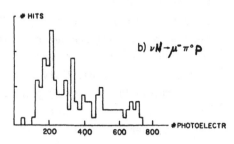

PULSE HEIGHT SPECTRUM

Acknowledgements
The author would like to thank all of the HPW collaborators who have
done most of the work reported here. Special thanks is due to
J. Gaidos, D. Cline, and D. Winn whose reported works provided much
of the form and content of this report. Finally I would like to
acknowledge the work of those people who are "in the trenches",
namely Tim Smart, and the miners L. North and M. Ryan who have
spent a good portion of the last year 600 meters underground.

SUPERSYMMETRY, GRAND UNIFICATION

AND PROTON DECAY

Serge Rudaz

School of Physics and Astronomy
University of Minnesota
Minneapolis, Minnesota 55455

ABSTRACT

In this talk, I review some phenomenological issues pertaining to supersymmetric
grand unified theories (SUSY GUTs), with
special emphasis on new features for proton
decay.

I. Supersymmetry and the Hierarchy Problem in Grand Unified Theories

Grand Unified Theories - GUTs for short, see [1] for recent reviews -
provide an appealing framework for a true unification of the strong and
electroweak interactions successfully described by the standard $SU(3)\times$
$SU(2)\times U(1)$ low energy theory. Theories of this type lead to unique new
insights into the systematics of quark and lepton masses and interactions. A most impressive result in GUTs is the accurate prediction of the
neutral current mixing angle $\sin^2\theta_W$ (see [2] for a review of SU(5) GUT
phenomenology). GUTs also offer an explanation for electric charge quantization and lead to the striking prediction that all baryonic matter is
unstable [1]. Still, GUTs leave some very basic questions largely unans-
wered: For example, why the replication of fermion generations ? Another difficulty arises as a result of the existence of two widely dispa-
rate symmetry breaking scales: In addition to the usual low-energy
electroweak scale $O(M_W)$, GUTs involve a superheavy scale M_X , characte-
rizing baryon number violating processes. Typically, one finds that
$M_W/M_X \lesssim 10^{-12}$. The possibility of any real quantitative predictive

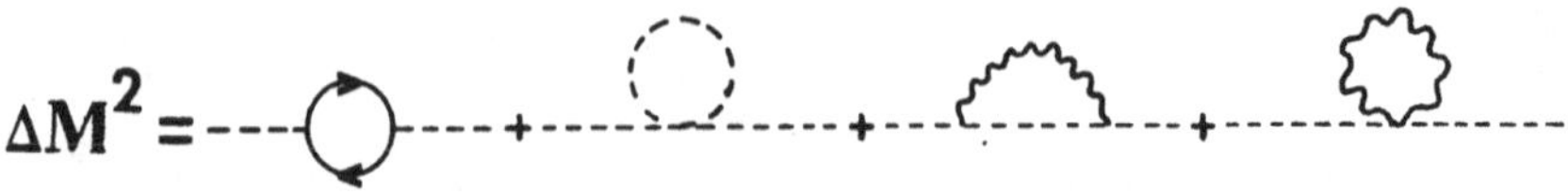

FIGURE 1   Contributions to the light scalar self-mass

power is predicated on there being a "Grand Desert" between these scales
in which coupling constants run unhindered by the influence of new particles and interactions with a priori unknown parameters. The existence
of these two widely different scales, and of the Desert in between, are
at the roots of the hierarchy problem.

To illustrate some aspects of the hierarchy problem, consider a
gauge theory with scalar fields ϕ and Φ with potential $V(\phi,\Phi)$. At tree
level, one can look for vacuum expectation values v and V for these
fields by solving a pair of minimization equations for the potential. If
we require that $v \ll V$, this amounts to two simultaneous equations for
what is, to $O(v^2/V^2)$, essentially one unknown V. Consistency will require a very fine tuning of the bare parameters in the potential, to accuracy $O(v^2/V^2)$: Suppose this is done. Now, however, the fact that the
low-energy theory (characterized by the scale v) is imbedded in a larger
one means that radiative corrections to its parameters are meaningful
and not just to be absorbed in counter-terms. As a result, the parameter v is dragged up to a value comparable to V. To see how this disas-
trous situation obtains, consider the light scalar particle mass, which
is closely related to v, and which receives radiative corrections from a
variety of graphs, as illustrated in Figure 1. One obtains an equation
for the scale-dependent scalar mass, of the form:

$$M^2(v^2) = M^2(V^2) + \sum_i A_i \alpha_i \int_{v^2}^{V^2} dq^2 + \sum_i B_i \alpha_i + O(\alpha^2) \quad . \tag{1}$$

Here, $M(v^2)$ is the effective mass at low energies, while $M(V^2)$ is the
"bare" mass, defined at the superheavy scale at which new physics appears.
The sums run over loops of virtual scalars, fermions and gauge particles,
with couplings α_i , and with coefficients $A_i \sim O(1)$ and $B_i \sim O(\log V^2/v^2)$.
Because of this quadratic dependendence of the scalar self-mass on the
high energy cut-off V, any relation between bare parameters arranged to
give $v \ll V$ will be spoiled by radiative corrections. There is also the
related issue that the standard Higgs doublet is associated with a color
triplet scalar in a GUT multiplet (e.g. a $\underline{5}$ of SU(5)) and that a fine

tuning of parameters in the potential is also required to give this color triplet a superheavy mass so as to prevent catastrophic proton decay.
This too is subject to radiative corrections.

To alleviate these difficulties requires an appeal to some new physics: One tack is to let the Desert bloom. For example, we may let the
interactions in the scalar sector only become strong, thus invalidating
the above perturbative analysis: This would lead to a new strong interaction sector around or above 1 TeV [3], while not necessarily spoiling
GUT predictions for gauge and Yukawa interactions. We may try to do
away completely with elementary scalars, as in the technicolor scenario
(see [4] for a recent review), where again there is new physics around
1 TeV: There are however serious difficulties in implementing this program in the context of GUTs.

A more attractive possibility is that a symmetry exists which ensures that $\sum A_i \alpha_i = 0$. This would entail cancellations between fermionic
and bosonic loop contributions in Figure 1: What is needed is a symmetry
relating fermions to bosons, and vice-versa, in other words, global
supersymmetry [5].

It has in fact been known for some time that globally supersymmetric
theories exhibit drastically softened quantum divergences as compared to
non-supersymmetric ones. An example of the so-called non-renormalization theorems of supersymmetric theories [6] is the absence of independent coupling constant or mass renormalization for matter fields, with
only logarithmically divergent wave-function and gauge coupling constant
renormalization.

As a consequence of supersymmetry (SUSY), bosons and fermions are
paired, arranged in supermultiplets, and their contributions to quadratic divergences in scalar self-masses cancel. This may provide a solution for both aspects of the radiative hierarchy problem: SUSY GUTs
would only require one "unnatural" adjustment of parameters at the superheavy scale, and this would be unaffected by radiative corrections, as
long as SUSY is unbroken.

Of course, the real world is conspicuously non-supersymmetric, with
no sign of degenerate multiplets of fermions and bosons in nature. On
the other hand, for SUSY to be useful in the solution of the hierarchy
problem, it can only be broken at a relatively low energy scale. To see
this, let SUSY be broken at a scale Λ_{SS}, leading to a lifting of the
degeneracy between say fermions f and their scalar partners sf: We can

then expect that $m_{sf}^2 - m_f^2 \sim O(\Lambda_{SS}^2)$, roughly. Then the cancellation of quadratic divergences will be imperfect with the result that $M^2(v^2) \sim O(\alpha(m_{sf}^2 - m_f^2)) \sim O(\alpha\Lambda_{SS}^2)$. Requiring that $M^2(v^2) \lesssim v^2$, we are led to the constraint $\Lambda_{SS}^2 \lesssim v^2/\alpha$ and with $v^2 \sim O(M_W^2/\alpha)$, finally to the conclusion that

$$\Lambda_{SS} \lesssim M_W/\alpha \sim O(1\text{-}10) \text{ TeV}$$

The hierarchy problem has thus been rephrased: The question is now why should SUSY be broken at such low energies ?

For a variety of reasons, supersymmetry in fact turns out to be quite resilient against breaking. Still, it may be broken either spontaneously à la O'Raifeartaigh or with a Fayet-Iliopoulos D-term [7], or it can be broken explicitly, but softly [8] (recently it has been suggested that gravitational effects could mimic explicit SUSY breaking[9]). Either way seems as good as the other, as the non-renormalization theorems hold in both cases [8]. We will assume here that SUSY is explicitly broken [10] and pursue the consequences of this assumption.

The difficulties in formulating a low-energy SU(3)×SU(2)×U(1) model with broken SUSY are formidable and while progress has recently been made towards this goal (some attempts are reviewed in [11]), a consistent model is still lacking. In spite of this, it seems worthwhile to analyze, in as general terms as possible, the properties expected of a SUSY GUT [10, 12, 13].

II. Phenomenology of a Minimal SUSY GUT: Particle Content, Grand Unification Scale, and All That.

We will consider the minimal SU(5) GUT, extended to include an unbroken N=1 global supersymmetry at the superheavy scale M_{SX}. The symmetry breaking pattern will then be

$$SU(5) \times (N{=}1 \text{ SUSY}) \xrightarrow{M_{SX}} SU(3){\times}SU(2){\times}U(1){\times}(N{=}1 \text{ SUSY})$$

$$\xrightarrow{\Lambda_{SS}} SU(3){\times}SU(2){\times}U(1)$$

$$\xrightarrow{M_W} SU(3){\times}U(1)_{em}$$

For definiteness, we will set $\Lambda_{SS} = M_W$, which will also then be the mass about which the unseen supersymmetric partners are clustered. The superfield content of this minimal Super-SU(5) [10,12,13] is as follows:

- Vector supermultiplet ($\underset{\sim}{24}$ of SU(5)) containing the SU(3), SU(2) and U(1) gauge bosons, as well as the superheavy leptoquark bosons X and Y, and their spin-1/2 superpartners, the gauginos, denoted by a tilde ($\tilde{W}$, $\tilde{B}$, $\tilde{g}$, $\tilde{X}$ and $\tilde{Y}$, winos, binos, gluinos,...)
- Three generations of left-handed ($\underset{\sim}{\overline{5}} + \underset{\sim}{10}$)$_L$ quark and lepton chiral superfields, containing the usual quarks and leptons and their spin-0 superpartners, the squarks (sq) and sleptons (sℓ).
- Two Higgs chiral superfields ($\underset{\sim}{5}_L$ and $\underset{\sim}{\overline{5}}'_L$), containing the scalars H_1 and H_2^* and their spin-1/2 partners (Higgsinos).
- A left-handed chiral supermultiplet of adjoint Higgs Φ_L, whose scalar components break SU(5) by acquiring a vacuum expectation value in the usual way down to SU(3)×SU(2)×U(1) [10].

Note that this minimal SUSY GUT has more than twice the usual number of states in standard SU(5). In particular, two distinct 5-plets of Higgs supermultiplets are required [10,12,13], for at least two reasons. First, to get a reasonable fermion spectrum, a point first emphasized by Sherry [14]: The new constraint of SUSY on Yukawa couplings forces the existence of separate $\underset{\sim}{5}_L$ and $\underset{\sim}{\overline{5}}'_L$ left-handed chiral superfields to give masses to up-type and down-type fermions respectively. Second, the couplings of spin-1/2 Higgsinos within a single chiral superfield $\underset{\sim}{5}$-plet will lead to anomalies, unless there is an accompanying $\underset{\sim}{\overline{5}}$ to cancel them.

Given this new particle spectrum, it is a simple matter to write down the appropriate renormalization group equations (taking higher order effects into account) to determine the evolution of the gauge coupling constants and the new grand unification scale M_{SX}. This is shown schematically in Figure 2 where we contrast coupling constant evolution in both the usual SU(5) GUT and its minimal supersymmetric extension.

Grand Unification Scale

The significant new result is that both the SUSY GUT unification scale and the grand unified coupling α_{SUM} are increased as compared to the corresponding quantities in minimal SU(5). This is easily understood to follow from the proliferation of new fermionic species whose presence modifies the evolution equations [15,16,17,18] so that the non-abelian couplings are less asymptotically free than before: As seen in Figure 2, the SU(2) coupling α_2 in fact increases with E in SUSY GUT. This defers grand unification to a higher scale, while making α_{SUM} larger than the corresponding α_{GUM}. Quantitatively, M_{SX} is extremely sensitive to the

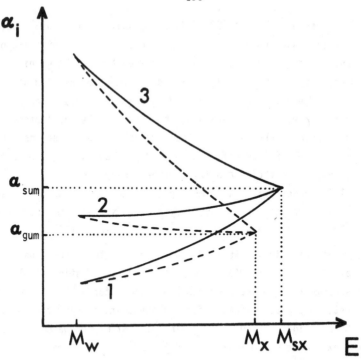

FIGURE 2 Schematic representation of coupling constant evolution
in Minimal SU(5) [dashed lines] and its supersymmetric
extension [solid lines]

number N_H of light Higgs doublets[15,16,17,18]:

$$M_{SX} \simeq 6 \times 10^{16} \; \Lambda_{\overline{MS}} \; [GeV] \; GeV \quad (\text{for } N_H = 2) \qquad (2)$$

$$\simeq 3 \times 10^{15} \; \Lambda_{\overline{MS}} \; [GeV] \; GeV \quad (\text{for } N_H = 4) , \qquad (2')$$

as compared to the usual SU(5) result [1,19]

$$M_X \simeq 1.5 \times 10^{15} \; \Lambda_{\overline{MS}} \; [GeV] \; GeV .$$

We note in both cases the familiar linear dependence on the QCD scale
parameter, here expressed in GeV. We can immediately deduce from previous experience the expected proton lifetime in SUSY GUT arising from
the exchange of superheavy gauge bosons (GEX): Using the scaling law
$\tau_p^{GEX} \sim (\text{const.}) \times M_X^4$ leads to the estimates

$$\tau_p^{GEX} \simeq 4 \times 10^{38\pm1} \times (\Lambda_{\overline{MS}} \; [GeV])^4 \; years \quad (\text{for } N_H=2) \qquad (3)$$

$$\simeq 2 \times 10^{33\pm1} \times (\Lambda_{\overline{MS}} \; [GeV])^4 \; years \quad (\text{for } N_H=4), \qquad (3')$$

with the usual guess of the uncertainties involved [1, 19]. We see that if GEX is the only contribution to proton decay, SUSY GUT with N_H = 2 in all likelihood leads to an unobservably long lifetime, which for the case N_H = 4 may be experimentally accessible, given a plausible value of $\Lambda_{\overline{MS}}$, say between 100 and 200 MeV: However, taking N_H = 4 also leads to an unacceptably large value of $\sin^2\theta_W$, as we will see. In any case, we will argue in the next section that there are in fact new contributions to proton decay arising from dimension-5 operators which unless forbidden by a symmetry will lead to an observable decay rate for the proton with N_H = 2.

The new value of the grand unified coupling should also be mentioned: One obtains that α_{SUM} = 1/24, to be compared with the usual α_{GUM} = 1/42.

Neutral Current Mixing Angle

It is also found that the predicted value of the neutral current mixing angle is increased in SUSY GUT [15,16,17,18] :

$$\sin^2\theta_W(M_W) \simeq 0.236 \pm 0.002 \quad \text{(for } N_H = 2) \tag{4}$$

$$\simeq 0.259 \pm 0.002 \quad \text{(for } N_H = 4) \tag{4'}$$

to be compared with the usual SU(5) prediction [2]

$$\sin^2\theta_W(M_W) = 0.208 + 0.006 \log (0.4 \text{ GeV}/\Lambda_{\overline{MS}})$$

$$= 0.214 \pm 0.002$$

On the other hand, the radiatively corrected experimental average value for this quantity is [2] :

$$\sin^2\theta_W{}^{exp} = 0.215 \pm 0.012$$

Thus while the case N_H = 2 merely looks uncomfortable, N_H = 4 seems to disagree with experiment.

Quark-Lepton Mass Ratios

The usual SU(5) predictions in the symmetry limit remain of course unchanged in SU(5) SUSY GUT. What is surprising is that when effective masses are renormalized down to physical values, the prediction m_b/m_τ = = 3.0 also remains unchanged in SUSY GUT. Needless to say, the same problems remain for light quark and lepton masses.

Raising the SUSY Breaking Scale

Finally, we may ask what happens if we slightly increase the SUSY breaking scale, to a value $\Lambda_{SS} > M_W$, thus displacing the SUSY partner thresholds into the Desert. It is found in this case that both M_{SX} and $\sin^2\theta_W(M_W)$ are slightly lowered [18]: for $N_H = 2$,

$$M_{SX} = (9\pm3) \times 10^{15} \text{ GeV} \qquad \text{(when } \Lambda_{SS} = M_W) \qquad (5)$$

becomes

$$M_{SX} = (4\pm2) \times 10^{15} \text{ GeV} \qquad \text{when } \Lambda_{SS} = 10 \text{ TeV,} \qquad (6)$$

while

$$\sin^2\theta_W = 0.236 \pm 0.002 \qquad \text{(when } \Lambda_{SS} = M_W)$$

becomes

$$\sin^2\theta_W = 0.233 \pm 0.002 \qquad \text{when } \Lambda_{SS} = 10 \text{ TeV.} \qquad (7)$$

We now turn to a description of a new mechanism for proton decay peculiar to SUSY GUT, through closed loops of supersymmetric partners of the usual matter and gauge fields. We will see that this may lead to proton decay with a peculiar signature and with an observable lifetime, the increase in the grand unification scale notwithstanding.

III. Super-Nucleon Decay: New Features

It is well-known that the phenomenological consequences at low-energy of a theory with new interactions at a superheavy mass scale M* are conveniently parametrized by an effective Lagrangian involving local operators transforming as singlets of the low-energy residual symmetry group [20]. In analyzing GUT predictions, for example, one writes an expansion

$$\mathcal{L}_{eff} = \sum_d A_d \, (M*)^{4-d} \, O^{(d)} \qquad (8)$$

in terms of SU(3)×SU(2)×U(1) symmetric operators $O^{(d)}$, of dimension d, and of the superheavy mass M*. For baryon number violation in standard GUTs, the minimal dimension operator must involve at least three quark fields to form a color singlet, and have an even number of fermions for Lorentz invariance: Thus it is of the form qqqℓ and has dimension six, and so the effective interaction is of the form

$$\mathcal{L}^{GUT}_{eff} = O(g^2)M_X^{-2} \; q \, q \, q \, \ell + \cdots \qquad (9)$$

Now it was pointed out by Weinberg [21] and by Sakai and Yanagida [22] that if the low energy theory is extended to be SU(3)×SU(2)×U(1)×SUSY symmetric, new baryon number violating operators will be generated. Given the superfield content of the minimal SUSY GUT, we can immediately write down the dimension-four operator

$$\int d^2\theta \; U_{R\alpha} \, D_{R\beta} \, D_{R\gamma} \; \epsilon^{\alpha\beta\gamma} \qquad (10)$$

with U_R, D_R right-handed SU(2) singlet quark superfields, and α,β,γ color indices. The operator in (10) gives rise to an effective interaction between two quarks and a squark, with no suppression by inverse powers of M_{SX}, leading to an extremely short proton lifetime and so it must be forbidden [21,22,23]. This can be achieved in a number of ways, the simplest being to impose a discrete family reflection symmetry, under which the quark and lepton superfields are replaced by their negatives. Such a symmetry could even be considered "natural" in the context of the simply supersymmetric grand unified theory considered here [10,13]

Under these conditions, there will still survive some dimension - five operators contributing to baryon decay, of the form:

$$\int d^2\theta \; Q_{Li\alpha} \, Q_{Lj\beta} \, Q_{Lk\gamma} \, L_{Ln} \; \epsilon^{\alpha\beta\gamma} \, \epsilon^{ij} \, \epsilon^{kn}$$

$$\int d^2\theta \; Q_{Li\alpha} \, Q_{Lj\beta} \, Q_{Lk\gamma} \, L_{Ln} \; \epsilon^{\alpha\beta\gamma} \, (\vec{\tau}\epsilon)^{ij} (\vec{\tau}\epsilon)^{kn} \qquad (11)$$

$$\int d^2\theta \; U_{R\alpha} \, U_{R\beta} \, D_{R\gamma} \, \ell_R \; \epsilon^{\alpha\beta\gamma} \qquad \cdots$$

where Q_L and L_L are left-handed doublet quark and lepton superfields, with $i,j,k,\ldots$ SU(2) indices, and ℓ_R is a right-handed singlet lepton superfield. Note the following important property: These operators must involve three distinct quark flavors, or vanish identically (generation indices are suppressed in (11)). To see this, note that when expressed in component form, these operators yield interactions involving for example two squarks and a quark and lepton. The requirement of color antisymmetrization will thus clash with that of Bose symmetry unless more than one generation of quarks are involved. As a consequence of this, we expect that in SUSY GUTs, strange particle final states will be dominant in proton decay [17,23].

A diagram giving rise to just such a dimension five operator is shown in Figure 3, leading to an effective baryon number violating inter-

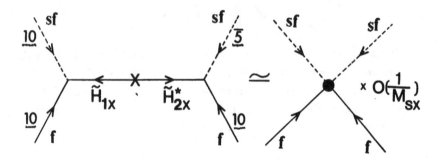

FIGURE 3 A diagram contributing to a dimension-five operator with baryon number violation.

action of the form:

$$\mathcal{L}_{eff} \sim \frac{g_H^2}{M_{\tilde{H}_X}} [(sq\ sq\ q\ \ell)\ or\ (sq\ s\ell\ q\ q)] \tag{12}$$

as a result of mass mixing between the spin-1/2 superpartners of the (superheavy) color triplet components of the $\underset{\sim}{5}$ and $\overline{5}'$ of Higgs scalars. We will in fact assume that $M_{\tilde{H}_X} \sim M_{SX}$. Also, g_H^2 will be a typical Higgs coupling, of the form (cf. Figure 3):

$$g_H^2 = \sqrt{2}\ G_F\ m_s m_u \times 2K \tag{13}$$

with $2K = v_0^2/v_1 v_2 = (v_1^2 + v_2^2)/v_1 v_2 > 1$, a numerical factor related to the Higgs v.e.v. in the standard model v_0 and to the v.e.v's v_1 and v_2 of the two Higgs doublets in the minimal SUSY GUT.

To get an operator actually resulting in proton decay, the scalars involved in the dimension five operators must still be transformed into fermions, quarks and leptons. This can be achieved via the loop diagram shown in Figure 4. The cross on the gaugino line ($\tilde{W}$, $\tilde{B}$ and $\tilde{g}$) represents a Majorana mass insertion which explicitly breaks supersymmetry: In fact, in the absence of such breaking, supersymmetry leads to a vanishing of such loop contributions [24] ! Naturally, the values of the various SUSY partners' masses are just so many free parameters, but we may consider some limiting cases. For example, if (as assumed in Figure 4) we let $M_{\tilde{W}} \gg m_{sf}$ we get an effective interaction for baryon decay,

$$\mathcal{L}_{eff} \sim C\ \frac{g_H^2\ g_2^2}{M_{\tilde{H}_X} M_{\tilde{W}}}\ q_L\ q_L\ q_L\ \ell_L \tag{14}$$

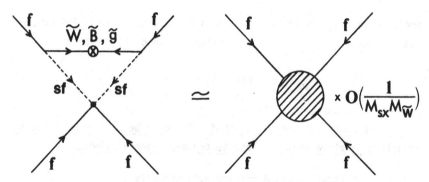

Figure 4 Loop diagram contributing to baryon decay. Here, as in
Figure 3, arrows denote helicities.

where C is an important numerical factor, to be compared with the usual
GUT result of Equation (9). One would at once suspect that since $M_{\widetilde{W}} \sim$
$\sim O(\Lambda_{SS}) \ll M_X$, such an interaction would lead to much too rapid proton
decay. A careful analysis shows, however, that this is not necessarily
the case [17,23]. For example, one finds for the coefficient C in Equation (14) the result

$$C = \frac{1}{32\pi^2} \log M_{\widetilde{W}}^2/m_{sf}^2 \tag{15}$$

If, on the other hand, we take say $m_{\widetilde{g}} \ll m_{sf}$, the entire coefficient
of $qqq\ell$ becomes

$$\frac{1}{16\pi^2} \cdot \frac{8g_3^2}{3} \cdot \frac{m_{\widetilde{g}}}{m_{sq}^2} \frac{g_H^2}{M_{\widetilde{H}_X}} \tag{16}$$

Using these interactions, and taking into account the renormalization
effects as usual (with suitable modifications!) we can compute the proton lifetime and obtain bounds on the various SUSY partners' masses consistent with existing limits from experiment. Taking as a benchmark the
value $M_X = 4{\times}10^{14}$ GeV for the usual GUT scale, we find a lifetime exceeding a lower bound of 10^{30} years or so, provided [17]

$$M_{\widetilde{W}} \cdot M_{\widetilde{H}_X} \geq (0.4 \text{ to } 4) \times 10^{18} \text{ GeV}^2 .$$

Assuming $N_H = 2$ and therefore $M_{H_X} \sim M_{SX} = (0.5 \text{ to } 2) \times 10^{16}$ GeV, this
yields the bound (applicable to the case $M_{\widetilde{W}} \gg m_{sf}$)

$$M_{\widetilde{W}} \geq (10 \text{ to } 10^3) \text{ GeV}, \tag{17}$$

or a factor of 20 larger if $N_H = 4$. This is clearly compatible with one's naive expectations for SUSY GUTs as outlined above. The other case, $m_{\tilde{g}} \ll m_{sq}$ (lighter gaugino exchange) leads to the bound

$$m_{sq}^2/m_{\tilde{g}} \geq 0(0.1 \text{ to } 6) \text{ TeV} \tag{18}$$

which is also not impossible to arrange.

These numbers suggest that SUSY GUTs would be most comfortable with a nucleon lifetime that is close to the experimental limit.

IV. Nucleon Decay Modes: A Test of Supersymmetry ?

The question of what precise decay modes will follow for protons in SUSY GUTs can only be partially answered, if relative model-indepedence is to be retained. We have already found out that strange final states will probably dominate: We also know that, for $SU(5) \times$ SUSY, B - L is conserved. This uniquely singles out the mode $n \to K^0 \bar{\nu}$ for neutrons bound in nuclei, while leaving as likely dominant modes $p \to K^0 \mu^+$ and $p \to K^+ \bar{\nu}$ for protons.

We may still want to take an admittedly model-dependent look at proton decay modes in minimal SUSY GUT, and make for example the following assumptions [17] :

(a) include all possible superheavy Higgs scalar and Higgsino exchanges, as well as all "light" gaugino exchanges;

(b) allow for three generations and all possible Higgs couplings (proportional to up-type and down-type quark mass matrices);

(c) assume that all mixing angles in the new scalar and fermionic sectors of the theory are equal to the usual Cabibbo-Kobayashi-Maskawa angles, and take the resulting GIM cancellations seriously.

This last assumption is clearly the strongest and may be somewhat questionable. However, it does lead to some amazing cancellations, resulting in the following hierarchy of decays (where K and π generically denote strange and non-strange final states respectively):

$$
\begin{array}{lll}
N \to \bar{\nu} K & \gg N \to \bar{\nu} \pi & (\sim 10\% \text{ level}) \\
& \gg N \to \mu^+ K & (\sim 10^{-3}) \\
& \gg N \to \mu^+ \pi & (\sim 10^{-4}) \\
& \gg N \to e^+ K & (\sim 10^{-5}) \\
& \gg N \to e^+ \pi & (\sim 10^{-6}) \quad,
\end{array}
$$

a pattern in sharp contrast with standard SU(5) expectations. In particular, note that many experiments are designed for maximum sensitivity to the $e^+\pi^0$ mode, which in the case of SUSY GUTs will be maximally suppressed (Yukawa couplings!) . For example, we may consider the IMB experiment with a projected benchmark sensitivity of 10^{33} years to the $p \rightarrow e^+\pi^0$ mode: Then, as a result of branching ratio effects and of less constrained kinematics, the sensitivities to the modes $p \rightarrow \bar{\nu} K^+$ and $n \rightarrow \bar{\nu} K^0$ are decreased to only $O(10^{31})$ years, whereas the mode $p \rightarrow \mu^+ K^0$, if appreciable, could be seen up to the $O(10^{32})$ years level [25]. *Caveat emptor* !

We may therefore summarise the possibilities for nucleon decay in SUSY GUTs:

1) <u>via superheavy gauge boson exchange</u>: As outlined above, this is likely to lead to an unobservably long partial lifetime (τ_p^{GEX} , Equations (3) and (3')) into the usual decay modes expected in SU(5);

2) <u>via loops of scalar fermions and gauginos</u>, with a very probably observable lifetime ($\tau_p \sim 10^{30\pm?}$ yrs), but with peculiar final states, dominated by kaons.

3) There remains the possibility, not mentioned above, of proton decay occurring <u>via the exchange of color-triplet Higgs scalars of mass around $O(10^{11})$ GeV</u>. This is what happens in the geometrical hierarchy model of Dimopoulos and Raby [26], and in a model of Nanopoulos and Tamvakis [27]. The main decay modes would again be dominated by strange final states, but with the modes involving muons and antineutrinos at roughly the same level. A lifetime of $O(10^{31})$ years is possible, but this decay mechanism could also conceivably happen in a standard GUT although such low masses (compared to M_X) for the color-triplet Higgs are unlikely in that case, especially when radiative corrections are present.

V. Conclusion

We have seen that, while there are many uncertainties tied to the large number of a priori unknown parameters, SUSY GUTs may yield an observable proton lifetime, arising from processes at the loop level, in spite of a larger unification scale, and with the promise of a solution of the hierarchy problem. It would indeed be ironic if the first indication of the importance of supersymmetry in Nature were to be the discovery of proton decay into predominantly strange final states !

Acknowledgements

I wish to thank J. Ellis and D. V. Nanopoulos for a most pleasant collaboration over a long period of time, leading to some of the results discussed here. My thanks also to P. Frampton and the organizers of TWOGU for the opportunity of presenting them at this workshop. The preparation of this talk was supported in part by the U.S. Department of Energy under Contract No. DE-AC02-82ER40051.

References

[1] J. Ellis, in Gauge Theories and Experiments at High Energies, K.C. Bowler and D.G. Sutherland, eds. (Scottish Universities Summer School in Physics, Edinburgh, 1981) p. 201; P. Langacker, Physics Reports 72C, 185 (1981).

[2] W. J. Marciano and A. Sirlin, in Proceedings of the Second Workshop on Grand Unification, J.P. Leveille, L.R. Sulak and D.G. Unger, eds. (Birhäuser, Boston, 1981) p. 151.

[3] M. Veltman, Acta Phys. Polon. B8, 475 (1977)

[4] E. Farhi and L. Susskind, Physics Reports 74C, 277 (1981)

[5] J. Wess and B. Zumino, Nucl. Phys. B70, 39 (1974); D. Volkov and V.P. Akulov, Phys. Lett. 46B, 109 (1973); Y.A. Golf'and and E.P. Likhtman, Pisma Zh. Eksp. Teor. Fiz. 13, 323 (1971)

[6] J. Wess and B. Zumino, Phys. Lett. 49B, 52 (1974); J. Iliopoulos and B. Zumino, Nucl. Phys. B76, 310 (1974); S. Ferrara, J. Iliopoulos and B. Zumino, Nucl. Phys. B77, 413 (1974) S. Ferrara and O. Piguet, Nucl. Phys. B93, 261 (1975); M.T. Grisaru, W. Siegel and M. Roček, Nucl. Phys. B159, 420 (1979).

[7] L. O'Raifeartaigh, Nucl. Phys. B96, 331 (1975); P. Fayet and J. Iliopoulos, Phys. Lett. 31B, 461 (1974)

[8] L. Girardello and M.T. Grisaru, Nucl. Phys. B194, 65 (1982)

[9] B.A. Ovrut and J. Wess, IAS Princeton preprint (1981); S. Weinberg, these Proceedings

[10] S. Dimopoulos and H. Georgi, Nucl. Phys. B193, 150 (1981); N. Sakai, Z. Phys. C11, 153 (1981); N.V. Dragon, Heidelberg preprint HD-THEP-82-3

[11] B.A. Ovrut, these Proceedings

[12] E. Witten, Nucl. Phys. B185, 513 (1981)

[13] H.P. Nilles and S. Raby, Nucl. Phys. B198, 102 (1982)

[14] T.N. Sherry, ICTP Trieste preprint IC/79/105, September 1979

[15] L. Ibáñez and G.G. Ross, Phys. Lett. 105B, 439 (1981)

[16] M.B. Einhorn and D.R.T. Jones, Nucl. Phys. B196, 475 (1982)

[17] J. Ellis, D.V. Nanopoulos and S. Rudaz, CERN preprint Ref.TH.3199, to be published in Nucl.Phys.

[18] W.J. Marciano and G. Senjanovic, BNL preprint BNL 30398, Nov. 1981

[19] J. Ellis, M.K. Gaillard, D.V. Nanopoulos and S. Rudaz, Nucl. Phys. B176, 61 (1980) and references therein.

[20] S. Weinberg, Phys. Rev. Lett. 43, 1566 (1979); F. Wilczek and A. Zee, Phys. Rev. Lett. 43, 1571 (1979).

[21] S. Weinberg, Harvard preprint HUTP-81/A017, to be published

[22] N. Sakai and T. Yanagida, Nucl. Phys. B197, 533 (1982)

[23] S. Dimopoulos, S. Raby and F. Wilczek, Phys. Lett. 112B, 133 (1982)

[24] W. Lang, Karlsruhe preprint, March 1982

[25] L. Sulak, private communication

[26] S. Dimopoulos and S. Raby, these Proceedings

[27] D.V. Nanopoulos and K. Tamvakis, CERN preprint Ref.TH.3255.

NEUTRINO OSCILLATION EXPERIMENTS AT ACCELERATORS[†]

Herbert H. Chen
Department of Physics, University of California
Irvine, California 92717

Abstract

The status of neutrino oscillation experiments at accelerators is reviewed. Present limits on neutrino oscillations are given. Several current oscillation experiments are described. The sensitivities of these, and other oscillation experiments in progress are summarized.

1. Introduction

Interest in searching for neutrino oscillation phenomena is very high at fixed target proton accelerators where neutrino experiments exist. In contrast to reactors where, due to the low neutrino energies, only the disappearance of $\bar{\nu}_e$'s can be studied, at accelerators, both the disappearance of a given neutrino type as well as the appearance of new neutrino types can be studied.

In searching for new effects, it is sufficient to consider initially the simplest phenomenology to parametrize that effect. Necessary conditions for neutrino oscillations to occur are that individual lepton numbers (L_i, i = e, μ, τ,...) are not separately conserved, and that at least two neutrino masses are non degenerate. The standard analysis for neutrino oscillations incorporates the minimal assumption of two neutrino mixing (as well as CP and CPT invariance). In this case, the observed neutrino types, e.g. ν_e and ν_μ, are quantum admixtures of neutrino mass eigenstates ν_1 and ν_2. This mixing is represented usually as follows:

[†]Talk presented at the Third Workshop on Grand Unification, University of North Carolina, Chapel Hill, North Carolina, April 15-17, 1982.

$$\nu_e = \nu_1 \cos \theta + \nu_2 \sin \theta$$

$$\nu_\mu = - \nu_1 \sin \theta + \nu_2 \cos \theta$$

where θ is the mixing angle. Observable consequences of such mixing would be neutrino oscillations, e.g. the appearance of ν_e's in an initially pure ν_μ beam, with the corresponding disappearance of ν_μ's.

The associated probability for the $\nu_\mu \rightarrow \nu_e$ appearance transition is $P(\nu_\mu \rightarrow \nu_e)$ which satisfies the following relations:

$$P(\nu_\mu \rightarrow \nu_e) = P(\bar{\nu}_\mu \rightarrow \bar{\nu}_e) \qquad\qquad \text{CP} \quad \text{invariance}$$

$$= P(\bar{\nu}_e \rightarrow \bar{\nu}_\mu) \qquad\qquad \text{CPT} \quad \text{invariance}$$

$$= P(\nu_e \rightarrow \nu_\mu) \qquad\qquad \text{T} \quad \text{invariance}$$

$$\equiv P$$

and

$$P = \sin^2(2\theta) \sin^2\left(\frac{1.27 \ L \ \delta m^2}{E_\nu}\right)$$

where
$$\delta m^2 = |m_1^2 - m_2^2| \qquad\qquad\qquad\qquad \text{(in eV}^2\text{)}$$
$$L = \text{neutrino source to detector distance} \quad \text{(in meters)}$$
$$E_\nu = \text{neutrino energy} \qquad\qquad\qquad\qquad \text{(in MeV)}$$

and the probability for initial ν_μ's remaining ν_μ's is:
$$P(\nu_\mu \rightarrow \nu_\mu) = 1 - P(\nu_\mu \rightarrow \nu_e)$$
$$= 1 - P$$

For experiments utilizing a single detector where one compares observation with expectation, it is convenient to examine $\sin^2(1.27 \ L \ \delta m^2/E_\nu)$ in the low and high δm^2 limits. In the low δm^2 limit, the sine function can be replaced by its argument to give,

$$\delta m^2 \sin(2\theta) = 0.8 \left(\frac{E_\nu}{L}\right) \sqrt{P} \qquad \text{(low } \delta m^2 \text{ limit)}$$

while in the high δm^2 limit, the sine squared function averages to $\frac{1}{2}$, and one gets,

$$\sin^2(2\theta) = 2P \qquad\qquad \text{(high } \delta m^2 \text{ limit)}$$

The low and high δm^2 limits c.n be shown graphically as straight lines in a log-log plot. Figure 1 shows these limits superposed on results published by Baker et al [1]. One sees that the low and high δm^2 limit lines provide a not unreasonable approximation to the detailed results shown.

2. Present Limits

Existing data from neutrino experiments at accelerators have been analyzed for oscillations. From this data, no firm evidence for oscillations has been found (beam dump experiments will be discussed in the next section), and many limits on the oscillation parameters have been published. The first accelerator results on oscillations came from Gargamelle. [2,3] The data was taken at the CERN-PS with a heavy freon (CF_3Br) fill. These early limits on $\nu_\mu \rightarrow \nu_e$ and $\bar{\nu}_\mu \rightarrow \bar{\nu}_e$ oscillations remain among the most stringent for such appearance experiments. The results on $\bar{\nu}_\mu \rightarrow \bar{\nu}_e$ oscillations was improved by an experiment at LAMPF which used a water (H_2O and D_2O) Cerenkov counter. [4] The LAMPF data was also used to set limits on $\nu_e \rightarrow \nu_e$ disappearance.

These "low energy" experiment have been supplemented in the past year by "high energy" experiments, i.e. Fermilab and CERN-SPS. With one exception, these results come from bubble chambers filled with high density liquids (15'BC and BEBC filled with a heavy Ne/H_2 mix, and Gargamelle). [1,5,6,7,8] The exception is a hybrid emulsion/spectrometer neutrino experiment designed for a search of charm particles. The fine spatial resolution in emulsion allows a direct search for the production and decay of the $\tau^\pm$ lepton from $\nu_\mu \rightarrow \nu_\tau$ oscillations, [9] while the bubble chamber searches rely on $\tau^\pm \rightarrow \nu_\tau \nu_e e^\pm$ decay (17% branching ratio) and the absence of anomalous electrons to set limits on $\nu_\mu \rightarrow \nu_\tau$ appearance. [1,6,7,8] The absence of anomalous electrons is also used to set limits on $\nu_\mu \rightarrow \nu_e$ appearance [1,6,7], and a comparison of the observed with the expected neutrino fluxes was used to set limits on $\nu_e \rightarrow \nu_e$ disappearance. [1,5,7] These results are summarized in Table I, and shown in Figures 2 through 4.

A limit on $\nu_\mu \rightarrow \nu_\mu$ disappearance is also given by Armenise et al [6], however, assumptions about the ν_μ cross section are required.

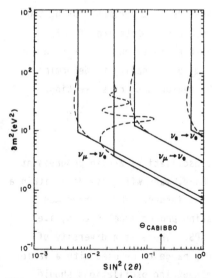

Figure 1. The low δm^2 and the high δm^2 limit lines on oscillation parameters are superposed on the limits from Baker et al. [1]

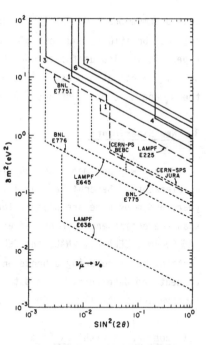

Figure 2. Present and proposed limits at accelerators on $\nu_\mu \rightarrow \nu_e$ appearance.

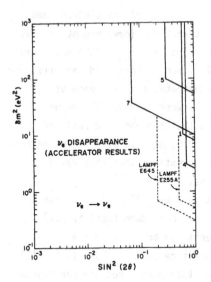

Figure 3. Present and proposed limits at accelerators on $\nu_e \rightarrow \nu_e$ disappearance.

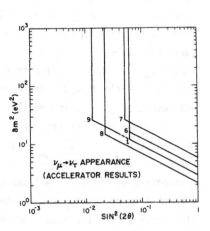

Figure 4. Present limits at accelerators on $\nu_\mu \rightarrow \nu_\tau$ appearance.

Such assumptions are clearly not necessary in experiments using two detector positions, and most of the new experiments are of this type. It is also useful to note that the limits on $\nu_\mu \rightarrow \nu_e$ and $\nu_\mu \rightarrow \nu_\tau$ oscillations assume ν_e, ν_μ, ν_τ universality to determine the ν_e and ν_τ cross sections from the measured ν_μ cross section.

3. New Experiments

Many new experiments on neutrino oscillations are in progress at accelerators. Several have data in hand, many will have data within a year, and many more are planned for the future. These experiments span the entire energy range of existing proton accelerators, i.e. LAMPF, BNL, CERN-PS, FNAL, and CERN-SPS. There is a diversity of neutrino beams covering a broad energy range to be used with a variety of neutrino detectors. Thus data on neutrino oscillations should improve substantially as these experiments are completed.

3.1 Experiments with New Data

3.1.1 Beam Dump Experiments The primary goal of the CERN-SPS and Fermilab beam dump experiments is to search for (new) prompt sources of neutrinos and/or new types of neutrinos, e.g. ν_τ. Data was taken at CERN-SPS in 1977 and 1979. The 1977 run demonstrated the existence of prompt ν_e's and the 1979 run demonstrated the existence of prompt ν_μ's. In each case new questions were raised. The major question raised from the 1979 run was that the observed ratio of prompt ν_e to prompt ν_μ flux is about 0.6, not 1.0 as anticipated from charmed particle decays. There are suggestions that this may be a manifestation of neutrino oscillations.

Data was taken at Fermilab last year by E613 [10] and a run was started recently at CERN-SPS using a new beam dump facility [11]. Additional details of these experiments can be found in the proceedings of the 1981 neutrino conference in Maui, Hawaii. [10,11] Irrespective of the outcome of these experiments, because the sources of neutrinos from the beam dump are not well known, no strong conclusions can be drawn directly about neutrino oscillations from such experiments.

3.1.2 <u>Large $\delta m^2 \nu_\mu$ Disappearance Experiments</u>. Two experiments using electronic neutrino detectors are currently in progress to search for ν_μ disappearance oscillations.

E701 at Fermilab is a Columbia/Caltech/Chicago/Fermilab/ Rochester/Rockefeller collaboration using narrow band ν_μ and $\bar{\nu}_\mu$ beams, and a reshuffled Caltech/Fermilab/Rochester/Rockefeller detector which is now located both in lab E and in the wonder building 337 meters upstream. [12] In this two detector geometry, the front detector measures the neutrino flux times cross section, while the rear detector searches for deviations from expectation. This experimental setup is shown in Figure 5, and the region of sensitivity to oscillations is shown in Figure 6. This experiment is being run now at Fermilab.

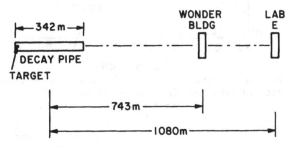

Figure 5. The experimental layout for E701 at Fermilab.

E734 at Brookhaven is a BNL/Brown/KEK/Osaka/Pennsylvania/Stony-brook/Tokyo collaboration which has recently built a totally live 170 ton electronic neutrino detector to study neutrino-electron and neutrino-proton elastic scattering. [13] This fine-grained detector consists of 118 alternating layers of $3\frac{1}{2}''$ thick liquid scintillation counters and x and y low mass proportional drift tubes. [14] The detector length is about 24 meters and it sits 115 meters from the target. This geometry is shown in Figure 7. A 10-week run was completed earlier this year. Data from the front and rear halves of this detector can be compared to search for large $\delta m^2 \nu_\mu$ oscil-lations. [15]

3.2 <u>Scheduled Experiments</u>

E225 at LAMPF is an Irvine/Los Alamos experiment using a 14 ton plastic scintillator/flash chamber detector for the study of neutrino-electron elastic scattering. [16] A schematic of this detector system

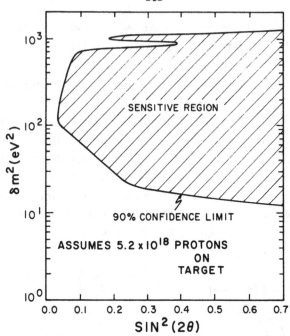

Figure 6.   Proposed limits of sensitivity on $\nu_\mu \rightarrow \nu_\mu$ disappearance
oscillations from E701 at Fermilab.

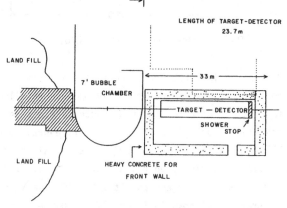

Figure 7.   The experimental layout for E734 at BNL.

is shown in Figure 8.   The neutrino source at LAMPF is the beam stop
where π^+'s are produced, stop and decay, and the resulting μ^+'s also
stop and decay.   The available neutrino types and spectra are shown in
Figure 9. π^-'s and μ^-'s are absorbed so that few μ^-'s decay.   Thus
the $\bar{\nu}_e$ flux is suppressed by a factor of at least 10^3.   This allows a
search for $\bar{\nu}_\mu \rightarrow \bar{\nu}_e$ oscillations via the inverse beta reaction

$\bar{\nu}_e$ p → e⁺n. This experiment is scheduled to begin a run in May.

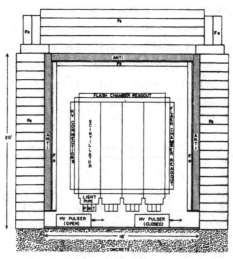

Figure 8. Schematic of E225 at LAMPF (end view).

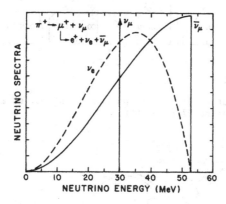

Figure 9. Types and spectra of neutrinos from the LAMPF beam dump.

E775 phase 1 (E775I) at BNL is an expansion of the E734 collaboration by an Irvine contingent. [15] The E734 detector system would be exposed to a two horn narrow band neutrino beam [17] to search for ν_μ → ν_e oscillations at relatively large δm^2 but with high sensitivity to small $\sin^2(2\theta)$. Figure 10 shows a schematic of the horn/collimator system used by Baltay et al [17]. The narrow band beam has a lower intrinsic ν_e background since these come from three body decays ($K^{\pm}$, $\mu^{\pm}$) which gives a broad band spectrum. This experiment is scheduled for a five week run this fall.

The three "low energy" neutrino oscillation experiments at the CERN PS utilizing CDHS, CHARM and BEBC have already been described in detail. [18,19] These experiments share a new neutrino beam line at the PS which is currently under construction. Figure 11 shows the layout of this beam at CERN. These experiments are schedule to run next spring.

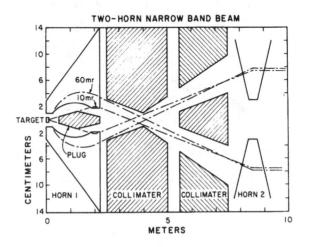

Figure 10. The narrow band neutrino beam of Baltay et al. [17] to be used by E775I at BNL.

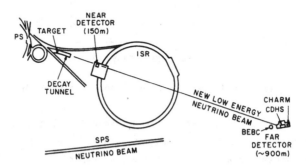

Figure 11. The layout of the new PS neutrino beam at CERN.

3.3 Planned Experiments

The planned neutrino oscillation experiments at LAMPF [20,21] and CERN [18,19] have been described in previous conferences, and proceedings from these conferences are already available. Therefore, an update is sufficient. The present situation at LAMPF and CERN is that E225A, E645 and E638 at LAMPF, and the SPS-Jura experiment at CERN are continuing.

Two experiments, E775 [15] and E776 [22] are being planned at BNL. Both experiments would search for $\nu_\mu \rightarrow \nu_e$ appearance as well as $\nu_\mu \rightarrow \nu_\mu$ disappearance. Both are two detector experiments and the detector locations at BNL is shown in Figure 12.

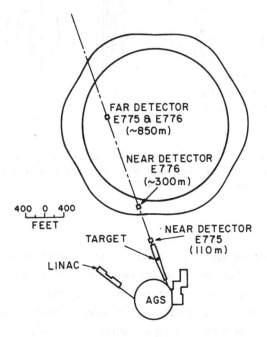

Figure 12. The layout for E775 and E776 at BNL.

E775 would relocate half the E734 detector to a new position 850 meters from the target. The existing high intensity wide band neutrino beam provides a sufficient rate so that one could examine NC/CC ratios, e.g. $\nu p \rightarrow \nu p / \nu_\mu n \rightarrow \mu^- p$ and $\nu N \rightarrow \nu X / \nu_\mu n \rightarrow \mu^- p$, as well as a comparison of the observed spectra and rate in the two detector geometry. Should an effect be seen, this variety of tests would provide confidence that the effect is real.

E776 would use a new narrow band neutrino beam having substantially higher intensity than the Baltay beam, and new massive (300 ton) sampling calorimeters as neutrino detectors. Figure 13 shows a schematic of their detector. The narrow band beam requires a massive detector to get a sufficient event rate, but allows a significantly less sensitive detector. Phase one of this experiment would use a single detector to search for $\nu_\mu \to \nu_e$ appearance.

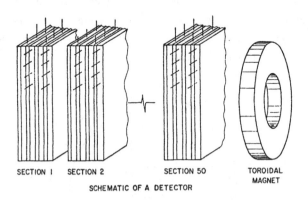

SECTION 1   SECTION 2   SECTION 50   TOROIDAL
                                                          MAGNET
              SCHEMATIC OF A DETECTOR

Figure 13. Schematic of BNL E776 which consists of alternating layers of passive absorber and proportional drift tubes followed by a muon spectrometer

4. Summary

Over the past year, much of the existing neutrino data at accelerators has been examined for neutrino oscillations. No definite effect has been observed either in the $\nu_\mu \to \nu_e$ and $\nu_\mu \to \nu_\tau$ appearance modes, or in the $\nu_e \to \nu_e$ disappearance mode. Assuming two neutrino oscillations in each case, a large region of the δm^2, $\sin^2(2\theta)$ parameter space can be ruled out.

Many new experiments are continuing to search for neutrino oscillations. Several experiments now have new data to analyze. Many are scheduled to run within the next year, and many more are planned. These experiments would have progressively higher sensitivity to oscillations. One anticipates that in about a year the sensitivity to small δm^2 will improve by about an order of magnitude, and that the presently planned experiments will eventually increase this sensitivity by another order of magnitude. Further improvements

by yet one more order of magnitude is possible if higher energy meson
factories are built. Such possibilities are being considered at
LAMPF, [23] SIN [24] and TRIUMF [25] e.g. LAMPF II, operating at 16
GeV with an average current of 100μA, is being actively studied. [26]
This path, progressing towards three orders of magnitude improvement
in sensitivity to neutrino oscillations, might well have surprises
along the way!

5. Acknowledgements

This work was supported in part by the Department of Energy and
the National Science Foundation.

References

[1] N. J. Baker et al, Phys. Rev. Letters 47, 1576 (1981).

[2] E. Bellotti et al, Nuovo Cim. Letters 17, 553 (1976).

[3] J. Blietschau et al, Nucl Phys. B133, 205 (1978).

[4] P. Nemethy et al, Phys. Rev. D23, 262 (1981);
    S. Willis et al, Phys. Rev. Letters 44, 522 (1980);
    45(E), 1370 (1980).

[5] H. Deden et al, Phys. Letters 98B, 310 (1981).

[6] N. Armenise et al, Phys. Letters 100B, 182 (1981).

[7] O. Erriquez et al, Phys. Letters 102B, 73 (1981).

[8] A. E. Asratyan et al, Phys. Letters 105B, 301 (1981).

[9] N. Ushida et al, Phys. Rev. Letters 47, 1694 (1981).

[10] R. C. Ball et al, Proc. Int. Neutrino Conf. Maui, Hawaii, 1981
     (ed. R. J. Cence et al) 1, 467.

[11] P. O. Hulth, Proc. Int. Neutrino Conf., Maui, Hawaii, 1981
     (ed. R. J. Cence et al) 1, 474.

[12] M. Shaevitz (Columbia), Spokesman, E701 Fermilab.

[13] D. H. White (BNL), Spokesman, E734 Brookhaven.

[14] A. Mann, Proc. Workshop on Nucl. and Particle Phys. at Energies
     up to 31 GeV: New and Future Aspects, Los Alamos, New Mexico,
     1981 (ed. J. D. Bowman et al) p. 79; Proc. Second Kaon Factory
     Phys. Workshop, Vancouver, B.C., Canada, 1981 (ed. R. M.
     Woloshyn and A. Strathdee) p. 54.

[15] M. Murtagh (BNL), Spokesman, E775 Brookhaven.

[16] H. H. Chen (Irvine), Spokesman, E225 LAMPF.

[17] C. Baltay et al, Phys. Rev. Letters 44, 916 (1980).

[18] J. Rothberg, BNL Neutrino Oscillation Workshop, BNL 51380
      (1981) p. 101.

[19] A. L. Grant, Proc. Int. Neutrino Conf. Maui, Hawaii, 1981
      (ed. R. J. Cence et al) 2, 214.

[20] R. L. Burman, BNL Neutrino Oscillation Workshop, BNL 51380
      (1981) p. 43.

[21] H. H. Chen, Proc. Int. Neutrino Conf. Maui, Hawaii, 1981
      (ed. R. J. Cence et al) 2, 183.

[22] W. Y. Lee (Columbia), Spokemsn, E776 Brookhaven.

[23] Proc. Workshop on Nucl.and Particle Phys. at Energies up to
       31 GeV: New and Future Aspects, Los Alamos, New Mexico, 1981
      (ed. J. D. Bowman et al).

[24] ASTOR project, SIN Jahresbericht 1981.

[25] Proc. Second Kaon Factory Physics Workshop, Vancouver, B.C.
      Canada, 1981 (ed. R. M. Woloshyn and A. Strathdee).

[26] H. A. Thiessen (LASL), LAMPF II Workshop, Los Alamos, New
      Mexico, February 1982.

[27] T. Y. Ling and T. Romanowski (Ohio) Spokesmen, E645 LAMPF.

[28] T. Dombeck (LASL), Spokesman, E638 LAMPF; Also see "A proposal
      to the Department of Energy for a High Intensity Los Alamos
      Neutrino Source", Los Alamos National Laboratory, Los Alamos,
      New Mexico, January 15, 1982.

TABLE I

Present limits at accelerators on neutrino oscillations

mode	experiment	$\delta m^2 \sin(2\theta)$ low δm^2 limit (eV^2)	$\sin^2(2\theta)$ high δm^2 limit
$\overset{(-)}{\nu_\mu} \rightarrow \overset{(-)}{\nu_e}$ appearance	Gargamelle-PS (Blietshau et al)[3]	1.0	0.002
	LAMPF (Nemethy et al)[4]	0.9	0.2
	15' BC ν_μ (Baker et al)[1]	0.6	0.006
	Gargamelle-SPS (Armenise et al)[6]	1.4	0.008
	BEBC-SPS (Erriquez et al)[7]	1.7	0.01
$\overset{(-)}{\nu_\mu} \rightarrow \overset{(-)}{\nu_\tau}$ appearance	15' BC ν_μ (Baker et al)[1]	3.0	0.06
	Gargamelle-SPS (Armenise et al)[6]	4.0	0.06
	BEBC-SPS (Erriquez et al)[7]	6.0	0.05
	15' BC $\bar{\nu}_\mu$ (Asratyan et al)[8]	2.2	0.022
	Emulsion/Spectrometer (Ushida et al)[9]	3.0	0.013
$\nu_e \rightarrow \nu_e$ disappearance	LAMPF (Nemethy et al)[4]	2.5	0.7
	15' BC ν_μ (Baker et al)[1]	8.	0.6
	BEBC-SPS(NB) (Deden et al)[5]	55.	0.3
	BEBC-SPS (Erriquez et al)[7]	10.	0.07

TABLE II
New Neutrino Oscillation Experiments at Accelerators

Experiments	Institutions Source / Detected	Oscillation Modes	L/E$_\nu$ (m/MeV)	δm^2 sin(2θ) low δm^2 limit (eV^2)	sin^2(2θ) high δm^2 limit
FNAL-E701	Columbia,Caltech Chicago,FNAL,Rochester, Rockefeller [12]	$\nu_\mu \rightarrow \nu_\mu$	0.02		$10 < \delta m^2 < 10^3$
BNL-E734	BNL,Brown,KEK,Osaka, Pennsylvania,Stony brook,Tokyo [13]	$\nu_\mu \rightarrow \nu_\mu$	0.1		$10 < \delta m^2 < 10^2$
BNL-E775I -E775	E734 + Irvine [15]	$\nu_\mu \rightarrow \nu_e$	0.03	0.6	0.0016
		$\nu_\mu \rightarrow \nu_e$	0.3 - 1.5	0.06	0.04
		$\nu_\mu \rightarrow \nu_\mu$		$0.3 < \delta m^2 < 40$	
BNL-E776	BNL,Columbia, Illinois,Johns Hopkins,NRL [22]	$\nu_\mu \rightarrow \nu_e$	0.5 - 1.5	0.035	0.002
		$\nu_\mu \rightarrow \nu_\mu$		$0.3 < \delta m^2 < 40$	
LAMPF-E225 -E225A	Irvine, Los Alamos [16]	$\bar{\nu}_\mu \rightarrow \bar{\nu}_e$	0.2	0.35	0.02
		$\bar{\nu}_\mu \rightarrow \bar{\nu}_e$	0.6	0.18	0.08
		$\nu_e \rightarrow \nu_e$		0.5	0.5
LAMPF-E645	Ohio,ANL,Caltech, Louisiana [27]	$\bar{\nu}_\mu \rightarrow \bar{\nu}_e$	0.5 - 1.5	0.06	0.008
		$\nu_e \rightarrow \nu_e$		0.3	0.2

TABLE II (Continued)

New Neutrino Oscillation Experiments at Accelerators

Experiments	Institutions	Oscillation Modes Source	Oscillation Modes Detected	L/E_ν (m/MeV)	$\delta m^2 \sin(2\theta)$ low δm^2 limit (eV^2)	$\sin^2(2\theta)$ high δm^2 limit
LAMPF-E638	Los Alamos,Maryland, New Mexico,Temple, UC Riverside [28]	ν_μ ν_μ	ν_e ν_μ	15	0.002 0.02	0.004 0.14 (min)
CERN-PS-BEBC	Padova,Pisa,Athens, Wisconsin	ν_μ	ν_e	0.3 – 2	0.09	0.03
CERN-PS-CHARM	CERN,Hamburg,Aachen, Rome,Moscow	ν_μ	ν_μ	0.6 – 2	0.3	0.15 (min)
CERN-PS-CDHS	Cern,Dortmund, Heidelberg,Saclay	ν_μ	ν_μ	0.4 – 2	0.25	0.13 (min)
CERN-SPS-Jura	Annecy,CERN, IC London,Oxford	ν_μ ν_μ ν_μ	ν_e ν_τ ν_μ	0.6 – 2	0.1 0.15 0.15	0.014 0.034 0.02 (min)

COSMOLOGICAL CONSTRAINTS ON WITTEN'S HIERARCHY MECHANISM

So-Young Pi

Harvard University

and

University of New Hampshire

Abstract

In Witten's hierarchical supersymmetric model the SU(5) symmetry
stays unbroken until the temperature becomes of order of the fundamental scale, M, of the model. This together with other phenomenological constraints suggests that M must satisfy the Weinberg bound
$M \gtrsim 10^{11}$ GeV. Baryon number generation and primordial monopole
density pose difficulties for the model.

I shall discuss the unusual early universe phase transition in the
SU(5) supersymmetric model which was proposed by Witten [1] as a
possible solution to the mass hierarchy problem [2]. In this model
the small mass scale is fundamental and the large one is generated
dynamically in the one-loop effective potential. Symmetry breaking in
this model is entirely determined at the tree level by the Higgs
field with small mass characteristics. The interesting early universe
phase transition, which I shall discuss, is a consequence of this
peculiar feature of the model: at finite temperature the symmetry-
changing phase transition is also governed by the Higgs field with
small mass characteristics, thus the SU(5) symmetry stays unbroken
until the universe cools to the temperature of the order of the small
mass scale.

The above symmetry behavior both at zero and finite temperatures,
seems to occur in the class of models in which the mass hierarchy
appears through Witten's mechanism.

First, I shall introduce the model and demonstrate the unusual finite temperature behavior. Then, I shall discuss the possibility of dynamical generation of baryon asymmetry and suppression of primordial monopole density. Baryon number generation may pose difficulties for this model. Initial monopole density may be suppressed only if the fundamental scale is taken to be the weak interaction scale.

In this model, there are two complex Higgs fields A^i_j and Y^i_j in the adjoint representation of SU(5), and one complex singlet Higgs field X. The Higgs fields in the fundamental representation are ignored for simplicity. The scalar potential is given by

$$V_0(A,Y,X) = g^2 \, \mathrm{Tr}|A^2-M^2|^2 + \lambda^2(\mathrm{Tr}\ A^2A^{*2} - \frac{1}{5}\ \mathrm{Tr}\ A^2\ \mathrm{Tr}\ A^{*2})$$

$$+\ \mathrm{Tr}\left|\lambda(AY+YA) - \frac{2}{5}\ \lambda\ \mathrm{Tr}\ AY + 2\ gX\ A\right|^2 +e^2\ \mathrm{Tr}(i[A,A^*] + i[Y,Y^*])^2.$$

$$(1)$$

where * denotes complex conjugate . M is the only mass scale in the theory, and it characterizes supersymmetry breaking. We shall take M to be the weak interaction scale or a scale much smaller than the unification scale.

To minimize the energy, one must have

$$\langle A\rangle = \frac{gM}{\sqrt{30\ g^2+\lambda^2}}\begin{pmatrix}2 & & & & \\ & 2 & & & \\ & & 2 & & \\ & & & -3 & \\ & & & & -3\end{pmatrix}.$$

$$(2)$$

Y is parallel to A at the minimum and is given by

$$\langle Y\rangle = \frac{g}{\lambda}\ \langle X\rangle\begin{pmatrix}2 & & & & \\ & 2 & & & \\ & & 2 & & \\ & & & -3 & \\ & & & & -3\end{pmatrix}.$$

$$(3)$$

However, $\langle X\rangle$ is undetermined at the tree level. The broken gauge symmetry is SU(3) x SU(2) x U(1). The one-loop effective potential must be calculated in order to determine X; it is given by [3]

$$v^0_1(\phi) = \sum_i \frac{(-1)^F}{64\pi^2}\ M^4_i(\phi)\ \ell n\ M^2_i(\phi)/\mu^2$$

$$(4)$$

where the sum runs over all helicity states, $M_i(\phi)$ is the field dependent mass of the ith such state, F=1 for fermions, F=0 for bosons, μ is a renormalization mass. Witten has shown [1] that for large X, X >> M, the effective potential evaluated at the minimum, given by Eqs. (2) and (3), has the following form, which includes the lowest order term.

$$V_1^0(X) = \frac{M^4 g^2 \lambda^2}{30g^2 + \lambda^2} \left| 1 + \frac{g^2}{g^2 + \frac{1}{30}\lambda^2} \frac{(29\lambda^2 - 50e^2)}{80\pi^2} \ln |X|^2/\mu^2 \right|$$

$$+ 0\left(\frac{1}{|X|^2}\right) \quad \text{for} \quad X \gg M. \qquad (5)$$

If $29\lambda^2 - 50e^2 < 0$, X increases without limit until perturbation theory breaks down. However, Witten argues that asymptotic freedom will force e^2, which really depends on X, to vanish with increasing X. Hence the effective coupling $29\lambda^2 - 50e^2$ changes sign at some large value X_0. Then a stable minimum will be produced near X_0, which can be interpreted as the unification mass scale ($\gtrsim 10^{14}$ GeV). Therefore, X_0 is independent of the fundamental scale M of the theory and is determined by the renormalization group equation. SU(5) is strongly broken by the vacuum expectation value of Y given in Eq. (3).

Now I shall discuss the unusual phase transition of this model. The finite temperature effective potential [4] tells us that usually, if the symmetry of a theory is broken at a mass scale m at zero-temperature, the symmetry-changing phase transition occurs at temperature $T_c \simeq 0(m)$. Since in this model SU(5) symmetry is broken to SU(3) x SU(2) x U(1) at the mass scale X_0, one would naively expect that the phase transition SU(5) $\rightarrow$ SU(3) x SU(2) x U(1) occurs at $T_c \simeq 0(X_0)$. However, as we have seen in Eq. (2), although SU(5) is strongly broken at X_0, the symmetry breaking pattern is entirely determined by the A field, with a small vacuum expectation value of $0(M)$. Consequently, the SU(5) symmetry stays unbroken until the field A gets a non-vanishing vacuum expectation value at temperature $0(M)$. I shall demonstrate that in fact <u>the large scale X_0 is induced by the non-vanishing vacuum expectation value of A</u>.

The finite temperature effective potential [4] in this model has the following form:

For $T \gg M_i(A,X,Y)$

$$V_{eff}(T,A,XY) = V_0(A,X,Y) - \frac{\pi^2}{90} N(T)T^4$$

$$+ \sigma_A T^2 \text{Tr} AA^* + \sigma_Y T^2 \text{Tr} YY^* + \sigma_X T^2 XX^* \qquad (6)$$

$V_0(A,X,Y)$ is as in Eq. (1); $N(T) = N_B(T) + 7/8 \, N_F(T)$ where $N_B(T)$ and $N_F(T)$ are the total number of distinct helicity states for bosons and fermions with mass $\ll T$; σ_A, σ_Y and σ_X are functions of e^2, λ^2, g^2 of order (g^2, λ^2, e^2) and are positive.

For $T \ll M_i(A,X,Y)$

$$V_{eff}(T,A,X,Y) = V_0(A,X,Y) + V_1^0(A,X,Y) - \frac{\pi^2}{90} N'(T)T^4 \qquad (7)$$

Here V_1^0 is the zero-temperature one-loop potential of Eq. (4) and the definition of $N'(T)$ is the same as $N(T)$ but for $T \ll M_i(A,X,Y)$.

In the remaining region we shall simply extrapolate the two limiting forms (6) and (7).

The above finite temperature effective potential shows that essentially, nothing happens until the temperature becomes the fundamental mass scale M of the theory. Let us discuss this in some detail: at very high temperature $T \gg X_0$, where X_0 is the large mass scale, an absolute minimum exists at $A = X = Y = 0$. The universe is in the SU(5) phase. As the temperature decreases, the A field may undergo a second order phase transition. The highest possible critical temperature is

$$T_A = \sqrt{2/\sigma_A} \ gM. \qquad (8)$$

This phase transition will actually occur at T_A if the minimum of X stays at X=0 until this temperature. Therefore, the minimum of A will be at $<A> = 0$ for the temperatures higher than T_A. In the following we argue that no phase transitions occur for $T \gtrsim T_A$.

(1) $T \gtrsim T_A$

First, let us consider the behavior of the X and Y fields for the temperatures $T \gtrsim T_A$ where $<A> = 0$. At very high temperatures an absolute minimum exists at $X = Y = 0$. This minimum will be present until the temperature becomes $\lesssim O(M)$. However, as the temperature decreases well below the large mass scale X_0, a non-zero minimum may develop at large X and Y. For the regions where $X, Y \gg T$, T is much less than the field dependent mass M_i and as shown in Eq. (7) the behavior of X and Y is dominated by the zero-temperature effective potential. When $<A> = 0$, unlike the zero-temperature situation, the classical potential does not determine the direction of Y nor the magnitudes of X and Y. These are fixed by the one-loop corrections. The calculation of V_1^0 for arbitrary Y and X at $<A> = 0$ becomes very simple due to the following: When $<A> = 0$, the A-Y mixing in the mass matrices disappears and V_1^0 splits into two parts.

$$V_1^0 = V_{1A}^0 + V_{1Y}^0. \tag{9}$$

V_{1A}^0 is the contributions from the fields A,X and their superpartners. The masses of these particles are independent of the gauge coupling e^2 when <A> = 0, and the coefficient of the logarithmic function in the effective potential is positive definite as in the O'Raifeartaigh model [5], where there are no gauge fields. V_{1Y}^0 is the contribution from the Y and gauge fields and their superpartners. Supersymmetry is unbroken among these particles when <A> = 0, and they have the same field dependent mass m(X,Y). Therefore, the one-loop correction V_{1Y}^0 vanishes.

$$V_{1Y}^0 = \Sigma(-1)^F m(X,Y)^4 \ln m^2(X,Y) = 0. \tag{10}$$

The sum is over Y, gauge fields and their superpartners. Hence, when <A> = 0, the effective potential V_1^0 is independent of e^2 and increases logarithmically for large X and Y in all SU(5) directions of Y. Therefore, the absolute minimum is at X = Y = 0 and SU(5) stays unbroken for $T \gtrsim T_A$. The field A will have an independent second order phase transition at $T_A \simeq O(M)$ given in Eq. (8).

(2) $T \lesssim T_A$

When $T \lesssim T_A$, A will have a non-zero minimum <A(T)> = <A(0)>$(1 - T^2/T_A^2)^{1/2}$ with <A(0)> given by Eq. (2). The gauge symmetry will be softly broken to SU(3) x SU(2) x U(1).

For large X and Y, where M_i(A,X,Y) >> T, the effective potential will have the form of Eq. (5) as <A(T)> becomes non-zero and a stable minimum will develop near X_0; the large scale is induced by the non-vanishing value of <A(T)>. The minimum at X=0 will become metastable as the temperature decreases and X_0 will become the true vacuum. Depending on the sign of $d^2V_1^0/dX^2|_{X=0}$ there are two possible ways for the transition from X=0 to X=X_0 to occur. In a wide range of parameters, for which the coefficient of $\ln X^2$ in Eq. (5) is negative, the sign of $d^2V_1^0/dX^2|_{X=0}$ tends to be negative. In this case $d^2V_{eff}(X,T)/dX^2|_{X=0}$ can be negative and the transition from X=0 to X_0 can proceed without tunneling. If the minimum at X=0 becomes metastable before $d^2V_{eff}(X,T)/dX^2|_{X=0}$ becomes negative, tunneling [6] will begin. However, this is a slow process, and the transition will actually occur after the curvature becomes negative at some temperature T_X. For $T \lesssim T_X$, X will smoothly move from X=0 to X_0. For some range of the parameters it may be possible to have $d^2V_1^0/dX^2\big|_{X=0} > 0$. In this case, $d^2V_{eff}(X,T)/dX^2\big|_{X=0}$ is always positive and the transition from X=0 to X_0 can occur only through tunneling. Since it is a slow process, the temperature at which the tunneling is completed will be well below T_A.

In both cases, as X increases, SU(3) x SU(2) x U(1) symmetry breaking will become strong and the masses of particles which depend on X will become heavy. However, we notice that the vacuum energy difference $V(X=0) - V(X=X_0)$ is only $O(M^4)$ and the phase transition temperature is also $O(M)$. This means that there is no available energy to make heavy particles and therefore, the particles that would become heavy decay as X increases, and no heavy particles are formed in the early universe. Of course, this decay cannot occur for magnetic monopoles and for them, an energy puzzle remains. Presumably they will slow down the X-transition.

Now I shall briefly discuss baryon number generation and monopole density, assuming a smooth X-transition $(d^2V_1^0/dX^2\big|_{X=0} < 0)$ which occurs for a wide range of parameters in the model.

BARYON NUMBER PRODUCTION

Let us consider the conventional out-of-equilibrium decay scenario for baryon generation through the decay of heavy bosons with constant mass M_b [7]. The decay rate Γ is given by $\Gamma = f\, M_b$ where $f=\alpha$, $(\alpha = e^2/4\pi)$ for a gauge boson decay and $f = $ (Yukawa coupling)^2 for a Higgs boson decay. Baryon number will arise only when the universe is out of eqiulibrium; this requires that the ratio $K \equiv \Gamma/H\big|_{T=M_b}$, where H is the universe's expansion rate, be less than 1. In Witten's model $K \simeq f\, M_p/M$, where M_p is Planck mass. For $M < f\, M_p$, which is the value of M which we are assuming, the universe is in thermal equilibrium and net baryon number is not produced.

The fact that M_b increases does not seem to alter the situation. As I have mentioned earlier, the heavy bosons will decay as their masses increase during the X-transition. If the rate of change in the equilibrium distribution of the heavy boson due to this decay, $\dot{n}_b/n_b$, were greater in magnitude than $\Gamma = f\, M_b(t)$, then thermal equilibrium would not be maintained and baryon asymmetry would arise. However, a rough calculation gives $\Gamma > |\dot{n}_b/n_b|$ at $T \lesssim M_b(t)$. A detailed calculation must be performed to reach a definite conclusion.

MONOPOLE PRODUCTION

When $SU(5)$ breaks down to $SU(3) \times SU(2) \times U(1)$ at $T=T_A$ light monopoles will be produced prolifically due to large fluctuations in the A field. These monopoles annihilate each other strongly when they are closely together. However, as the universe expands, the annihilation becomes negligible. Preskill's [8] estimate for the monopole density to entropy ratio, at temperatures well below the critical temperature, is given by

$$\frac{n_m}{T^3} \simeq 10^{-6} \frac{M_m}{M_p} \qquad (11)$$

where M_m is the mass of the monopoles. In standard $SU(5)$ theory, where $M_m \simeq 10^{16}$ GeV, Preskill's calculation tells us that the monopole annihilation can reduce n_m/T^3 only to $0(10^{-10})$. However, the standard scenario of helium synthesis requires $n_m/T^3 \lesssim 0(10^{-19})$ for $M_m \simeq 10^{16}$ GeV. The bound on the present value of n_m/T^3 imposed by the observed Hubble constant and deceleration parameter is $\lesssim 10^{-24}$ for $M_m \simeq 10^{16}$ GeV. This is the well known problem of initial monopole production in the standard $SU(5)$ theory [9].

Eq. (11) tells us that monopole density is proportional to the mass of monopoles. In the scenario we are considering, the monopole mass is $0(M/\alpha)$ for the temperature $T \gtrsim T_X$, and it increases with temperature for $T \lesssim T_X$. A rough estimate shows that if $T_X \lesssim 10^{-3} T_A$, the annihilation can reduce the monopole density to $n_m/T^3 \simeq 0(10^{-20})$ if M is taken to be the weak interaction scale. If these monopoles become heavy as X increases to X_0, their mass density will be within the bound at the time of helium synthesis.

Finally, I shall comment on some phenomenological difficulties that arise when the small mass scale M is identified with the weak interaction scale, as in Witten's model. When the Higgs fields in the fundamental representation are included in the model, Witten finds that $SU(5)$ is strongly broken to the unphysical group $SU(3) \times U(1) \times U(1)$ at the large scale X_0 [1]. Another problem arises from the fact that the mass splittings between bosons and fermions are much less than M. This is unacceptable if M is taken to be the weak interaction scale [10].

Further difficulties with equating M to the weak scale present themselves as a consequence of the finite temperature behavior I have discussed. In asymptotically free theories (which I am assuming, although no asymptotically free Witten-type model exists at present) when SU(5) stays unbroken until low temperatures, the gauge coupling becomes large and eventually blow up at scale Λ_5, which in super-symmetric theories is typically 10^9 GeV. Therefore, if we take M as the weak scale ($\lesssim 10^3$ GeV), the SU(5) gauge coupling will become singular. One possibility that may arise due to the strong forces is that SU(5) is broken dynamically. However, we do not have any under-standing of such strong coupling phenomena. Therefore, a well defined theory, in the absence of the dynamical symmetry breaking mechanism, must not identify M with the weak interaction scale, rather must be greater than Λ_5 ($\simeq 10^9$ GeV).

Recent analysis by Weinberg [11] shows that the supersymmetry breaking scale must satisfy the bounds $M < 10^6$ GeV (obtained by Pagels and Primack [12]) or $M \gtrsim 10^{11}$ GeV. This bound and the above discussion suggests that in Witten-type models the desirable value of M exceeds 10^{11} GeV. Witten suggested that M be $0(10^{11}$ GeV) and the weak scale be $0(M^2/X_0)$, where X_0 is the large scale of 0(Planck mass). Recently, a model with this feature was constructed by Dimopoulos and Raby [13]. However, as long as $M \simeq 10^{11}$ GeV not only baryon asymmetry but also primordial monopole density pose difficulties.

REFERENCES

[1]   E. Witten, Phys. Lett. 105B, 267 (1981).

[2]   S. Weinberg, Phys. Rev. D 13, 974 (1976); D 19, 1277 (1978); E. Gildner and S. Weinberg, Phys. Rev. D 13, 3333 (1976).

[3]   S. Coleman and E. Weinberg, Phys. Rev. D 7, 788 (1973).

[4]   L. Dolan and R. Jackiw, Phys. Rev. D 9, 3320 (1974); S. Weinberg, Phys. Rev. D 9, 3357 (1974).

[5]   L. O'Raifeartaigh, Nucl. Phys. B96, 331 (1975).

[6]   S. Coleman, Phys. Rev. D 15, 2929 (1977).

[7]   Yu. Ignatiev, N.V. Frasnikov, V.A. Kuzmin, and A.N. Tarhelidge, Phys. Lett. 76B, 436 (1978); M. Yoshimura, Phys. Rev. Lett. 41, 281 (1978); 42, 146 (E)(1979); S. Dimopoulos and L. Susskind, Phys. Rev. D 18, 4500 (1978); B. Toussaint, S.B. Treiman, F. Wilczek and A. Zee, ibid, 19, 1036 (1979); J. Ellis, M.K. Gaillard, and D.V. Nanopoulos, Phys. Lett. 80B, 360 (1979); S. Weinberg, Phys. Rev. Lett. 42, 850 (1979); A. Yildiz and P.H. Cox, Phys. Rev. D 21, 906 (1980);. For a review, see M. Yoshimura, National Laboratory for High Energy Physics preprint (1981) and references therein.

[8]   J. Preskill, Phys. Rev. Lett. 43, 1365 (1979).

[9] Some papers on the suppression of initial monopole production in grand unified theories are: A. Guth and H. Tye, Phys. Rev. Lett. 44, 631 (1980); P. Langacker and S.Y. Pi, Phys. Rev. Lett. 45, 1 (1980); G. Lazarides and Q. Shafi, Phys. Lett. 94B, 149 (1980).

[10] E. Witten, Princeton University preprint (1982).

[11] S. Weinberg, University of Texas preprint (1982).

[12] H. Pagels and J. Primack, Rockefeller University preprint (1982).

[13] S. Raby and S. Dimopoulos, Harvard University preprint (1982).

THE LEPTON ASYMMETRY OF THE UNIVERSE

Paul Langacker, Gino Segrè, Sanjeev Soni
Department of Physics, University of Pennsylvania
Philadelphia, Pennsylvania 19104, USA

The Lepton Asymmetry

The standard hot big bang cosmological model has been extremely successful in explaining the $T_\gamma \simeq 2.7\,^\circ K$ microwave radiation and the relative abundance of primordial helium and deuterium [1] . When combined with the ideas of grand unification [2] it may also give a plausible dynamical explanation of the small baryon asymmetry $B \equiv (n_B - n_{\bar B})/n_\gamma \simeq 10^{-10 \pm 1}$ observed in the present universe.

The hot big bang model predicts the existence of cosmological relic neutrinos analogous to the microwave radiation. The neutrinos would have stayed in equilibrium via the weak interactions until they decoupled at a temperature of about 1 MeV. The momentum distribution of the decoupled neutrinos would have been subsequently redshifted, so that at present the relic neutrinos would have a distribution of relativistic thermal form characterized by an effective temperature

$$T_\nu = \left(\frac{4}{11}\right)^{\frac{1}{3}} T_\gamma \simeq 1.9\,^\circ K, \tag{1}$$

corresponding to a number density $n_{\nu_i} \simeq n_{\bar\nu_i} \simeq 50/cm^3$.

This scenario involves one largely untested assumption, however, viz. it assumes that the lepton asymmetry

$$L_i \equiv \frac{n_{e_i^-} - n_{e_i^+} + n_{\nu_i} - n_{\bar\nu_i}}{n_\gamma} \simeq \frac{n_{\nu_i} - n_{\bar\nu_i}}{n_\gamma} \tag{2}$$

of the ith lepton family is negligibly small. In fact the existing limits on L_i are very weak. It is useful to define the dimensionless parameters $\xi_i \equiv \mu_i/T_\nu$ where μ_i is the chemical potential of the ith species. (The ξ_i remain constant for an adiabatically expanding universe if there are no lepton number violating interactions). Then the asymmetry is

$$L_i = \frac{2}{3} \left(\frac{T_\nu}{T_\gamma}\right)^3 (\xi_i + \frac{1}{\pi^2} \xi_i^3) \tag{3}$$

and the energy density associated with ν_i and $\bar\nu_i$ is

$$\rho_{\nu_i} + \rho_{\bar\nu_i} = \frac{\pi^2}{15} T_\nu^4 \left[\frac{7}{8} + \frac{15}{4\pi^2} \xi_i^2 + \frac{15}{8\pi^4} \xi_i^4\right] \tag{4}$$

The only significant direct limit on the ξ_i is that the neutrino energy density not exceed $\rho_0 = 8 \times 10^{-29} \text{gm/cm}^3$, the upper limit on the total energy density of the universe [1], [3]. For massless neutrinos this implies

$$[\sum_i \xi_i^4]^{\frac{1}{4}} < 80 \Rightarrow |L_i| < 4 \times 10^4$$

i.e. an enormous asymmetry is allowed.

For massive neutrinos the situation is somewhat improved, but large asymmetries are still allowed [4]. For $m_{\nu_i} \simeq 20$ eV, for example, one obtains $|\xi_i| < 6$ or $|L_i| < 20$.

Additional constraints are imposed by nucleosynthesis. In the standard model ($\xi_i = 0$) the abundances of primordial ^4He and D are successfully given [5] and the number of neutrino species limited; furthermore, the weak dependence on the baryonic density requires a low baryon density universe, $0.01 < \Omega_N < 0.10$, corresponding to $n_B/n_\gamma = (4\pm1) \times 10^{-10}$, where Ω_N is the ratio of baryon density to critical density.

It has been pointed out by many authors [6], however, that the situation is completely changed if $\xi_i \neq 0$. An asymmetry in the electron neutrino directly affects the equilibrium fraction of neutrons to nucleons

$$X_n \approx \frac{\lambda(p \to n)}{\lambda(p \to n) + \lambda(n \to p)} = \frac{1}{1 + \exp\left[\xi_{\nu_e} + \frac{M_n - M_p}{T}\right]} \tag{5}$$

where $\lambda(p \to n)$ is the reaction rate for processes where a proton is converted into a neutron. We see that $\xi_{\nu_e} > 0$ (<0) implies fewer (more) neutrons and hence less (more) ^4He. Asymmetries in ν_μ or ν_τ increase the expansion rate of the universe, implying a higher freezeout temperature for X_n and therefore more ^4He. The observed element abundance allows $|\xi_{\nu_e}| \lesssim 0.2$ or $|\xi_i| \lesssim 2$ ($i = \nu_\mu$ or ν_τ) as perturbations around the standard model. In addition, David and Reeves [7] have found a continuum of new solutions to nucleosynthesis in which ξ_{ν_e} increases from 0 to $\simeq 1.2$; the effect of ξ_{ν_e} is balanced by an increased expansion rate, which could be due to asymmetries in ν_μ or ν_τ (with $0 \lesssim |\xi_i| \lesssim 80$), additional neutrino species, anisotropic shear, magnetic monopoles, etc. The required Ω_N increases with ξ_{ν_e}.

Let us conclude this introduction by mentioning the theoretical expectations for the lepton asymmetry. A priori, any initial value of L_i at the time of the big bang is possible. If lepton number is conserved then L_i (or more accurately, the ratio of $n_{\nu_i} - n_{\bar{\nu}_i}$ to the entropy) is unchanged by an adiabatic expansion of the universe.

It is generally assumed [8] that the lepton number violating interactions in GUTs would dilute an initial large asymmetry to a negli-
gibly small value ($|L_i| \approx n_B/n_\gamma \approx 10^{-10}$) long before nucleosynthesis.
However, Harvey and Kolb [3] have shown that $|L_i| \gg n_B/n_\gamma$ is possible
even in a GUT provided there are (a) a large initial asymmetry in some
quantum number and (b) approximately conserved global quantum numbers.
They have constructed an explicit SO_{10} model with an arbitrarily large
value of $|L_i|$.

The theoretical and observational constraints on the lepton asymmetry are therefore very weak. Furthermore, large asymmetries could
significantly alter the usual nucleosynthesis scenario and therefore
the determination of n_B/n_γ and the limits on the number of neutrino
species. In this paper we will consider the possibility that for some
(unexplained) reason the asymmetry in one or more neutrino species was
large ($\xi_i > 1$) subsequent to the epoch of grand unification (or at the
big bang if one does not believe in GUTs). We consider the role of the
lepton number violation associated with Majorana neutrino masses in the
10 eV range in reducing an initial large lepton asymmetry. We find [9]
that in most models with explicit hard lepton number violation any initial lepton asymmetry would be reduced to an insignificant level long
before nucleosynthesis. For models in which lepton number is violated
spontaneously, we naively expect the symmetry to be restored at high
temperatures and therefore the mechanism for lepton asymmetry erasure
to be inoperative at high temperatures. Following an argument of
Linde [10] , we find this to not be the case, i.e. a large enough asymmetry $L = \sum_i L_i > L_c$ acts on the Higgs potential to prevent restoration
of symmetry. In fact we expect, for L not too much larger than L_c,
that the relevant vacuum expectation values (v.e.v.'s) v(T) behave as

$$\left(\frac{v(T)}{T}\right)^2 \approx \frac{L^2(T)}{L_c^2} - 1 \qquad (6)$$

and for $L < L_c$, $v(T) = 0$.

We consider starting with a large $|L_i|$ as in the SO_{10} model [3]
(in practice the requirement that this model lead to a realistic cos-
mology implies $|L_i| \lesssim 10^{-1}$ [9]). We then discuss the two currently
popular models for spontaneously breaking lepton number; the first [11]
introduces a Majorana neutrino N_R coupled to an SU_2 singlet field Φ
(the so-called Majoron) which acquires a very large v.e.v. N_R also

couples to ordinary neutrinos via the Higgs doublet field which has a
v.e.v. of $v_\phi(T)$. The N-ν mixing leads to a neutrino mass
$m_\nu(T) \sim \dfrac{h^2 v_\phi^2(T)}{M_N}$ where h is an appropriate Yukawa coupling. The second
model introduces a Higgs triplet X with a direct coupling to lepton
doublets. The neutrinos acquire Majorana masses proportional to $v_x(T)$
of the form

$$f_{ij} \overline{\nu}^c_{R,i} \nu_{L,j} \, v_x(T) + h.c. \tag{7}$$

so in this case $m_\nu(T) \sim f \, v_x(T)$.

The lepton asymmetry erasure mechanisms occur through scatterings
such as $\nu + a \rightarrow \bar{\nu} + a'$, where a is a target. These reactions violate
lepton number conservation and are therefore proportional to $m_\nu(T)$, the
characteristic magnitude of the violation. Studying the Boltzmann equation for L, we find, for velocity u, density of target n_a and lepton
number violating cross section σ that

$$\frac{1}{L} \frac{dL}{dT} \sim - 2 \, n_a(T) \, \langle u\sigma \rangle \tag{8}$$

with $\sigma \propto m_\nu^2(T)/T^4$. Using equation (6) and the fact that $n_a \sim T^3$, we
find for the Higgs singlet model that $\dot{L}/L \sim T^3$ and in the Higgs triplet
model that $\dfrac{\dot{L}}{L} \sim T$. The difference is caused by the fact that $m_\nu \sim v_\phi^2(T)$
in the first case and like $v_x(T)$ in the second. We compare the result-
ing equation with the expansion rate of the universe

$$\frac{\dot{R}}{R} \sim 1.66 \, g* \, \frac{T^2}{M_p} \tag{9}$$

where $M_p \simeq 1.2 \times 10^{19}$ Gev and g* is the number of relativistic species
of particles at temperature T (appropriately modified if there are
large chemical potentials).

For both models, in a large class of initial conditions, we find
that $L_i \rightarrow 0$ for all but the species with the lightest mass neutrino,
which we call ν_1. L_1 tends to a fixed point value L_c which is of order
unity. Intuitively this is due to the driving mechanism of a large
lepton number violating interaction proportional to $v(T)$, which however
is turned off as $L(T) \rightarrow L_c \sim O(1)$.

Our conclusion therefore is that, for a wide range of models and
parameters, we arrive at neutrino asymmetries of the order of magnitude
to significantly affect nucleosynthesis. We believe therefore that this
possibility should be seriously entertained.

References

1. S. Weinberg, "Gravitation and Cosmology"
   John Wiley and Sons, publishers (N.Y. 1972).

2. P. Langacker, Phys. Reports 72, 4, p. 185 (1981).

3. J. Harvey and E. W. Kolb, Phys. Rev. D24, 2090 (1981).

4. J. M. Cohen and P. Langacker, to be published.

5. For a recent paper on the subject see K. A. Olive, D. N. Schramm
   G. Steigman, M. S. Turner and J. Yang, Astrophys. Jour. 246,
   557 (1981).

6. R. V. Wagoner, W. A. Fowler and F. Hoyle, Astrophys. Jour. 148, 3
   (1967), A. Yahil and G. Baudet, Astrophys. Jour. 206, 261 (1976),
   A. Linde, Phys. Lett. 83B, 311 (1979).

7. Y. David and H. Reeves, Phil. Trans. R. Soc. Lond. A296, 415 (1980).

8. S. Dimopoulos and G. Feinberg, Phys. Rev. D20, 1283 (1979);
   D. Schramm and G. Steigman, Phys. Lett. 87B, 141 (1979);
   D. V. Nanopoulos, D. Sutherland, and A. Yildiz, Lett. Nuovo Cimento
   28, 205 (1980); M. S. Turner, Phys. Lett. 98B, 145 (1981).

9. The details of these calculations will be presented elsewhere,
   P. Langacker, G. Segrè and S. Soni, to be published.

10. A. Linde, Rep. Prog. Phys. 42, 25 (1979).

11. Y. Chikashige, R. N. Mohapatra and R. Peccei, Phys. Lett. 98B, 265
    (1981); Phys. Rev. Lett. 45, 1926 (1980).

12. G. B. Gelmini and M. Roncadelli, Phys. Lett. 99B, 411 (1981);
    H. M. Georgi, S. L. Glashow, and S. Nussinov, Nucl. Phys. B193, 297
    (1981).

STATUS OF THE IRVINE-MICHIGAN-BROOKHAVEN NUCLEON DECAY SEARCH

W. R. Kropp
for the
Irvine-Michigan-Brookhaven Collaboration

R. M. Bionta[2], G. Blewitt[4], C. B. Bratton[5], B. G. Cortez[2,a],
S. Errede[2], G. W. Foster[2,a], W. Gajewski[1], M. Goldhaber[3],
J. Greenberg[2], T. W. Jones[2,7], W. R. Kropp[1], J. Learned[6], E. Lehmann[4],
J. M. LoSecco[4], H. S. Park[2], P. V. Ramana Murthy[1,2,b], F. Reines[1],
J. Schultz[1], E. Shumard[2], D. Sinclair[2], D. W. Smith[1], H. Sobel[1],
J. L. Stone[2], L. R. Sulak[2], R. Svodboda[6], J. C. van der Velde[2], and
C. Wuest[1].

(1) The University of California at Irvine
     Irvine, California 92717

(2) The University of Michigan
     Ann Arbor, Michigan 48109

(3) Brookhaven National Laboratory
     Upton, New York 11973

(4) California Institute of Technology
     Pasadena, California 91125

(5) Cleveland State University
     Cleveland, Ohio 44115

(6) The University of Hawaii
     Honolulu, Hawaii 96822

(7) University College
     London, U. K.

(a) Also at Harvard University

(b) Permanent address: Tata Institute of
     Fundamental Research, Bombay, India

Abstract

Progress in the construction and operation of the IMB ~ 10,000 ton water Cerenkov detector is discussed. Results from a brief (~ 40 hr) cosmic ray exposure are presented.

Introduction

The Irvine-Michigan-Brookhaven (IMB) collaboration has constructed a ~ 10,000 ton water Cerenkov detector to search for nucleon decay. Our detector is located at a depth of 1570 mwe (~ 600m) in a salt mine [1] near Cleveland, Ohio. At this depth, the cosmic radiation is reduced by ~ 4×10^4 from that observed at a surface laboratory. The detector (Fig. 1) is a roughly cubical volume viewed by an array of 2048, 5" hemispherical photomultiplier tubes (PMT). The PMT's cover the 6 faces of the cube on a ~ 1 meter grid giving a coverage of ~1.2%. Water is chosen as the detector medium because it is inexpensive, benign, and easily purified for good clarity. The water is contained in a dual liner system, each layer being composed of 2.5 mm thick, high density polyethylene.

The active volume (22.5m × 17m × 18m) is ~ $7000m^3$. The fiducial volume is variable via software. Our Monte Carlo calculations suggest that this volume can begin ~ 2m inside the PMT planes giving a fiducial mass of ~ 4000 metric tons or ~ 2.5×10^{33} nucleons. This mass, together with the estimated background for two body decay modes of ~1 yr $^{-1}$, establishes the maximum sensitivity of the detector (~ 3×10^{33} yrs).

The general properties of the detector, as predicted by Monte Carlo studies, are given in Table 1. Figure 2 shows the general plan of the laboratory which has been constructed to support the detector.

A more extensive description of the detector and its design criteria, the laboratory construction, the data acquisition system, software, etc. can be found in our report to the Second Workshop on Grand Unification. [2]

The activities of the group since the last workshop have centered around (1) completion of the laboratory and detector, including the installation of the water purification system, (2) finalization and installation of the electronic instrumentation, (3) completion of the PMT testing, including deep water testing of each PMT in its housing, (4) further development and evaluation of on-line and off-line software, and (5) improving the water containment system.

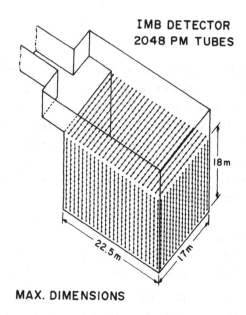

**IMB DETECTOR
2048 PM TUBES**

18m

22.5m   17m

MAX. DIMENSIONS

Figure 1.   The IMB Detector.

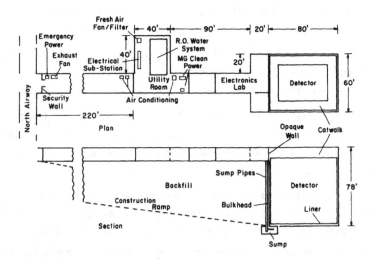

Figure 2.   Plan and cross section views of the Proton Decay
Underground Laboratory.

TABLE 1

Summary of IMB Detector Properties

Energy Resolution:

500 MeV shower:	σ = 11%
300-500 MeV/c $\pi^{\pm}$, $\mu^{\pm}$:	σ = 10-15%
p → e^{+}π^{0}:	σ = 8%
Trigger threshold:	~ 30 MeV (24 MeV measured)

Vertex Localiztion

Nucleon decay:	σ = 75 cm
~ 500 MeV single tracks:	σ = 75 cm

Track Direction:

No ambiguity

Angular Resolution

Showering tracks:	σ = 11^{0}
300-500 MeV/c $\pi^{\pm}$,$\mu^{\pm}$:	σ = 5^{0}
Straight-through muons:	σ = 2^{0}

μ → e Detection:

ε ~ 75% after trigger
Several μ → e ok
Position correlation ~ 1 m
e direction: σ = 25^{0}
e energy: σ = 30-40%

As of September, '81, the complete set of PMT's were in the underground laboratory, ready for submersion in the detector, and the data acquisition system was on-line.

In November, the water purification system became operational and the initial filling of the detector began. Minor leaks, well within the design capacity of the water handling system, were found and repaired.

A more serious leakage problem developed in December when the water depth was ~ 5m, requiring us to empty the pool. Repair of the liner is in progress with completion expected in mid-May. We expect to begin filling the detector again in early June.

Detector Studies

The approximately one-month period during which there was water in the detector afforded us the opportunity to evaluate the performance of the detector and its electronics and software with cosmic rays, and to test the water purification equipment. The remainder of this report will be devoted to these topics.

Water Purification

During this past year, a major effort was the installation and testing of the reverse osmosis water purification system. Its initial operation has been very satisfactory. More than a million gallons of high quality water were produced. The product water typically has an absorption length of 40 m at 440nm. Our Monte Carlo calculations suggest that this absorption length - which exceeds the maximum dimension of the detector - is more than sufficient for the successful operation of the detector. Thus, production of quality water in adequate quantity is not a problem.

Maintenance of such high quality water in large volume and for long periods in the detector has been questioned. During our December, '81 fill, in situ tests of light attenuation were made using a nitrogen laser (337nm) and a green LED array (550nm). Figure 3 shows the results of these tests superimposed on spectrophotometer measurements of the 40m product water. No deterioration in light

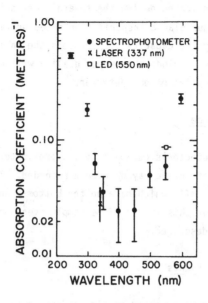

Figure 3. Spectrophotometer measurement of 40m (440nm) product water from the reverse osmosis water purification system. Also shown are the results of in situ attenuation measurements with a N_2 laser and an LED array.

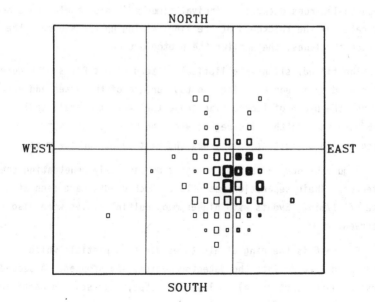

Figure 4. Detector response to a single penetrating muon.

transmission is indicated during the several week storage in the detector. It is our plan to repurify the detector water as required. We have the capacity to reprocess the entire volume in about 35 days. We conclude that water quality will not be a limiting factor in the operation of our detector.

Detector Performance

To test the detector's capabilities, approximately 40 hours (~ 250 K events) of cosmic ray data were recorded. The full complement of 384 PMT's which make up the bottom plane of the detector was operational for this study. Some sample events and preliminary analysis will be described.

Sample Cosmic Ray Events

The full range of expected cosmic ray events was observed. These included single and multiple muons, showers, and stopping muons. Figures 4 to 6 are examples. The displays represent the bottom plane of PMT's divided into 6 "patches" of 64 PMT each. The small squares are "hit" tubes. The area of the square is proportional to the number of photoelectrons detected. Arrival times of the photoelectrons are indicated by the thickness of the lines making up the squares; the thicker the lines, the earlier the photoelectron.

The filled, slightly elliptical pattern of hit PMT's of Figure 4 was caused by a muon entering the top surface of the water and exiting through the plane of tubes. The trajectory was at a small angle (~ 5°) to the zenith. The region with the highest intensity of Cerenkov light (largest boxes) was the exit point for the muon.

Figure 5 shows a pair of muons, simultaneously penetrating the detector. Their separation was ~ 15m. Such events were seen at a rate of ~1/min. Events with higher muon multiplicities were also observed.

Figure 6 is the ring of hit tubes due to a particle which penetrated 1 to 2m into the detector medium, and stopped. A second burst of light (not shown), delayed by 0.6μs, suggests the particle was a muon. The difference between short tracks (Fig. 6),

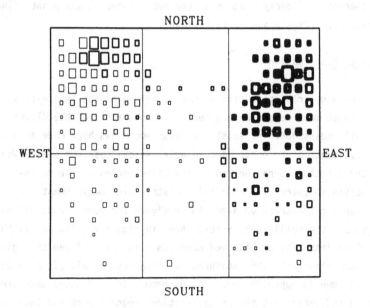

Figure 5. Detector response to a pair of simultaneously penetrating muons.

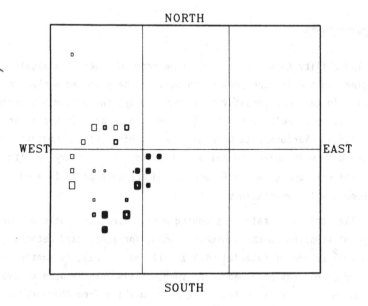

Figure 6. Detector response to a short path length stopping particle.

chacteristic of decay fragments from nucleon decay, and penetrating
cosmic rays (Figs 4 and 5) is striking.

Angular Distribution

The angular distribution of cosmic ray muons has been extensively
studied at many depths underground. The character of the distribution
is well understood. At least one measurement has been made at a depth
nearly identical to ours. Accordingly, we can test our understanding
of the detector operation and of the efficiencies of the reconstruction software with an angular distribution measurement. A crude
but rapid reconstruction algorithm designed for muon tracks was used
to obtain the preliminary result shown in Figure 7. Plotted is the
absolute intensity of observed muons as a function of zenith angle.
For comparison, the 1952 measurement of Barrett et al. [3] at a depth
of 1574 mwe is superimposed. The agreement in both shape and normali-
zation indicates that the reconstruction programs work and that the
efficiencies are reasonably well understood. (The structure at
$\sim 42^0$ - the Cerenkov angle in water - shows that some refinement in
software is still required.)

Stopping Muons

The ability to detect the electron from muon decay is vital for
tagging muons which are present in many of the proposed nucleon decay
modes. To test our sensitivity to muon decay, the data was searched
for events with delayed signals (> 4 PMT in 60 ns), in the interval
0 to 7.5µs. Various cuts were employed to minimize the contamination
of the data with noise and PMT afterpulsing. Preliminary results of
this analysis are given in Figure 8, and are seen to be in good
agreement with expectations.

The normalized rate of stopping muons, i.e., the ratio of the
rates of stopping to through-going muons, for a standard detector (a
100g cm^{-2} sphere of material with Z = 11 and A = 22), is another well-
studied cosmic ray phenomena. As such, it also can be used to evaluate our efficiency estimates. Figure 9, adapted from the work of
Cassidy, et. al, [4] gives the results of a theoretical analysis of
the depth dependence of the normalized stopping rate, together with

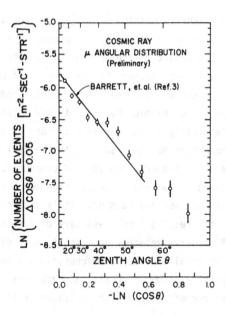

Figure 7. Angular distribution of cosmic ray muons observed with
the IMB detector. Also shown is the 1952 results of
reference 3.

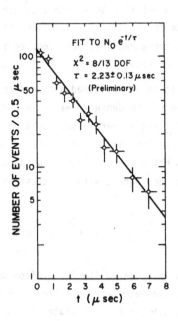

Figure 8. Decay spectrum of stopping events.

several relevant experimental points. [5 - 7] Two previous measurements [6,7] made in a tunnel adjacent to the proton decay laboratory are included, as is the result of the present work. The theoretical curves show the various contributions to the normalized stopping rate: R_a is the component due to penetrating muons, produced in the atmosphere and stopping in the detector. R_μ is the contribution from pions and muons made locally in nuclear interactions by virtual photons secondary to the primary muons. R_γ is the real photon contribution, while R_ν is the neutrino induced component. [8] At our depth, approximately equal numbers of penetrating muons and locally produced muons should be observed to stop. It is difficult to make an exact comparison of the results from these very different detectors and the theoretical standard detector because of their differing sensitivies to the vertically peaked atmospheric muons and the more nearly isotropic, locally produced muons. Nevertheless, the agreement is adequate to provide confidence in our efficiency estimates.

Trigger Requirements and Energy Thresholds

A 6 PMT trigger was employed during the cosmic ray studies. The trigger structure was 3 PMT's in 50ns from a single 8 PMT x 8 PMT patch and 2 such patch triggers in 150ns. These times, 50 ns and 150ns are the maximum times required for light to traverse one patch, and the detector respectively.

An energy calibration was obtained by reconstructing single muon tracks which traversed the 5m depth of water in the detector. Only muons with small zenith angles were used. This procedure led to the value of 1 photoelectron for each 4 MeV deposited, giving a threshold of ~ 24 MeV for the 6 PMT trigger. This value is comfortably below proton decay energies, and agrees with our previous measurements of PMT sensitivities.

Summary

We have successfully operated the IMB nucleon decay detector as a cosmic ray detector. The detector was filled to approximately one-third of its design depth and had one-sixth of its PMT's operating. The data acquisition system, and the on-line and off-line software has

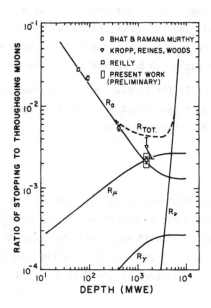

Figure 9. Theoretical ratios of stopping to through-going muons, with representative experimental results.

been tested and found to work as expected. Our ability to identify muons via their decay has been verified. The water purification system operates well and has demonstrated its ability to produce high clarity water in the quantities we require. We have also shown that the quality of the water can be maintained in the detector.

Our schedule calls for completion of the liner repairs in May (1982) with filling of the detector to start in June. Data collection should commence ~ 2 months thereafter.

Acknowledgements

The IMB collaboration expresses its thanks to the Morton-Norwich Products, Inc., owners and operators of the Fairport Mine, for their gracious hospitality. We wish especially to thank Mr. J. Davis, B. Lewis, J. Stoffer, and all other members of the Fairport Mine staff. This work is supported in part by the United States Department of Energy.

References

[1] The Fairport Mine, Grand River, Ohio, is owned and operated by Morton-Norwich Products, Inc.

[2] D. Sinclair and T. Jones in The Second Workshop on Grand Unification, p. 84, University of Michigan (April 1981).

[3] P. H. Barrett, et. al., Rev. Mod. Phys. 24, 133 (1952).

[4] G. L. Cassidy, J. W. Keuffel and J. A. Thompson, Phys. Rev. 7D, 2022 (1973).

[5] P. N. Bhat and P. V. Ramana Murthy, J. Phys. 6A, 1961 (1973).

[6] W. R. Kropp, F. Reines, and R. M. Woods, Phys. Rev. Lett. 20, 1451 (1968).

[7] T. Reilly, Ph.D. Thesis, Case Western Reserve University, 1970 (unpublished).

[8] C. Grupen, A. W. Wolfendale, and E. C. M. Young, Il Nuovo Cimento, 10, 144 (1972)

Fractionally Charged Color Singlets

Michael T. Vaughn

Department of Physics

Northeastern University

Boston, Massachusetts 02115

1. Introduction

Larue, Phillips, and Fairbank have reported the observation of excess charges $\pm 1/3$ e on cold Nb spheres[1,2]. Since every schoolboy now knows that color is permanently confined (except perhaps at the very high temperatures of the early universe), this observation implies the existence of a new kind of particle, a fractionally charged color singlet which we shall generically call a "fracton".

Although the observations of Ref. 1 require further experimental confirmation, and are in fact inconsistent with the interpretation of the "monopole" event reported by Cabrera[3] at this conference unless there is a profound modification of the quantization conditions for monopole charges, I wish to discuss various implications of the existence of "fractons" in the context of grand unified gauge theories.

First, I will review the incorporation of fractons into grand unified gauge theories. The minimal theory of this type is an SU(7) model[4,5], which requires a rich structure of new physics in the energy region 100–1000 GeV. This SU(7) model is almost ruled out by present low-energy phenomenology, and some attempted extensions of the model[6,7] will be described.

Then I will outline an interesting study by Goldberg[8] of the cosmological implications of the low relative abundance of fractons in the present universe. In the standard big bang cosmology, this low abundance, coupled with experimental bounds on the cosmic neutrino flux, implies the absence of primordial hydrogen in the earth's environs (local hydrogen has been part of, even if not

249

recycled through, early supernovas). Moreover, first generation stars must have been reasonably heavy (≥ 20 M$_\odot$) to allow the nearly complete annihilation of fractons.

Finally, I will have some brief concluding remarks.

2.   Grand Unification with Fractons – minimal SU(7) model[4,5]

In order to find a grand unified gauge group G which can accomodate fractionally charged color singlet fermions, we list the constraints which must be satisfied by the fermion representation in G.   We   look   for   complex,   flavor–chiral,   anomaly–free representations[9,10] which contain only 1^c, 3^c, and 3^{*c} fermions. This restriction is made primarily for reasons of simplicity, and can be removed (although in a supersymmetric theory, which necessarily contains color octet gluinos, such a restriction on the matter multiplets is still reasonable), but it will certainly be appropriate for the minimal model.

Then we require

(i)   a vector–like representation with respect to the unbroken subgroup $SU_c(3) \otimes U_{e-m}(1)$.

(ii) $\Delta Q = \pm 1$ weak currents, which require ($\ell \nu_\ell$) doublets to occur in single irreducible representations.

(iii)   only $Q = 0$, $\pm 1/3$, $\pm 2/3$, ± 1 in the fundamental representation (this is necessary to have a reasonable value of $\sin^2 \theta_w$).

(iv)   color singlets with fractional charge.

These constraints lead to the unique minimal group[4,5] SU(7), with basic fermion representation

$$7 + 21^* + 35$$

derived from the spinor 64 of SO(14) in the same way that the $5 + 10^*$ of SU(5) is derived from the spinor 16 of SO(10).

With charge assignment in the fundamental representation of SU(7) given by

$$Q = \text{diag} \ (1/3, \ 1/3, \ 1/3, \ 0, \ -1, \ q, \ -q)$$

the unrenormalized weak mixing angle at the unification mass is given by

$$\sin^2\theta_w(\text{GUM}) = \frac{3}{8 + 4q^2} = \frac{9}{28} \quad (q = 1/3)$$

$$\doteq \frac{9}{40} \quad (q = 2/3)$$

For $q = 2/3$, $\sin^2\theta_w$ is simply too small, so we limit our attention to $q = 1/3$ (even here it will be difficult to obtain a reasonable renormalized value of $\sin^2\theta_w$).

The fermion content of $\underset{\sim}{7} + \underset{\sim\sim}{21^*} + \underset{\sim\sim}{35}$ can be summarized as follows:

$$3^c: \quad \begin{pmatrix} u \\ \\ d \end{pmatrix}, \quad \begin{pmatrix} u' \\ \\ d' \end{pmatrix}, \quad \begin{pmatrix} Q^1 \\ \\ Q^0 \end{pmatrix}, \quad \begin{pmatrix} Q^{1/3} \\ \\ Q^{-1/3} \end{pmatrix}$$

(two normal quark doublets, two charge-shifted quark doublets) accompanied by the corresponding $\underset{\sim}{3^{*c}}$ antiquarks, and

$$1^c: \quad \begin{pmatrix} \ell^- \\ \\ \nu_\ell \end{pmatrix}, \quad \begin{pmatrix} \ell'^- \\ \\ \nu_\ell \end{pmatrix}, \quad \begin{pmatrix} \Lambda^{-2/3} \\ \\ \Lambda^{1/3} \end{pmatrix}, \quad \begin{pmatrix} \Lambda^{-4/3} \\ \\ \Lambda^{-1/3} \end{pmatrix}$$

(two normal lepton doublets, two charge-shifted lepton doublets) accompanied by the antileptons corresponding to the charged leptons only. In addition, there are SU(2) singlet leptons $L^{1/3}$, $L^{-1/3}$, L^0.

In order to accomodate the b quark and the τ lepton, two copies of this representation are required, which implies a <u>fourth</u> generation of ordinary quarks and leptons, in addition to many exotically charged quarks and leptons.

The pattern of symmetry breaking in the SU(7) model requires a rather complicated collection of Higgs scalars (or dynamical equivalents). An adjoint ($\underset{\sim\sim}{48}$) of Higgs scalars only gives a breaking

$$SU(7) \rightarrow SU(4) \otimes SU(3) \otimes U(1)$$

and the breaking to SU(3) ⊗ SU(2) ⊗ U(1) requires a tensor with many indices in order to lose rank (rank 6 → rank 4) and to give the right charge assignments. If we define a "normal" charge

$$Q_N \equiv \text{diag} \; (1/3, \; 1/3, \; 1/3, \; 0, \; -1, \; 0, \; 0)$$

and an "exotic" charge

$$Q_E \equiv \text{diag} \; (0, \; 0, \; 0, \; 0, \; 0, \; q, \; -q)$$

then we need a scalar with $Q_N \neq 0$, $Q_E \neq 0$ but $Q_N + Q_E = 0$ (otherwise Q_N, Q_E will be separately conserved).

Possible scalar vacuum expectation values which break the separate conservation of Q_N and Q_E are given in the table below.

Component	Representation	Dynkin label
$H_{77}^{[1236]}$	756	(201000)
$H_{66}^{[457]}$	840	(200100)
$H_{777}^{[12367]}$	1008	(310000)
$H_{666}^{[45]}$	1575	(300010)

(here [ ] denotes antisymmetrization of tensor indices). These vacuum expectation values must be of the order of the unification scale since the color triplet components of these multiplets can mediate proton decay, but the color singlet components may be light (although this requires the usual mysterious fine tuning).

The breaking of SU(2) ⊗ U(1) to U_{e-m} (1) can be achieved with a 7 of scalars; however, this multiplet alone leads to undesirable mass relations such as

$$m(u) = m \; (\Lambda^{1/3})$$

In order to evade this difficulty, it is necessary to have a scalar multiplet $H_c^{[ab]}$ (dimension 140) in order to keep normal fermions light and exotic fermions with masses of order m_W.

In addition to the unaesthetic proliferation of scalar multiplets, there are phenomenological difficulties with the minimal SU(7) model.

(i)   With a grand unification mass $M \sim 10^{14}$ GeV, $\sin^2\theta_W$ is renormalized too much! Including only fermions in the one-loop renormalization group (RG) equations gives

$$\sin^2\theta_W(m_W) = 0.15 \qquad \text{(fermions only)}$$

Including effects of light color-singlet scalars from 7 and 140 leads to

$$\sin^2\theta_W(m_W) = 0.19 \qquad \text{(with light scalars)}$$

which is still low, but possibly acceptable in a one-loop calculation.

(ii) With 16 quark flavors, and a large number of charged leptons and scalars, the SU(2) $\otimes$ U(1) coupling constants grow large between m_W and M in a one-loop RG calculation (in fact the Landau singularity in these coupling constants can occur below M), as was noted in Ref 4. If one naively insists upon using the one-loop RG equations[6], there is no way out without lowering M to give too rapid proton decay, but a careful two-loop calculation including the effects of Yukawa couplings[11] may lead to a tightly constrained resolution of the problem..

(iii) Since $SU^c(3)$ is barely asymptotically free with 16 quark flavors, $\alpha_s(\mu)$ is large, and the mass ratio m_b/m_τ is too large at the one-loop level. Again, higher order corrections may, or may not, alter this conclusion.

3. Refinements of the Minimal Model

There are two recent extensions of the minimal SU(7) model which I should mention briefly here.

Frampton and Kephart[7] have examined supersymmetric GUTs with fractional charges, and, in particular, a supersymmetric model with only a gauge multiplet, the minimal fermion multiplet, together with the supersymmetric partners, and two light Higgs doublets coming from an unspecified supersymmetric multiplet (perhaps as supersymmetric partners of the observed fermions). The difficulties with $\sin^2\theta_w$, and m_b/m_τ of the minimal non-supersymmetric model can be evaded, with a unification mass $M \sim 3 \times 10^{13}$ GeV, which may be OK. But a new problem exists – there is only one fermion multiplet and hence two ordinary generations of leptons and quarks. The b and τ cannot be ordinary quarks and leptons – can they be SU(7) gluinos?

Extending the minimal SU(7) model to SO(14) has been considered briefly by Li and Wilczek[5], and more recently by Li and Unger[6]. This allows the introduction of an intermediate mass scale at which baryon number is still naturally conserved[6], so that proton decay can be put off to higher mass scales while solving other problems at intermediate mass scales. This model has less predictive power than the SU(7) version, but (i) at least the predictions do not conflict with experiment, and (ii) there are already many undetermined parameters associated with the fermion mass matrix and Higgs potential in the minimal model.

4. Cosmological Implications of Fractons[8]

Goldberg[8] has recently studied the cosmological question of how the abundance of fractionally charged leptons can be reduced from order unity (relative to protons) at the big bang, down to $O(10^{-19})$ or less suggested by the Stanford experiment[1]. The scenario for this reduction has four stages:

Stage I: big bang → freezeout. The annihilation of L^+L^- into qq, $\ell\bar{\ell}$ and W^+W^- proceeds through γ and Z^0 intermediate states. With $L^\pm$ identified as leptons with charge $\pm1/3$, mass of order 100 GeV, this leads to a relative abundance

$$\frac{n_\pm(t)}{n_p(t)} \cong 10^{-4}$$

at freezeout.

Stage II: expansion and cooling. There is no further annihilation, but some L^- combine with protons and He^4 leading to concentration

$$(pL)^{+2/3} \sim 7\% \; , \quad (He \; L)^{+2/3} \sim 4\%.$$

Stage III: protostar formation. The remaining H and He^3 are neutral and cannot combine with L^- due to barrier penetration factors, until the temperature reaches $10^{4\circ}K$, when the neutral H ionizes, and the L^- again combine with protons to form $(pL)^{+2/3}$ ions (with binding energy $\sim$ 3keV), and there is no further annihilation with L^+ until the $(pL)^{+2/3}$ ionizes. At this point, the star is very dense, and γ-rays from the annihilation do not escape.

After annihilation in the protostellar environment, the abundance of $L^{\pm}$ is reduced to

$$\frac{n_{\pm}}{n_p} \sim 10^{-11} \text{ to } 10^{-12}$$

(for a star with mass $25M_\odot$) which the star is in radiative equilibrium; further annihilation before hydrogen burning starts leads to

$$\frac{n_{\pm}}{n_p} \sim 10^{-12} \text{ to } 10^{-14}$$

At this point, the $(He \; L)^{+2/3}$ has now ionized as well, and the annihilation is further enhanced by the greater density of $L^{\pm}$ in the core of the star due to gravity.

Stage IV: nuclear burning → supernova. Further annihilation in the core now reduces the abundance to

$$\left(\frac{n_{\pm}}{n_p}\right)_{final} \sim 10^{-19}$$

for a star with mass $25M_\odot$

Note that

(i)   the density of $L^{\pm}$ must be reduced before nuclear burning to keep the star from exploding due to the $L^{\pm}$ annihilation.

(ii) $25M_\odot$ is a typical mass expected for a first generation star.

(iii)   independent support for the belief that $L^{\pm}$ must be annihilated in first generation stars comes from cosmic ray neutrino flux – the ν_μ flux from $L^\pm$ annihilation (through the Z^0 diagram) must be strongly redshifted ($z \gtrsim 20$) in order to avoid producing μ's significantly in excess of the observed atmospheric background.

5. Conclusions

(i) Heavy fractionally charged color singlet fermions with masses ~100 GeV can be accomodated in an SU(7) or SO(14) grand unified theory, but only with a rich structure of fractionally charged leptons and hadrons in this mass range.

(ii) The standard GUT phenomenology of $\sin^2\theta_w$, α_s/α_{em}, m_b/m_τ must be substantially modified. There are certainly significant two-loop effects, as well as effects due to Yukawa and scalar quartic couplings, and there may also be intermediate mass scales which do not affect proton decay directly, but modify the RG analysis.

(iii) An abundance of $L^\pm$, relative to protons, of 10^{-19} (consistent with the Stanford experiment) can be obtained only after annihilation in first generation stars.

I am indebted to P. Frampton, H. Goldberg, T. W. Kephart, and D. Unger for useful discussions of the topics covered in this talk.

References

1. G. S. LaRue, J. D. Phillips, W. M. Fairbank, Phys. Rev. Lett. **46**, 967 (1981).

2. In a recent comment, J. Schiffer [Phys. Rev. Lett. **48**, 213 (1982)] has suggested that free charged +1/3 e can diffuse and evaporate readily at room temperature, so that such charges can be for observation only on spheres held at low temperature, and that the data of Ref.1 are consistent with a random distribution of charges +1/3 e over the Nb spheres in the sample.

3. B. Cabrera, these proceedings.

4. H. Goldberg, T. W. Kephart, M. T. Vaughn, Phys. Rev. Lett. **47**, 1429 (1981).

5. L.-F. Li and F. Wilczek, Phys. Lett. **107B**, 64 (1981).

6.   L.-F. Li and D. Unger, Carnegie-Mellon preprint (1982).

7.   P.H. Frampton and T. W. Kephart, University of North Carolina preprint (1982).

8.   H. Goldberg, Phys. Rev. Lett. _48_, 1518 (1982).

9.   M. Gell-Mann, P. Ramond, R. Slansky, Rev. Mod. Phys. _50_, 721 (1978).

10.   M. T. Vaughn, J.Phys. _G5_, 1317 (1979).

11.   Two-loop RG equations including effects of Yukawa and scalar quartic coupling constants are discussed by M. Fischler and J. Oliensis (to be published) and by M. E. Machacek and M. T. Vaughn (to be published).

NEUTRINO MASSES FROM β END-POINT MEASUREMENTS

J. J. Simpson

1. Introduction

One of the most challenging problems facing experimental physicists today is that of determining whether or not neutrinos have non-zero rest masses. The most accurate measurements have been made in attempts to determine the mass of the neutrino emitted in β^--decay, and almost invariably these experiments involve measuring accurately the shape of the continuous β-spectrum of tritium near its high energy end. It is the purpose of this paper to describe the methods that have been used to achieve the lowest upper limits on the mass of the neutrino from tritium decay (and in one case a non-zero lower limit), to discuss what new experimental conditions should be realized in order to increase the credibility of a specific non-zero value, and briefly to discuss a search for the emission of more massive neutrinos in tritium decay.

2. Theoretical Considerations

In the Fermi theory of allowed β-decay, the spectrum as a function of the β-energy E is given by [1]

$$N_\beta(E,Z) \propto pE[(Q - E)^2 - m_\nu^2]^{1/2}(Q - E)F(E,Z) \tag{1}$$

or, in terms of the momentum p of the β-particle, by

$$N_\beta(p,Z) \propto p^2[(Q - E)^2 - m_\nu^2]^{1/2}(Q - E)F(E,Z) \tag{2}$$

where Q is the total energy available in the decay and F(E,Z) is the Fermi function to account for the effect on the electron wave of the

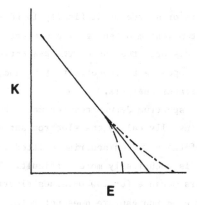

Figure 19.1. Kurie plot for allowed β-decay. The smooth curve is
for m_ν = 0, the dashed curve is for a non-zero m_ν and the dot-dash
curve shows the effect of finite resolution.

charge Z of the daughter nucleus. The β-spectrum can be displayed in
the form of the Kurie plot

$$ K = \left[\frac{N(E,Z)}{pEF(E,Z)} \right]^{1/2} \tag{3} $$

which is a linear function of E extrapolating to an end point energy Q
(fig. 19.1) in the case of zero neutrino mass. For a non-zero
neutrino mass the Kurie plot is still nearly linear far from the high
energy end, the linear portion effectively extrapolating to the same
end-point energy Q, but near the end the Kurie plot curves downward to
terminate at an energy m_ν below Q (fig. 19.1).

As can be seen, the most dramatic change occurs in an energy
interval ΔE below Q which is at most a few times the neutrino mass.
This explains much of the difficulty in the mass measurement — the
fraction of the number of β-particles having energies greater than Q-
ΔE (for m_ν = 0) is extremely small. As an example, for $\Delta E/T_m$ = 0.01
(where T_m is the maximum kinetic energy of the β-particles
corresponding to an energy Q) only a few millionths of the β-particles
have energies greater than T_m-ΔE. Furthermore the number of β-
particles in this energy region varies as ΔE^3. This fact strongly
influences the choice of radioactive source since it is desirable to
have one with a low Q-value and a reasonably short half-life to obtain
a high β-emission rate. The choice has usually fallen on tritium
[2-5] which has an end-point energy of 18.6 keV and a half-life of
12.3 years.

The other major source of difficulty in the neutrino mass determination is that momentum or energy measurements have a finite resolution, and this distorts the end of the β-spectrum or Kurie plot (fig. 19.1) somewhat opposite to the effect of m_ν and, as we shall see, seriously complicates analysis.

The tritium spectrum determines the mass of the neutrino emitted in β⁻-decay (usually called the electron anti-neutrino). Determing the mass of the electron neutrino emitted in positron decay and electron capture is technically more difficult. There are few low-energy β⁺ emitters because for low Q-values electron capture dominates. Ordinary electron capture does not permit a neutrino mass measurement easily, but occasionally a bremsstrahlung photon is emitted with the ν giving rise to a continuous spectrum much like that of β-decay and this can be used in a similar way to determine the neutrino mass [6]. However, bremsstrahlung-accompanied electron capture is a fairly rare event.

Throughout this paper the word neutrino will be used for neutrinos associated with β⁻-decay.

Recently, there has been speculation that neutrinos emitted in the weak interaction are not mass eigenstates but are linear combinations of primitive, massive neutrinos. In the event that the masses of these primitive neutrinos are small (smaller than the resolution function of the detecting system) the measurement determines the probability-averaged mass of these neutrinos; if only one mass is small and the others are much larger, the measurement determines the mass of this lightest neutrino. In principle the existence, masses and mixing parameters of the heavier neutrinos can be determined by looking for kinks in the Kurie plot associated with the emission of these heavier neutrinos [7-10].

3.   Experimental Method

The measurement of the tritium spectrum has provided the lowest upper limits, and in one case a non-zero lower limit, on the mass of the neutrino. There are two methods that have been used for studying the tritium spectrum, β-spectrometry and calorimetry.

3.1   β-spectrometry

The most recent and successful applications of this method [2,3,5] have involved using a magnetic spectrometer to measure the

momentum distribution of the electrons. The characteristics of the
two instruments which have achieved the lowest mass limits are listed
in table 19.1. The $\pi\sqrt{2}$ iron-cored spectrometer of Bergkvist [2] is
seen to have a very large luminosity but also a high background, in
contrast to the lower luminosity but lower background of the four-pass
air-cored toroidal spectrometer at ITEP [3,5]. Generally a solid
source containing tritium is placed in the object focal plane and some
sort of gas detector, either Geiger counter or proportional counter,
at the image plane. Because the magnetic field must be changed step-
wise, a fairly time consuming process, only a relatively restricted
region of momenta encompassing the end-point is measured. In general,
the resolution of these spectrometers is very good – they have
effective resolutions (including source thickness) under the operating
conditions used equivalent to about 50 eV for 18.6 keV electrons.

When a free tritium atom decays, the daughter $^3He^+$ ion is
left about 70% of the time in the 1s (ground) state, 25% in the 2s
state, and 5% of the time in higher electronic s-states[2]. The
energy separation between the 1s and 2s states is about 40 eV, so that
in effect a β-spectrometer with a source of free tritium atoms would
be measuring the sum of two β-spectra of differing end-point energies.
This has serious implications, for it means that one must know the
fraction of various $^3He^+$ ionic final states and possibly about

Table 19.1   Characteristics of β-spectrometry measurements

	Bergkvist[2]	ITEP[3]
Type	$\pi\sqrt{2}$ iron-cored 50 cm radius	air-cored toroidal $4 \times 180^\circ$ deflection
Luminosity(cm^2)	1	7×10^{-2}
Source	T in Aℓ	Valine-T on Aℓ
Resolution (FWHM at 18.6 keV)	55 eV	45 eV
Background (counts/minute)	300	6
Detector	Geiger	1)Geiger 2)Proportional

relative energy shifts of final states for a solid, sometimes
molecular source (which may be different from the free atom case).

Furthermore, the final states act to spread out the region of the "missing" counts associated with finite neutrino mass over an energy region of the order of 40 eV, or equivalently, to cause an effective increase in resolution from about 50 to 70 eV[2].

3.2 Calorimetry

Historically, the earliest method of studying tritium decay for determining $m_{\bar{\nu}}$ measured electron energies from a trace of tritium gas added to a proportional counter [11,12]. Recently, the energy spectrum of tritium β-rays implanted in a solid state detector has been measured[4]. The latter method has features which are very different from the spectrometer experiments. For one, the entire spectrum is obtained. For another, the $^3He^+$ ions in excited states de-excite in the detector, adding their energy to the β-energy, and hence there is only one end-point. The chief drawback of the method is that the energy resolution is worse (~ 250-300 eV) than that of the magnetic spectrometers. However, the fact that the entire β-spectrum is measured allows the determination of the end-point accurately which can compensate for the disadvantage of worse resolution in extracting the neutrino mass.

4. Discussion of Results

Table 19.2   Comparison of neutrino mass results

Reference	$m_{\bar{\nu}}$ (eV)	
Bergkvist [2]	<55	(90% C.L.)
Tretyakov et al [3]	<35	(90% C.L.)
Simpson [4]	<60	(90% C.L.)
Lyubimov et al [5]	34 ± 4	(1 S.D.)
	14 < $m_{\bar{\nu}}$ < 46	(99% C.L.)

Table 19.2 presents the most accurate results of recent years. A general conclusion one might draw is that it seems unlikely that some systematic effect could have hidden a neutrino mass larger than the upper limit of these results, about 60 eV, because of the variety of experimental techniques involved. There is one result which stands out because of its importance - that of the second ITEP

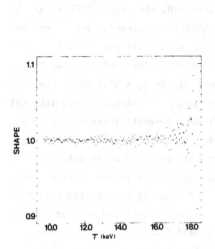

Figure 19.2. A typical β-shape spectrum for tritium [4], which is the ratio of the experimental to the theoretical spectrum. T is the β-ray kinetic energy.

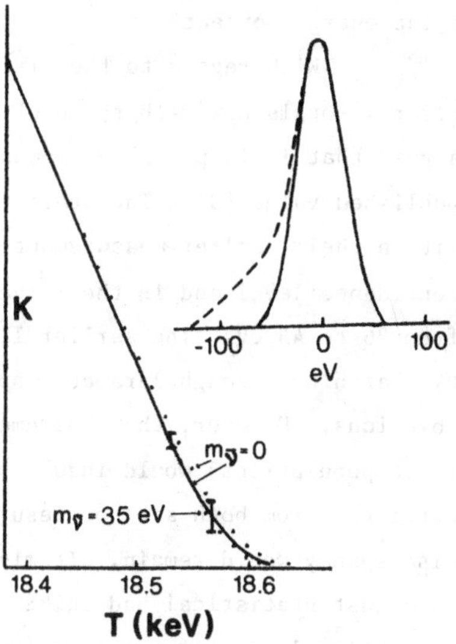

Figure 19.3. Effect on the Kurie plot of different resolution functions (inset). The data (smooth curve and with typical error bars) corresponding to Q = 18575 eV and $m_\nu = 0$ is assumed to have been acquired with the narrow resolution function. The best fit for the assumed broader resolution function (dots) requires Q=18589eV and m_ν=35eV. For $m_\nu = 0$ the best fit for the latter resolution is also shown (Q = 18590 eV).

paper [5] giving a lower bound to the neutrino mass. It should be pointed out that the 99% confidence interval quoted by the authors includes in it the probably unlikely situation that the $^3He^+$ ion is left in only one final state, and that if one assumes that it is more likely that the final states will be closer to the free atom case the lower bound would be raised to about 26 eV. (If only one $^3He^+$ ionic final state is indeed involved this would make the valine-T source

used in reference [3,5] the source of preference for these experiments
- better even than atomic tritium.)

What internal checks are available to assess the reliability
of neutrino mass measurements? One might ask at least three questions
of an experiment. Is the mass consistent with previous mass limits?
Is the spectrum shape consistent with the Fermi theory? Is the end-
point energy correct?

With regard to the first question, the recent ITEP result is
clearly consistent with those of Bergkvist and Simpson, but it can be
argued that it is possibly inconsistent with their own earlier
published value [3]. The statistical limit determined from the best
fit in their earlier measurement was given as $m_\nu < 23.5$ eV at the 95%
confidence level and in their recent work as a 99% confidence interval
from 26 to 46 eV. The earlier limit was, however, increased to 35 eV
by searching through parameter space for additional acceptable
solutions. However, these parameter changes (such as changing final
state populations) would induce parallel changes in the neutrino mass
extracted from both sets of results and hence it is expected that the
discrepancy would remain. It might be argued that the earlier results
were just statistical bad luck; it should be noted, however, that a
proportional counter was substituted for the Geiger counter in the
recent work.

Unfortunately, magnetic spectrometer measurements usually
cover such a short portion of the β-spectrum that a comparison of the
spectrum with Fermi theory is not a very accurate test of experimental
reliability, such as linearity of the spectrometer. In the case of
experiments with implanted detectors, the whole spectrum is obtained
and this is a vital test of the linearity of the system. Figure 19.2
shows an example of the β-shape spectrum which is the experimental
spectrum divided by the theoretical, and which should give unity as a
function of β-energy, from reference [4].

Although it is true that a knowledge of the extrapolated
end-point energy is not strictly needed for a determination of the
neutrino mass, the measurements which have been carried out to date
are high precision measurements of this end-point energy (and
ultimately the T-^{3}He atomic mass difference) and disagreement between
the end-point energy and the "true" value could be a signal of a
systematic error in the measurements, perhaps in a knowledge of the
effective resolution function. Figure 19.3 shows the effect on the

end-point energy and neutrino mass if the assumed resolution function
used for analysis purposes were wider and with more low energy tail
than the actual resolution function. There is a significant change in
end-point energy in this case, and thus agreement with the correct
end-point energy might be an important check on the resolution
function. However as table 19.3 shows there is at present
considerable spread in empirical values of the T-^{3}He atomic mass
difference;   clearly it would be useful to have an accurate value for
it.

　　　　Now that there is some indication that the mass of the
neutrino might be around 35 eV, what criterion should be imposed on
the next experiments to make their results believable? Recalling
figure 19.1 for the moment, we see that the neutrino mass manifests
itself by an increasing magnitude of slope of the Kurie plot, whereas
the resolution function causes the magnitude of the slope to decrease.
The derivative of the Kurie plot, $|\Delta K|$, is plotted in figure 19.4 and
demonstrates dramatically the importance of resolution. For a

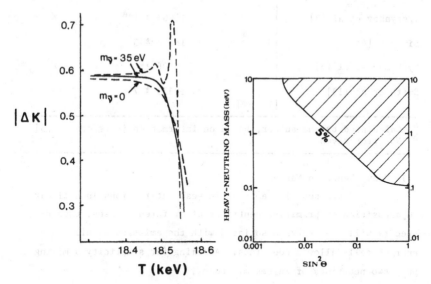

Figure 19.4. Differential effect on
Kurie plot of different resolution
functions. The dashed curves
correspond to a resolution function
of 20 eV, the smooth curve to one of
55 eV. The calculations assume two
^{3}He$^+$ final states separated
by 43 eV and with 30% population in
the upper state.

Figure 19.5. Limits on the
emission of a heavy neutrino
from tritium as a function of
mixing strength. The shaded
area is excluded at the 95%
confidence level.

resolution function of width 20 eV there is a profound difference between $|\Delta K|$ for $m_\nu = 0$ and $|\Delta K|$ for $m_\nu = 35$ eV. (One can also see the presence of two atomic final states separated by 43 eV). On the other hand, with a resolution function of width 55 eV the function $|\Delta K|$ just monotonically decreases in magnitude near the end with $m_\nu = 35$eV. Clearly a β-spectrum with 20-eV resolution would give an unambiguous signal of a 35 eV mass (assuming sufficient statistical accuracy). Furthermore, the result would be independent of the knowledge of final state effects (in fact might even serve to measure them) and would be relatively insensitive to uncertainties in the width of the resolution function.

Table 19.3   T-^{3}He atomic mass differences

Reference	T-^{3}He Atomic Mass Difference (eV)
Bergkvist [2]	18651 ± 16
Tretyakov et al [3]	18603 ± 13[a]
Simpson [4]	18567 ± 5
Lyubimov et al [5]	18606[a,b]
Smith et al [13]	18573 ± 7

[a] Calculated by the author, based on information in reference [3]
[b] No error given

4.1 Heavy Neutrino Masses

If the neutrino emitted in weak interactions is a linear superposition of primitive neutrinos of different masses, then β-spectra will show kinks associated with the emission of all energetically allowed neutrinos. Assuming for simplicity a mixing of only two neutrinos of masses m_1 and m_2,

$$\nu_e = \nu_1 \cos\theta + \nu_2 \sin\theta, \tag{4}$$

then the β-spectrum will be

$$N_\beta(E,Z) = N_\beta(E,Z,m_1)\cos^2\theta + N_\beta(E,Z,m_2)\sin^2\theta \tag{5}$$

where $N_\beta(E,Z,m)$ is given by eq. 1. The tritium spectrum from an implanted detector has been analyzed for a heavier neutrino between about 0.1 keV and 10 keV, and figure 19.5 shows the region excluded at the 95% confidence level [10].

5. Conclusions

The study of the β-spectrum of tritium has provided the tightest limits to date on the mass of any neutrino. One recent experiment has provided evidence that the mass is about 35 eV and this provides a challenge for future experiments. It has been argued that to make a measurement of a 35 eV mass truly credible the experiment should aim to demonstrate unequivocally the increasing magnitude of the slope of the Kurie plot caused by this neutrino, and this requires an effective resolution in the neighbourhood of 20 eV. These experiments are probably feasible today with large magnetic β-spectrometers such as at Chalk River [14]. It has also been argued that it would be useful to have an accurate end-point energy for tritium decay as a check on certain systematic effects (especially if the neutrino mass should turn out to be considerably less than 35 eV). An experiment to measure this is now in progress [15].

Finally, the study of β-spectra can be used to search for mixing of massive neutrinos in the neutrino wavefunction, and the tritium spectrum has been examined for emission of a heavy neutrino of mass between about 0.1 and 10 keV.

This work was supported by a grant from the Natural Sciences and Engineering Research Council of Canada.

University of Guelph
Guelph, Ontario, Canada

[1] H. F. Schopper, "Weak Interactions and Nuclear Beta Decay" (North-Holland, Amsterdam, 1966).
[2] K. E. Bergkvist, Nucl. Phys. B39, 317 (1972); B39, 371 (1972); Phys. Scr. 4, 23 (1971).
[3] E. F. Tretyakov, N. F. Myasoedov, A. M. Apalikov, V. F. Konyaev, V. A. Lyubimov, and E. G. Novikov, Bull. Acad. Sci. USSR, Phys. Ser. 40, 1 (1976).
[4] J.J. Simpson, Phys. Rev. D23, 649 (1981).
[5] V. A. Lyubimov, E. G. Novikov, V. Z. Nozik, E. F. Tret'yakov, and V. S. Kosik, Phys. Lett. B94, 266 (1980).
[6] A. DeRújula, CERN Preprint TH. 3045-CERN.

[7] M. Nakagawa, H. Okonagi, S. Sakata, and A. Toyoda, Prog. Theor. Phys. 30, 258 (1963).

[8] R. E. Shrock, Phys. Lett. 96B, 159 (1980).

[9] B. H. J. McKellar, Phys. Lett. 97B, 93 (1980).

[10] J. J. Simpson, Phys. Rev. D24, 2971 (1981).

[11] G. C. Hanna and B. Pontecorvo, Phys. Rev. 75, 983 (1949).

[12] S. C. Curran, J. Angus and A. L. Cockcroft, Phys. Rev. 76, 853 (1949).

[13] L. G. Smith, E. Koets, and A. H. Wapstra, Phys. Lett. 102B, 114 (1981).

[14] R. L. Graham, private communication.

[15] J. J. Simpson, W. R. Dixon and R. S. Storey, private communication.

FIRST STEPS TOWARDS THE MEASUREMENT OF
NEUTRINO MASSES IN ELECTRON CAPTURE

A. De Rújula

CERN -- Geneva

FOREWORD

I have been asked by Paul Frampton, organizer of this TWOGU, to
report on the progress of experiments aiming at the measurement of $m(\nu_e)$
in electron capture. It is very easy for a theorist to report on experiments: one need not understand what one is saying, and the audience
reacts with a motherly patience of an unusual kind. The people responsible for the work I will talk about are: J.V. Andersen, D.F. Anderson,
G.J. Beyer, G. Charpak, B. Elbek, H.Å. Gustafsson, P.G. Hansen,
B. Jonson, P. Knudsen, E. Laegsgaard, J. Pedersen and H.L. Ravn.

INTRODUCTION

"Research is when one does not know what one is doing"[1]. In this
sense, the attempts to measure neutrino masses are among the purest
forms of research: we have no serious prejudices about the answer. All
one can say, and this is a very positive statement, is that the arguments
defending massless neutrinos are very weak. Our present understanding
of elementary particles and their interactions is based on unified
gauge theories. In their realm, everything not explicitly verboten is
allowed, and happens. Non-vanishing neutrino masses and mixings, in
all models but one [minimal $SU(5)$[2]], are not forbidden. The mass spectrum of all particles, not only neutrinos, remains an absolute mystery,
perhaps the biggest single challenge to contemporary particle theory.
Measurements, or even upper limits, on neutrino masses could conceivably
constitute an important clue.

The most restrictive experiments on electron (anti-) neutrino masses deal with Tritium beta decay, and they are briefly discussed in the next section. The goal is to devise alternative methods to measure or constrain the electron neutrino mass with comparable or superior precision. We shall see that we are still far from this difficult goal, and that hope is also far from being lost.

TRITIUM BETA DECAY

The best upper limits on electron (anti-) neutrino masses

$$m(\bar{\nu}_e) < 60 \text{ eV} \quad (90\% \text{ confidence}) \quad \left[\text{Bergkvist}^{3)}\right] \tag{1}$$

$$m(\bar{\nu}_e) < 65 \text{ eV} \quad (95\% \text{ confidence}) \quad \left[\text{Simpson}^{4)}\right] \tag{2}$$

and the celebrated ITEP result giving a lower limit:

$$14 \text{ eV} < m(\bar{\nu}_e) < 46 \text{ eV} \quad (99\% \text{ confidence}) \quad \left[\text{Lubimov et al.}^{5)}\right] \tag{3}$$

result from the analysis of the shape of the electron energy spectrum[6] in Tritium beta decay near its endpoint, E_{max}. Tritium is a substance of choice because $E_{max} \sim 18.6$ keV is a particularly low value, the complete spectrum is particularly "short" and the fraction of events in a fixed interval near the endpoint is particularly large. For $m_\nu = 0$, the fraction of events in the window $E_{max} - 30 \text{ eV} < E_e < E_{max}$ would be $g \sim 10^{-8}$ in Tritium decay. We will in all cases define this fraction of events (normalized to the total decay rate) as the theoretical figure of merit, g, in a given process.

In ^{3}H β-decay the nuclear charge changes by a factor of two. The electron in the ^{3}H atom, with $Z = 1$ and principal quantum number $n = 1$, does not have the correct wave function to feel fully at home in the daughter atom with $n' = 1$, $Z = 2$. Thus, the decay of atomic Tritium will proceed with several branches with different n'

$$^3\text{H} \rightarrow {}^3\text{He}^+(n') + e^- + \bar{\nu}_e$$

The mass differences between the differently excited $^3\text{He}^+$ ions are of the same order of magnitude, or smaller, than the resolution of existing experiments $\left[\text{i.e., He}^+ (n'=2) - \text{He}^+ (n'=1) \sim 41 \text{ eV}\right]$. Moreover the expected branching ratios for different n' do not differ by orders of magnitude $\left[70\%\text{ for n}' = 1,\ 25\%\text{ for n}' = 2,\ ...,\text{ in atomic tritium}^{7)}\right]$. Thus, the shape of the spectrum tens of eV's from the endpoint, where one hopes to see the neutrino mass effects, is fairly complicated. There are two complementary ways out of this problem:

  i) Trust the theory of the spectral shape
  ii) Improve the resolution to such a level that the different decay
     branches are experimentally observable.

Clearly, the very difficult second way out would be the most satisfactory, particularly for experiments whose source is not atomic Tritium or a beam of tritons. I will not further discuss the "molecular" problems in the decay of Tritium containing molecules[5] or Tritium-coated surfaces[3], and refer the reader to the recent literature on the subject[8]. My feeling is that we have no serious reason to distrust the ITEP result, and every reason to attempt to reproduce it independently, as well as to develop the theoretical understanding of endpoint shapes in different environments.

It must be emphasized that the problems associated with the several branches of Tritium decay do not apply to inclusive or calorimetric experiments. This type of experiment measures all the energy of Tritium decay (but that of the escaping neutrino) rather than the energy of a single electron. Since J.J. Simpson has reviewed this approach at this Workshop, I will not further elaborate upon it.

ELECTRON CAPTURE; ^{163}Ho DECAY

In β-decay, the sensitivity to a neutrino mass occurs for electron energies close to their maximum kinematically allowed value. There, the neutrino energy is smallest and the effects of $m_\nu \neq 0$ would be largest. No further thought is required to conclude that any process involving very low energy neutrinos is of interest; provided its rate is not ridiculous. It is very easy to imagine many such low energy processes and somewhat more difficult to overcome the proviso concerning their rate. An example: ^{187}Re beta-decays with $E_{max} \sim 2.6$ keV, but its lifetime is $4 \cdot 10^{10}$ years. Even if other very low energy β-decays existed, their fate would be similar: β-decay rates go down fast with the available energy. There is a form of inverse β-decay: orbital electron capture, where the problem of diminishing rate with diminishing phase space is less serious. In electron capture a nucleus (Z,A) transmutes to (Z-1,A) as an orbital electron is captured and converts into a neutrino. The $\ell = 0$ electron wave functions at the origin grow with $Z^{3/2}$, somewhat **compensating** in large Z materials for an eventually small energy release. The isotope ^{163}Ho decays by electron capture to ^{163}Dy, with a record low Q value (mass difference between the neutral atoms in their ground state). For this reason ^{163}Ho, one way or another, is a good thing to have at hand, in view of possible measurements of neutrino masses. [Another reason, concerning internal bremsstrahlung and electron ejection in ^{163}Ho decay, will be presently discussed.]

Up to a few months ago, the data concerning ^{163}Ho were contradictory. Nuclear reaction data[9] gave a Q-value of -2.9 ± 3.1 keV. Electron capture by ^{163}Ho had been observed from the M shell but not from the L shell[10],[11], bracketing the Q value between the corresponding Dy ionization energies: 9.1 keV > Q > 2.05 keV. Wapstra and Bos[9] suggested a readjustment of the capture result[10] to 6 ± 3 keV, and combined it with the nuclear reaction data, to give Q = 2.6 ± 3 keV. The nuclear matrix elements in ^{163}Ho decay can be estimated with confidence by comparison with ^{161}Ho decay[12],[13]. This allows one to turn the limit 9.1 keV > Q into a lower limit of 150 years for ^{163}Ho decay[1],[12],[13]. The half-life of ^{163}Ho, as measured in 1968, was 33 ± 23 years[10]. New attempts to clarify this perversely controversial body of data seemed mandatory.

An Aarhus-CERN-Risø collaboration has recently measured[13] the M-capture half-life of ^{163}Ho and the (^{163}Ho,^{163}Dy) Q-value to be

$$T^M_{1/2} = (4.0 \pm 1.2) \; 10^4 \text{ years} \tag{4}$$

$$Q = 2.3 \pm 1.0 \text{ keV} \tag{5}$$

As we proceed to explain, this implies a half-life $T_{1/2}(^{163}\text{Ho}) = (7 \pm 2)10^3$ y. This preliminary experiment can also be used to set an upper limit $m(\nu_e) < 1.3$ keV[13].

A considerable initial difficulty in experiments involving very low energy radiation is the preparation of the radioactive source. Any radioactive contaminants constitute a higher energy background, and must be eliminated to a very high degree. This is particularly true for internal bremsstrahlung experiments (to be discussed in the next chapter) where one is searching for a very tiny signal: the levels of radioactive decontamination must reach one part in 10^{14} or 10^{15}. This decontamination level has been achieved in the preliminary ^{163}Ho experiment and the method[13] deserves a little pause.

A 2.4μA beam of 600 MeV protons from the CERN PS is shot into a Tantalum target. Rare-earth elements from the hot target evaporate, are surface ionized and mass separated in the ISOLDE facility. The mass 163 elements are collected and counted in a Faraday cup. A brief "cooling" time allows them to decay into the element of interest ^{163}Ho [or into stable ^{163}Dy]:

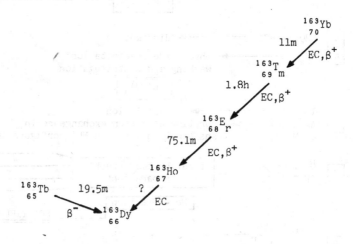

The on-line separator is not perfect, after some weeks of cooling a gamma ray spectroscopy analysis of the source reveals a contamination of ^{147}Gd (at the 1 per thousand level) and ^{167}Tm, ^{169}Yb (both at the $3 \cdot 10^{-6}$ level). There remain eleven or twelve orders of magnitude to go in the level of decontamination. This extra purification is achieved by a somewhat complex wet chemical procedure involving a triple ion-exchange chromatography. The method is summarized in next page. For tracing purposes a small amount of short lived ^{160}Ho is added; measurement of its activity results in a yield of 44% for the chemical clean-up. The purified stock of radioactivity is vacuum evaporated onto a backing, with ^{160}Ho still a tracer, to check the evaporation yield. After all this fiendishly complicated gymnastics one may or may not trust the estimated number of Ho atoms in the source. This number is measured independently by back scattering of α-particles (from the Aarhus single stage Van der Graaf accelerator) on the source. A typical recoil energy spectrum at 170°, from incident α's of 3.5 MeV, is shown in Fig. 1. Peaks in the spectrum reflect surface impurities, bulk

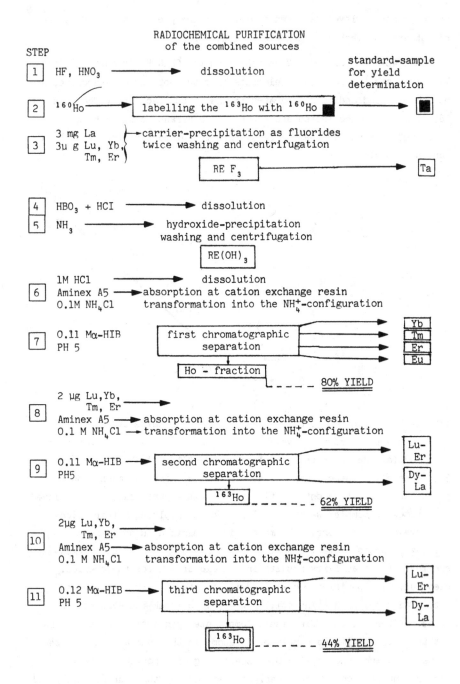

RADIOCHEMICAL PURIFICATION
of the combined sources

STEP

1   HF, HNO₃ ──────────► dissolution   standard-sample
                                               for yield
                                               determination

2   ¹⁶⁰Ho ──► labelling the ¹⁶³Ho with ¹⁶⁰Ho ──────► ■

3   3 mg La   ┌►carrier-precipitation as fluorides
     3u g Lu, Yb,   │ twice washing and centrifugation
         Tm, Er   ┘
                         RE F₃ ────────────────────► Ta

4   HBO₃ + HCl ──────────► dissolution
5   NH₃ ──────────► hydroxide-precipitation
                     washing and centrifugation
                         RE(OH)₃

6   1M HCl ──────────► dissolution
     Aminex A5 ──► absorption at cation exchange resin
     0.1M NH₄Cl   transformation into the NH₄⁺-configuration

7   0.11 Mα-HIB   first chromatographic ──────► Yb
     PH 5   separation   ──────► Tm
                                         ──────► Er
                         Ho - fraction   ──────► Eu
                         └ ─ ─ ─ 80% YIELD

8   2 µg Lu,Yb, ──────►
         Tm, Er
     Aminex A5 ──► absorption at cation exchange resin
     0.1 M NH₄Cl ──► transformation into the NH₄⁺-configuration

9   0.11 Mα-HIB ──► second chromatographic ──────► Lu–Er
     PH5   separation   Dy–La
                         ¹⁶³Ho ─ ─ ─ ─ 62% YIELD

10   2µg Lu,Yb, ──────►
         Tm, Er
     Aminex A5 ──► absorption at cation exchange resin
     0.1 M NH₄Cl   transformation into the NH₄⁺-configuration

11   0.12 Mα-HIB ──► third chromatographic ──────► Lu–Er
     PH 5   separation   Dy–La
                         ¹⁶³Ho ─ ─ ─ ─ ─ 44% YIELD

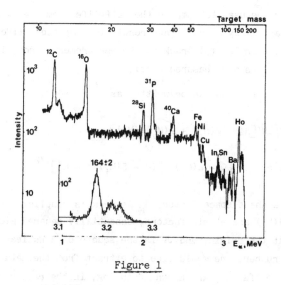

Figure 1

impurities give rise to step functions. If the well-resolved peak at
mass 164 ± 2 is interpreted as ^{163}Ho, its integrated number of atoms
($5 \cdot 10^{13}$ for one of the sources) is in excellent agreement with the
number obtained from the Faraday cup counting, chemical separation and
subsequent evaporation. Thus the method of making small but very pure
sources seems to be under control. No major problem in the purification
of mightier sources is foreseen.

When an electron is captured by the nucleus of ^{163}Ho, a hole is
left in the daughter ^{163}Dy atom, which subsequently emits Auger electrons and X-rays (preferentially the former for holes in M or higher
levels). The radiation from M holes is the more energetic and readily
detectable [the ionization energies in Dysprosium are $E(M_1)$ = 2.05 keV,
$E(M_2)$ = 1.84 keV, $E(N_1)$ = 0.416 keV, $E(N_2)$ = 0.332 keV, capture from
levels of higher angular momentum is negligible]. The M X-ray spectrum
has been measured with considerable resolution by a group in the USA[14].
The Aarhus-CERN-Risø collaboration[13] has measured M X-rays and Auger
electrons from the carefully prepared sources described in the previous
paragraph. The detectors in this preliminary experiment (a multi-wire
proportional chamber and a gas flow proportional counter) had a very
coarse resolution, but were good enough to observe the M Auger and X-
ray signals at the expected energy. Since the number of atoms in the
source is known, this allows one to determine the half-life for M-capture, already quoted in Eq. (5). A theoretical M-Auger fluorescence

yield[15], must be used to compute the half-life from the X-ray yield. The M-Auger and X-ray measurements result in compatible lifetimes. The error given in Eq. (5) brackets the measurements from different sources, detectors and measured particles.

The M capture rate can be written as

$$\omega = K \, G_F^2 \, \mathscr{N} \left[A(M_1)\phi(M_1) + A(M_2)\phi(M_2) \right]$$

$$\phi(M_i) = p_\nu E_\nu = \left[Q - E(M_i) \right] \left[(Q - E(M_i))^2 - m_\nu^2 \right]^{1/2}$$

where ϕ is a phase space factor, G_F is Fermi's coupling constant, K is a number, $A(M_i)$ are atomic factors, essentially squared electron wave functions at the origin, and $\mathscr{N}$ is the square of a nuclear matrix element. The numbers one would like to extract from the data are Q and m_ν. All other factors are known including, in the rather exceptional case of ^{163}Ho, the nuclear matrix element[12],[16],[17]. I will not indulge here in the details of the semi-empirical determination of $\mathscr{N}[^{163}\text{Ho}]$, that are given at length in Ref. 13). The point is that the phase space factor can be determined. For the dominant M-capture term, the result is:

$$\phi(M_1) = (Q - 2.05 \text{ keV}) \left[(Q - 2.05 \text{ keV})^2 - m_\nu^2 \right]^{1/2} =$$

$$= (0.53 \pm 0.10 \text{ keV})^2$$

(6)

which does not fix Q and m_ν but a combination theoreof. If m_ν is set to zero, the Q-value is

$$Q[EC, m_\nu = 0] = 2.58 \pm 0.10 \text{ keV}$$

(7)

Clearly a value of m_ν at the level of tens of eV would only affect this result by a negligible correction of order $1 + (m_\nu/530 \text{ eV})^2/2$. The Q-value of Eq. (7) is only 530 eV above $E(M_1, \text{Dy}) = 2.05$ keV. This means that N-capture will beat M-capture, thanks to its much bigger phase space, and in spite of a suppressed wave function at the origin. Thus the half-life of ^{163}Ho is much shorter than the measured M-half-life. A calculation with capture from all levels taken into account gives the result $T_{1/2} = (7 \pm 2)$ millenia, quoted before.

The soul is not entirely satisfied in setting $m_\nu = 0$ to obtain the result of Eq. (7). For this reason, the Aarhus-CERN-Risø collaboration[13] has directly measured Q in an independent nuclear reaction experiment, having nothing to do with neutrino masses. The central idea is to determine Q-values (mass differences) from the kinematics of the reactions

$$^2H + {}^{163}Dy \rightarrow {}^{162}Dy + {}^3H \tag{8a}$$

$$^3He + {}^{162}Dy \rightarrow {}^{163}Ho + {}^2H \tag{8b}$$

Reference lines are given by the elastic scattering of ^2H and ^3He; singly ionized ^3He was used, to provide a nearby reference line for tritons. A very preliminary experiment of this kind has been run with the Niels Bohr Institute tandem accelerator, and its pretty details are outlined in Ref. 13). The $(^{163}Ho, {}^{163}Dy)$ Q-value is obtained from the combined results of Eqs. (8a,b) and has already been quoted in Eq. (5). The error bars in this experiment, and the limits on m_ν to be presently discussed, are expected to become considerably tighter in the near future.

Given the nuclear reaction Q-value of Eq. (5) and the electron capture phase space measurement of Eq. (6), a domain of possible values of Q and m_ν can be determined. This is graphically done in Fig. 2, where $\phi^{1/2}(M_1)$ is plotted against Q. Curves for the function $\phi^{1/2}(Q,m_\nu)$ of Eq. (6) are given; as functions of Q for different values of m_ν (in keV). The horizontal strip is the experimental value of $\phi^{1/2}$. The vertical strip is the nuclear-reaction experiment value of Q. They overlap in the dotted region, giving a rough upper limit $m_\nu \lesssim 1.3$ keV.

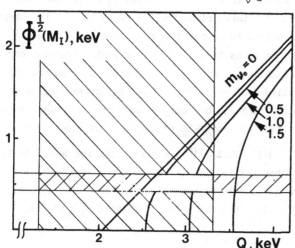

Figure 2

This limit will improve as the error bars, particularly from the nuclear reaction experiment, narrow down. Should they converge towards the lowest allowed Q-values, the electron neutrino is a tachyon ($m_\nu^2 < 0$).

THEORY OF INTERNAL BREMSSTRAHLUNG IN ELECTRON CAPTURE: (IBEC)

A photon may be emitted as a radiative correction in an electron capture process

$$Z \rightarrow (Z - 1)^H + \gamma + \nu \tag{9}$$

Here Z and $(Z - 1)^H$ refer to atoms, the daughter one being left with a hole H in the orbital from which the electron is captured. Photons, on the other hand, are emitted in electron capture as X-rays that "follow" the dominant non-radiative capture process

$$Z \rightarrow (Z - 1)^{\tilde{H}} + \nu \tag{10a}$$
$$\rightarrow (Z - 1)^H + \gamma \tag{10b}$$

The photon spectrum in Eq. (9) is referred to as an "Internal bremsstrahlung spectrum", while the approximately monochromatic photons of Eq. (10b) are referred to as X-rays. Since photons are not unlike each other, it is not difficult to find situations where the distinction makes no sense[18]. As an example, consider the case H = nP; the hole in Eqs. (9) and (10b) is in an angular momentum L = 1 state. In a non-relativistic approximation, an L = 1 wave function vanishes at the origin and "direct" capture from such an orbital is forbidden. The way the process of Eq. (9) takes place for H = nP is via the two steps Eqs. (10a,b). The electron is captured from a level $\tilde{H} = \tilde{n}S$ and the photon originates in the transition $\tilde{H} \rightarrow H\gamma$. For photon energies corresponding to X-ray lines, $E_\gamma = E_H - E_{\tilde{H}}$, the process is resonant and the photon can be called an X-ray. At energies not corresponding to X-ray lines the photon yield can be regarded as a bremsstrahlung spectrum or as a sum of tails of Breit-Wigner shaped X-ray lines.

The Q-value (Q = M(Z) - M(Z-1)) in electron capture decays is typically much bigger than the (positive) ionization energy E(H) of the missing electron in the daughter atom. The energetics of a typical three body decay, Eq. (9), is shown in the left-hand side of Fig. 3.

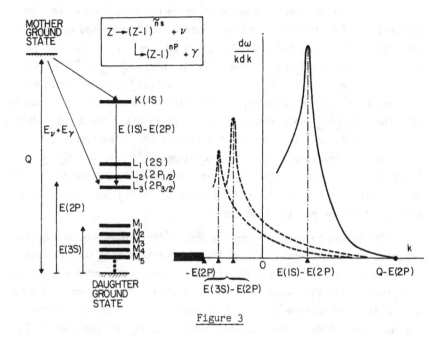

Figure 3

The mass difference between parent and daughter atoms is $Q - E[H]$.
Recoil effects are negligible and $E_\gamma + E_\nu = Q - E[H]$, in an obvious
notation. The maximum photon energy, upon neglect of the neutrino mass,
is also $Q - E[H]$. A typical $H = 2P$ photon spectrum is shown in the
right-hand side of Fig. 3. The process occurs via capture from $\tilde{n}S$
states, as in Eqs. (10). For $\tilde{n}S = 1S$, there is a photon energy $E_\gamma =$
$E[1S] - E[2P]$ at which the spectrum peaks at an X-ray resonance, as in
the figure. Virtual capture from $\tilde{n} > 1$ states also contributes to the
matrix element, giving poles at virtual (negative) photon energies:
$E_\gamma = E[nS] - E[2P]$, also symbolically shown in the figure. The endpoint
of the photon spectrum in the three body decay Eq. (9) is sensitive to a
neutrino mass in the same way as the endpoint of the electron spectrum
in β-decay. But the photon spectrum matrix element at the endpoint is
typically very small: $E_\gamma \simeq Q - E[2P]$ is far away from the X-ray resonances where the matrix element is large. An example: the figure of
merit for the dominant (2S) spectrum in the favourable case of ^{55}Fe
decay (number of events in the last 30 eV divided by the total decay
rate) is $g \sim 4 \cdot 10^{-16}$, to be compared with $g(^3H) \sim 10^{-8}$. A neutrino mass
hunt with ^{55}Fe would be out of the question: rates are too ridiculous.

The problem of tiny IBEC endpoint rates may be solvable, and the possibility of neutrino mass measurements in IBEC processes may be resurrected. Refer for definiteness of the 2P spectrum of Fig. 3. The endpoint energy Q - E[2P] and the resonant energy E[1S] - E[2P] are independent quantities. For isotopes with Q ≈ E[1S] the rate would be enhanced at the endpoint energy of interest to the neutrino mass hunter. Even for Q < E[1S] (no accidental degeneracy within a few X-ray widths), there is a considerable enhancement factor of the bremsstrahlung spectrum[19]. There is a short list of elements[19] with Q < E[1S], ^{193}Pt is an example. The arguments can be repeated, and the figures of merit further increased, for elements with Q < E[2S] (here the list is short: ^{163}Ho and nothing else known).

^{193}Pt decays by electron capture into ^{193}Ir. The decay energetics is shown in the left-hand side of Fig. 4. Capture from the 1S level is energetically forbidden. The shapes of the H = 2P and H = 3P spectra are given in the right-hand side of the figure. The resonances due to intermediate 1S capture now lie above the endpoint, but their presence enhances the matrix element at that point. In the 3P spectrum the 2S → → 3P resonance is real, the zero above it reflects the interference of this resonance with the virtual 1S → 3P one. The figure of merit for the 2P spectrum in ^{193}Pt is g ∿ $4 \cdot 10^{-13}$, a three orders of magnitude improvement over the typical case of ^{55}Fe, but still small enough to make experiments attempting to measure neutrino masses at the ∿ 30 eV level extremely difficult. ^{163}Ho may be the only hope.

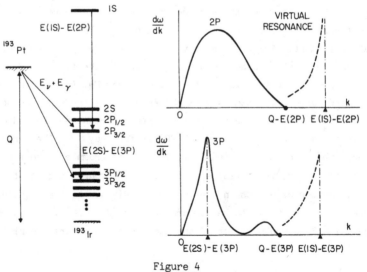

Figure 4

The 2P spectrum ends at $Q - E[2P]$, and the spectra for capture from higher levels end up at a higher energy: $Q - E[3P]$ for the case of 3P capture. The endpoint of the 2P spectrum is therefore submerged in a large background of photons from other captures. To see a clean end-point one must disentangle the different spectra. This may be done in a coincidence experiment. The 2P spectrum, for instance, may be observed via the chain

$$Z \rightarrow (Z - 1)^{2P} + \gamma + \nu \qquad \qquad \text{(11a)}$$

$$\longrightarrow (Z - 1)^{3S} + \gamma_X \qquad \qquad \text{(11b)}$$

where the 2P → 3S X-ray signs the 2P hole. Coincidence experiments of this type have been performed in the past, but not, to our knowledge, with the measurement of neutrino masses in the back of the mind.

Atomic and molecular interplay problems are much less serious in IBEC coincidence experiments than in Tritium beta decay[19]. In electron capture the nuclear charge changes from Z to $Z - 1$, a small fractional change in comparison to ^3H decay. Moreover, it is not the charge, but only the charge distribution in the inner atom that changes. The outer atomic electrons, that may change level by a few eV, are practically insensitive to the small rearrangement of charge in the inner atom.

COINCIDENCE IBEC EXPERIMENTS IN ^{193}Pt

The Aarhus-CERN-Risø collaboration[20] has made a preliminary mea-surement of the bremsstrahlung 2P spectrum in coincidence with the sub-sequent X-rays, as in Eqs. (11). The immediate aims are to check the theory of resonant enhancements, described in the previous chapter, and to demonstrate the coincidence method in a search for neutrino mass limits. Confidence in the theory may pump enthusiasm into the effort of pushing the ^{163}Ho measurements to their status of the art limit, where a hope exists to measure neutrino masses in the range of interest (Tritium-like or better).

To have an early start, the ^{193}Pt source was not made by isotope separation at ISOLDE. An estimated 10^{16} atoms of ^{193}Pt per typical running period should have survived in old lead targets used for the pro-duction of mercury isotopes. About 1kg of lead from these targets was sublimated and treated with a complicated sequence of liquid extractions and chromatographic purifications, not unlike the ^{163}Ho ones. About $5 \cdot 10^{14}$ ^{193}Pt atoms were obtained this way, with a decontamination factor of order 10^{12} to 10^{13}.

Bremsstrahlung photons from this source were measured with a Ge detector and L X-rays in coincidence were measured with a Na I detector. The data I will present are from a mere nine days of running plus nine days of background counting. Data from longer running periods and better sources and detectors should be available in the near future.

The single photon bremsstrahlung spectrum is shown in Fig. 5. The theoretical prediction[19] is shown as the full line. Both theory and experiment are given in an absolute scale. The data have not been deconvoluted from the detector response, which is not bell-shaped, but has a low energy plateau and an escape peak. This quantitatively explains the data excess at the lower energies. Notice the suppressed energy scale: enormous L X-ray peaks at E < 20 keV are not shown.

The bremsstrahlung spectrum in coincidence with L X-rays is shown in Fig. 6, and compared with theory[19] (full line) in an absolute scale. The L X-ray peaks that should not be present in these data are now enormously suppressed, but still there. This is a limited time-resolution effect, what remains are accidental coincidences of radiation from two different atoms.

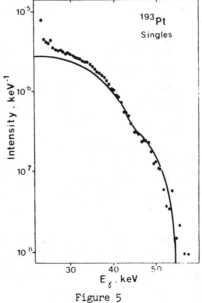

Figure 5

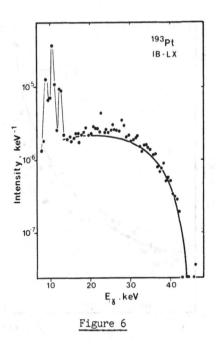

Figure 6

The agreement between theory and data is better than expected. The
number of atoms in the source is not known with absolute precision. The
theory ignores relativistic and screening effects, and is only semi-
empirically improved to give the resonance positions at their correct
values. Better data could only make the agreement worse. The agreement
corroborates the presence of the resonance enhancement factor in ^{193}Pt
IBEC. In this preliminary experiment the X-ray detector was not good
enough to distinguish 2S from 2P X-rays, and to disentangle the corresponding IBEC spectra. Therefore, we should not yet have convinced the
reader that the predicted enhancement of the 2P IBEC spectrum over the
2S IBEC spectrum (by a factor of $\sim$100) is what is being observed. This
can be done by a more detailed look at the shape of the data. The 2S
spectrum is predicted to have a shape given essentially by phase space:
$k(k_{MAX} - k)^2$, where k is the photon energy. In the 2P spectrum this
shape is modulated by a matrix element $Q^2_{2P}(k)$ that has a virtual pole
at the position of the 1S $\rightarrow$ 2P resonance (19 keV above the endpoint).
The number of events, divided by phase space, is shown in Fig. 7. The
data agree in shape and magnitude with the theoretical 2P matrix element
Q^2_{2P}. In the same absolute scale the 2S prediction would be a horizontal
line $\sim$0.01 above the abscissa.

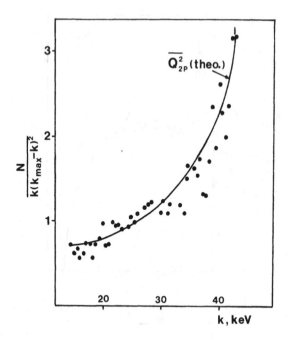

Figure 7

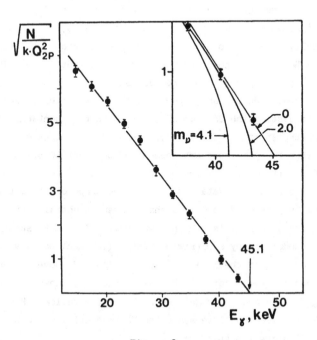

Figure 8

A typical neutrino mass hunt plot for the 2P spectrum is shown in Fig. 8. To get a straight line one must divide the number of events by photon momentum and by the non-trivial matrix element Q_{2P}^2, then take the square root: this is neither a Kurie nor a Jordan plot, but the idea is the same. The endpoint energy is ~ 45.1 keV, giving a $Q[^{193}Pt, ^{193}Ir]$ value of 56.8 ± 1 keV (to be compared with the previously measured value[21] of 61 ± 3 keV). The insert in the figure compares the data with the expected shapes for $m_\nu = 0, 2, 4.1$ keV. Neutrino masses above ~ 1 keV are excluded by this preliminary experiment. The peculiar figure of 4.1 keV does not reflect some peculiarity in the characters of the authors, but the fact that the best limit on the electron neutrino mass (as opposed to the electron antineutrino mass) is 4.1 keV[22].

THE FUTURE ?

The theory of resonant enhanced IBEC has been checked in ^{193}Pt. This may be something to write to Physics Letters about, but certainly not home, since the theory is essentially non-relativistic QED. The Q value in ^{163}Ho decay is measured to be very small. This implies a relatively very large figure of merit for the Ho IBEC spectra. To give an example, the number of events in the last 30 eV of the 4P capture spectrum, normalized to the total decay rate, is $g \sim 2 \cdot 10^{-10}$, a six orders of magnitude improvement over the favourable typical case of a K-capturing isotope, ^{55}Fe.

Presumably the Aarhus-CERN-Risø collaboration will attempt to measure, among others, the following bremsstrahlung photon-de-excitation-electron coincidences:

$$^{163}Ho \rightarrow {}^{163}Dy^{M_2} + \gamma + \nu \quad (k_{MAX} \sim 740 \text{ eV}) \tag{12a}$$
$$\rightarrow {}^{163}Dy^{M_4};^{N_6,7} + e^- \quad (E_e \sim 500 \text{ eV, BR} \sim 38\%[23]) \tag{12b}$$

$$^{163}Ho \rightarrow {}^{163}Dy^{M_3} + \gamma + \nu \quad (k_{MAX} \sim 900 \text{ eV}) \tag{13a}$$
$$\rightarrow {}^{163}Dy^{M_5};^{N_6,7} + e^- \quad (E_e \sim 380 \text{ eV, BR} \sim 37\%[23]) \tag{13b}$$

$$^{163}Ho \rightarrow {}^{163}Dy^{N_2} + \gamma + \nu \quad (k_{MAX} \sim 2250 \text{ eV}) \tag{14a}$$
$$\rightarrow {}^{163}Dy^{N_4};^{N_6,7} + e^- \quad (E_e \sim 175 \text{ eV, BR} \sim 57\%[23]) \tag{14b}$$

$$^{163}\text{Ho} \rightarrow {}^{163}\text{Dy}^{N_3} + \gamma + \nu \ (k_{MAX} \backsim 2290 \text{ eV}) \tag{15a}$$

$$^{163}\text{Dy}^{N_5;N_{6,7}} + e^- \ (E_e \backsim 135 \text{ eV, BR} \backsim 61\%)^{23)}) \tag{15b}$$

The techniques to measure photon energies in the keV range with efficiency and precision still have a long way to be developed. This extra
challenge only makes the experiment more tempting. A discussion of the
prospects of an IBEC ^{163}Ho experiment: rates, backgrounds, etc., is
now possible, given that the Q-value is known; and should appear
soon[24].

The transitions in Eqs. (12a), (13a), (14a) and (15a) are dominated
by M_1-capture, that makes them resonant a mere 530 eV below their end-
points. The fluorescence yield of an M_1 hole is very small: the atom
much prefers to de-excite by the emission of Auger and Coster-Kronig
electrons. This is not only true for "monochromatic" X-ray or electron
de-excitations, but also for their tails[25]: the IBEC spectrum and
the spectrum of electron ejection in electron capture (EEEC). This
means that the rate of the three body-process of electron ejection

$$^{163}\text{Ho} \rightarrow {}^{163}\text{Dy}^{H_1,H_2} + e^- + \nu \tag{16}$$

is much bigger at the endpoint than the corresponding IBEC rates. Once
again the kinematics must be specified by the detection of a signature
for the H_1, H_2 holes:

$$^{163}\text{Dy}^{H_1,H_2} \rightarrow {}^{163}\text{Dy}^{H_1',H_1'',H_2} + e^- \tag{17a}$$

$$\hookrightarrow {}^{163}\text{Dy}^{H_1',H_1'',H_2',H_1''} + e^- \tag{17b}$$

The possibility to measure neutrino masses in the three-electron coincidence experiments, Eqs. (16) and (17) has been discussed in Ref. 25).
The energy of electrons in the range of hundreds of eV's can be measured
with great precision with electrostatic methods. This may make this
kind of experiment very tempting.

REFERENCES

1) P.G. Hansen, private communication.

2) H. Georgi and S.L. Glashow, Phys. Rev. Lett. 32 (1974) 438.

3) K.E. Bergkvist, Nucl. Phys. B39 (1972) 317.

4) J.J. Simpson, Phys. Rev. D23 (1981) 849.

5) V.A. Lubimov et al., Phys. Lett. 94B (1980) 266.

6) E. Fermi, Nuovo Cim. 11 (1934) 1;
   F. Perrin, Comptes Rendus 197 (1933) 1625.

7) K.E. Bergkvist, Phys. Scripta 4 (1971) 23.

8) I.G. Kaplan, V.N. Smutny and G.V. Smelov, Karpov Institute of
   Physical Chemistry report, Moscow (1982);
   Ching Chengrui, Ho Tsohsiu and Chao Shaolin, Academia Sinica Report
   AS-ITP-81-22 (1981).

9) A.H. Wapstra and K. Bos, Atomic Data and Nuclear Data Tables 19
   (1977) 175.

10) R.A. Naumann, M.C. Michel and J.C. Power, J. Inorg. Nucl. Chem. 15
    (1960) 195.

11) P.K. Hopke, J.S. Evans and R.A. Naumann, Phys. Rev. 171 (1968) 1290.

12) R. Bengtsson and I. Ragnarsson, private communication (1981).

13) J.V. Andersen et al., CERN EP/82-50, April 1982, to be published
    in Phys. Lett.

14) C.L. Bennet et al., Princeton Univ. preprint (1982).

15) W. Bambynek et al., Revs. Mod. Phys. 49 (1977) 77.

16) A. Bohr and B.R. Mottelson, Nucl. Structure (W.A. Benjamin Inc.
    Reading, Mass. 1975), Vol. II, pp. 245, 296 and 306.

17) J. Zylicz, P.G. Hansen, H.L. Nielsen and K. Wilsky, Arkiv für Fysik
    36 (1967) 643.

18) R.J. Glauber and P.C. Martin, Phys. Rev. 104 (1956) 158.

19) A. De Rújula, Nucl. Phys. B188 (1981) 414.

20) J.V. Andersen et al., Status report CERN/PSCC/82-7, April 1, 1982.

21) P.K. Hopke and R.A. Naumann, Phys. Rev. 185 (1969) 1565.

22) E. Beck and H. Daniel, Z. Phys. 216 (1968) 229.

23) E.J. Mc. Guire, Phys. Rev. A5 (1972) 1043, 1052; Phys. Rev. A9
    (1974) 1840; Sandia Lab. reports SC-RR-710835, SAND-750443.

24) A. De Rújula and M. Lusignoli, to be published.

25) A. De Rújula and M. Lusignoli, CERN preprint TH.3300, (1982); to be published in the proceedings of the International Conference on Unified Theories and their Experimental Tests, Venice 16-18 March 1982.

PLANNING FOR THE NEXT GENERATION OF PROTON DECAY EXPERIMENTS IN THE UNITED STATES

D. S. Ayres

High Energy Physics Division, Argonne National Laboratory
Argonne, Illinois 60439

Abstract

There are now three well-developed proposals for new proton decay detectors to be built in the United States. These are the 1000-5000-ton Soudan 2 tracking calorimeter, the 1400-ton Homestake II liquid scintillator Tracking Spectrometer, and the 2500-ton University of Pennsylvania liquid-scintillator - proportional-drift-cell calorimeter. These proposals were reviewed by the Department of Energy Technical Assessment Panel on Proton Decay in February 1982. I shall describe the Soudan and Pennsylvania proposals, present the latest results from the 31-ton Soudan 1 experiment, and discuss the recommendations of the DOE Panel. Following these recommendations, a one-week workshop, to be held at Argonne in June, will focus on the optimization of techniques for future experiments.

1. Introduction

In February 1982 the U.S. Department of Energy Technical Assessment Panel on Proton Decay met to review the status of the U.S. proton decay experiments which are in operation or construction, and to consider proposals for new experiments. The experiments make use of three general techniques, each with special capabilities and limitations: the water Cerenkov detectors, the fine-grained sampling calorimeters (typically built of iron and track chambers), and the

289

totally active calorimeters (typically using liquid scintillator). Ideas for more advanced detectors were also presented: the liquid-argon drift calorimeter by the HPW collaboration, and the R&D program at Berkeley on a very high pressure argon time projection chamber.

In its recommendations [1], the panel advised that no new experiments be approved but that the situation be reviewed after a year, when results might be known from the water Cerenkov detectors and when R&D work in support of the calorimeter proposals would be completed. The panel also suggested that a workshop be held soon to discuss the optimum design and deployment of the next U.S. experiment. This workshop will be held at Argonne National Laboratory in June. Finally, the panel suggested that a national underground laboratory should be considered, to support both proton decay experiments and other underground physics research. A workshop on this subject will be held at Los Alamos National Laboratory in September.

I shall review the two proposals for new experiments which were presented to the panel and have not already been described at this meeting (Soudan 2 and the Penn Design Study), and also present the latest results from the Soudan 1 experiment. A synopsis of the issues to be resolved for the U.S. program is presented in the last section.

2.   Results from Soudan 1

The Soudan nucleon decay program [2-6] is being carried out in the Soudan iron mine in northeastern Minnesota, at a depth of 2000 m of water equivalent. A 31-ton prototype experiment, Soudan 1, has been built and is now being operated by a University of Minnesota - Argonne National Laboratory collaboration. The Soudan 1 detector consists of a 3 m × 3 m × 2 m block of taconite(iron ore)-loaded concrete, which is instrumented with 3456 gas proportional tubes in a crossed array, and has an average density of 1.85 g/cm^3. The detector began routine data acquisition in August 1981; turnon of the scintillation-counter active shield in October 1981 made the experiment fully operational. This shield surrounds the detector on the top and four sides, and signals the presence of charged particles from cosmic-ray interactions in the rock around the experiment. Such events can produce low-energy neutrals which resemble nucleon decays when they interact in the detector. Soudan 1 is providing valuable data on the rates and characteristics of such events, which are a potential source of background to nucleon decay in the proposed Soudan 2 experiment.

Preliminary analysis has now been completed on a total of 63 days of data from Soudan 1, yielding the following sample of events:

190,000	cosmic-ray muon events
1,100	multiple parallel muon events ($2 < N_\mu < 12$)
395	stopping muons (0.21% of the total)
70	observed μ^+ decays (giving 35±4% detection efficiency)
1	upgoing muon candidate (from a ν + rock interaction?)
0	neutrino interactions in the detector (~ 0.5 expected)
0	nucleon decay candidates
0	slow magnetic monopole candidates

Events with multiple parallel muon tracks can yield information on the nuclear composition of the very high energy cosmic-ray primaries which interact in the upper atmosphere to produce bundles of decay muons. Figure 1 shows the multiplicity distribution of these events. In addition, the very high-energy muons which traverse the detector occasionally produce spectacular interactions, demonstrating the excellent pattern-recognition capabilities of Soudan 1. An example is shown in Figure 2.

The lack of nucleon decay candidates gives the lower limits on the nucleon lifetime listed below. The fiducial-mass values are from Monte-Carlo event-containment studies, using the same cuts which were applied to the data. The detection efficiencies include a nuclear absorption loss of 20% per pion in the parent nucleus (a remaining e^+ is seen with 50% efficiency), and the > 8-tube cut on the number of hit proportional tubes.

Decay mode	Fiducial mass	Detection efficiency	τ/branching ratio (90% C.L.)
$p \rightarrow e^+\pi^\circ$	11 tons	90%	4.5×10^{29} years
$n \rightarrow e^+\pi^-$	13 tons	90%	5.4×10^{29} years
$n \rightarrow \mu^+\pi^-$	11 tons	90%	4.5×10^{29} years
$p \rightarrow \nu\pi^+$	16 tons	70%	5.2×10^{29} years
$n \rightarrow \nu\pi^\circ$	15 tons	80%	5.5×10^{29} years
$p \rightarrow \nu K^+$	16 tons	35%	2.6×10^{29} years

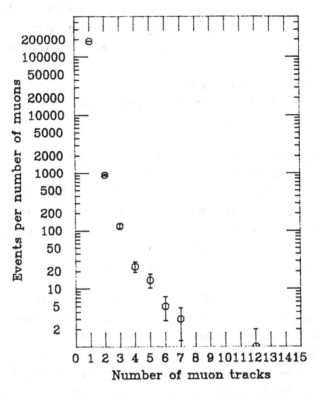

Figure 1. Muon multiplicity distribution from Soudan 1 for single muons and for events with two or more parallel tracks.

The scintillation-counter cosmic-ray veto shield has proved to be quite powerful in rejecting events induced by cosmic-ray interactions in the rock. Figure 3 shows an example of such an event, which strongly resembles a nucleon decay except for the hits in six veto counters.

A Monte-Carlo simulation has been used to calculate the Soudan 1 acceptance for slow, ultraheavy, "GUT" magnetic monopoles. The presence of these particles would be apparent in the timing data from the proportional tubes, which are sensitive for a 7 µsec period around the trigger time. For monopoles which ionize like muons, the velocity range in which Soudan 1 is sensitive is $2 \times 10^{-3} < \beta < 1.5 \times 10^{-2}$. The lack of candidates gives a 90% confidence-level lower limit on the monopole flux of $2.2 \times 10^{-5}/cm^2$ sr yr. This result is compared below

```
48  ...........:...........:...........:...........:...........:...........:...
46  ...........:...........:...........,...........:...........:...........:...
44  .....:...2...:...........:...........:...........:...........:...........:...
42  .....2:.....'...........:...........:...........:...........:...........:...
40  ......2.:...........:...........:...........:...........:...........:...
38  ......71.:A...4..:.....2.:...........:...........:...........:...........:...
36  ......6..*E654...94.3.....G.............:...........:...........:...........:...
34  ......7.K**MA8.942....5.............:...........:...........:...........:...
32  ....9.:..4.48G*****GH77...45..5.............:...........:...........:...
30  ........:....'87******IF87.1G.'.:...6...........6.............:...........:...
28  .......:..52G8CGR*****HGAD.5.9.:...........:...........:...........:...
26  .......:..F3I4FJR****SHTJ.8.43.............:...........:...........:...
24  .......:.B5...4CL***''''PEC..6.:...........:...........:...........:...
22  .......:8...A436CLJL****G5D679.:...........:...........:...........:...
20  ........:....52F.9A"7NGGHAE6...:2.............:...........:...........:...
18  ........:...T..7..2AB92B9..8.:...........:...........:...........:...
16  ........:...:2..6..:....1..:5.............:...........:...........:...
14  ........:...3:..7.F..5..43.:...........:...........:...........:...
12  ........:...8:....B.:..58.:...........:...........:...........:...
10  ........:........:....4...6.:...........:...........:...........:...
 8  ...........:...........:...........:...........:...........:...........:...
 6  ........:....5:...........:...........:...........:...........:...........:...
 4  ...........:...........:...........:...........:...........:...........:...
 2  ...........:...........:...........:...........:...........:...........:...


47  ...........:...........:...........:...........:...........:...........:...
45  ...........:...........:...........:...........:...........:...........:...
43  ...........:...........:...........:...........:...........:...........:...
41  ......:...4:...........:...........:...........:...........:...........:...
39  .........:...1D5.:...5.............:...........:...........:...........:...
37  .33...:....8.:...D..3.............:....3.:..5.............:...........:...
35  .1..4.**7IHB6I39.5..88.5.35.44.:"....7.............:...........:...
33  ...2.5C8R****L*HA9FCC.7C475..674.4.:...........:...........:...
31  ......:.BK2H********RD*MS7DH5.95H34...B7.....:....4:...........:...
29  ......:.4..297D*D*******RF*KB7CC82.:...4..3.:...........:...
27  .........:....:****''''"T"""8644...E...6.............:...........:...
25  ......35.14.5AJEA9"F*****N"I858BFD....2..2.:..:..3...:...........:...
23  ......5..F..KKEJ""L*"********..F"FB.:83.............:...........:...
21  .......7.:...92.7B.7.F8867ACFG8E"H.FGG8.1.1.:...........:...
19  ......:.9...8.1>545CD28.BD7DKFP....4.:......I.............:...
17  ......41.:.21..2421.368.:...7.76:...........:...........:...
15  .......:.....:....2:....3.34.6..53:..7....526..2B.8.............:...
13  .......:....:....5...3.:......22:...........:...........:...
11  ......5.:....8:14...5.:..2....:....432:..54....4
 9  ......2:...........:...........:...........:...........:...........:...
 7  ......7.............6.............:...........:...........:...
 5  ...........:...........:...........:...........:...........:...........:...
 3  ...........:...........:...........:...........:...........:...........:...
 1  ...........:...........:...........:...........:...........:...........:...
```

Figure 2. Two views of a fully contained electromagnetic shower in the Soudan 1 detector. In the lower (north-south) view, the incident muon enters from the top left and exits at the bottom right. The total shower energy is roughly 200 GeV. Numbers, letters and asterisks indicate observed pulse height, where pulse height increases as 1,2,...9,A,...,Z,*.

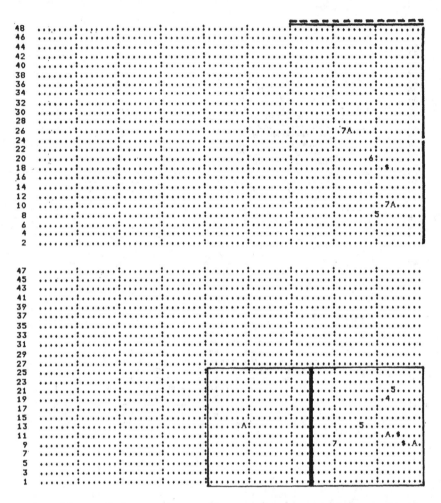

Figure 3. Two views of a nucleon decay background event in the Soudan 1 detector. Numbers, letters, and dollar signs show proportional tubes which were hit. This event is clearly identified as being due to a cosmic-ray interaction in the nearby rock by the six veto counters (indicated by lines and rectangles) which were hit.

with some other recent limits on the flux of GUT monopoles. Not shown here are even more restrictive bounds from the Kolar Gold Fields experiment and from the Baksan Underground Scintillation Telescope, which I understand will become available in the near future [7]. It seems clear that, if the monopole candidate recorded by Cabrera [8] is real, these particles ionize much less than relativistic light particles, are generally slower than $\beta = 10^{-3}$, or will soon be detected by other methods.

Experiment	Flux $(m^{-2}sr^{-1}day^{-1})$	dE/dx cut $(I_{min}$ units)	β Range	Method
B. Cabrera [8]	<0.5	none	any	superconducting loop
J. Ullman [9]	<0.05	>2	3×10^{-4} to 1.2×10^{-3}	proportional tubes
R. Bonarelli, et al. [10]	<0.002	>25	7×10^{-3} to 0.6	scintillation counters
Soudan 1 (preliminary)	<0.0006	≥0.3	2×10^{-3} to 1.5×10^{-2}	proportional tubes in deep mine

A 700-lb facsimile of the Soudan 1 detector has been exposed to low energy $\pi^{\pm}$, $\mu^{\pm}$, and $e^{\pm}$ in a test beam at the Argonne Rapid Cycling Synchrotron. The results have been compared to predictions of the Monte-Carlo simulation of the detector and the agreement is excellent. Figure 4 shows the average number of hit tubes for negative particles as a function of beam momentum.

3. The Soudan 2 Proposal

The Soudan 2 proposal [4] to build a 1000-ton tracking calorimeter detector was submitted to the U.S. Department of Energy and the U.K. Science and Engineering Research Council in September 1981 by the Minnesota-Argonne-Oxford University collaboration. Recently, the collaboration has suggested that the detector be made expandable, by installing it in an underground cavity large enough for

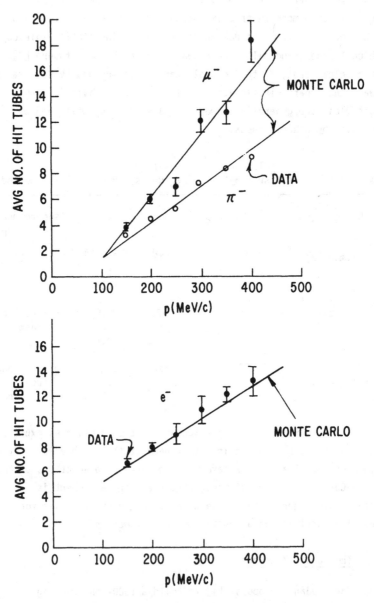

Figure 4. Comparison of Soudan 1 test-beam data and Monte-Carlo predictions. The average number of hit proportional tubes is shown as a function of beam momentum for π^-, μ^-, and e^-.

an eventual 5000 ton experiment, which would be built after experience
was gained with the initial 1000 tons. The 1000-ton Soudan 2 could be
operating in 1985, and the 5000-ton expansion in 1987.

Soudan 2 achieves fine-grained ionization sampling and tracking
at a reasonable cost by drifting ionization in long gas channels
within the iron calorimeter material. The ionization is detected by
planes of crossed proportional wires and cathode strips on the
5 m × 5 m faces of 50-ton movable modules. Planar drift-chambers with
50 cm drift distances have been developed to implement this
idea [6]. Figure 5 shows schematically how the detector would be
assembled from such chambers. Development work is continuing on
schemes which use drifting in 1 to 2 cm diameter cylindrical columns
to achieve more isotropic ionization sampling.

The Soudan 2 detector is characterized by excellent spatial
resolution and pattern-recognition ability. Its bubble-chamber-like
"pictures" of events provide a powerful tool for dealing with
unexpected background processes or unexpected physics. For $p \rightarrow e^+\pi^\circ$
decays, the detector records typically 60 drift-chamber-cell hits, and
achieves an energy resolution σ_E/E of about 20%. Since μ^- are
captured in the iron and do not decay visibly, the observation of μ^+
decays gives the ability to determine muon charge. Detection of $\pi^+ \rightarrow$
$\mu^+ \rightarrow e^+$ gives similar capability for $\pi^\pm$ charge determination. As for
most dense detectors, Soudan 2 has excellent event-containment
properties, which allow the detailed study of small modules in
accelerator test beams.

Sensitivity to three-prong nucleon decay modes in Soudan 2
averages 64% after cuts to reject background events. Using only the
three-prong modes, and SU(5) branching ratios, the 1000-ton experiment
would identify about one decay per year (after cuts) if the nucleon
lifetime were 10^{32} years. Monte-Carlo studies have recently shown how
to use the measurements of ionization along particle trajectories to
determine the direction of particle motion. Uncertainty on this point
had previously prevented the calculation of sensitivity to two-prong
nucleon decay modes which, in the absence of directionality
information, are topologically very similar cosmic-ray neutrino
interactions. Test-beam data from the Soudan 1 facsimile module have
been used to confirm the Monte-Carlo predictions. Figure 6 shows the
test-beam results for 300 MeV/c μ^- tracks, 84% of which deposit more

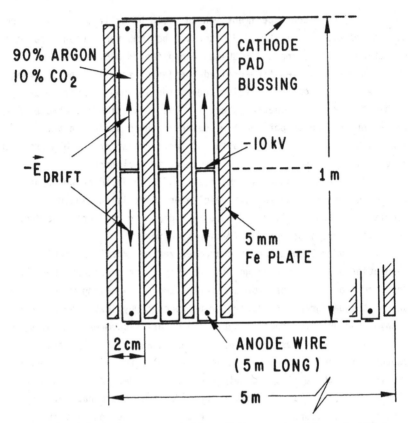

Figure 5. Schematic representation of the construction of a 50-ton Soudan 2 module from 5-mm thick iron plates and 5 m × 0.5 m × 1 cm planar drift chambers. Note that the horizontal and vertical scales are different. The module contains 500 chambers and is read out by 500 anode wires and 500 cathode strips on the two external 5 m × 5 m faces. The average density is 2 g/cm^3.

ionization in the last third of their tracks than in the first third. For 300 MeV/c π^-, the direction of motion is correctly determined for 72% of the tracks. The finer sampling in Soudan 2 gives particle direction for an even greater fraction of the tracks.

Monte-Carlo calculations show that neutrino-induced backgrounds to three-prong nucleon-decay modes is < 0.5 event/year in Soudan 2. Backgrounds from cosmic-ray induced interactions in the surrounding rock are expected to be < 0.1 event/year. Since this rate is already substantially less than the neutrino background rate, little would be gained by going deeper than the proposed depth of 2000 m of water

300 MEV/C MU MINUS

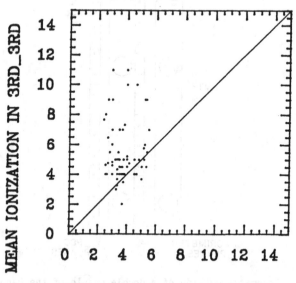

MEAN IONIZATION IN 1ST_3RD

Figure 6. Test-beam data from the Soudan 1 700-1b facsimile module. The data show that the direction of motion of 84% of 300 MeV/c μ^- can be determined by comparing the ionization deposited in the first third and the last third of the tracks. Monte-Carlo simulations predict even better directionality determination in Soudan 2.

equivalent. Continuing Monte-Carlo studies are aimed at optimizing background rejection in Soudan 2 before the calorimeter design is finalized later this year.

4. The Penn Design Study

The proposal to build the 2500-ton Penn Nucleon Decay Detector [11,12] was submitted to the Department of Energy in the fall of 1981 as a design study by a group from the University of Pennsylvania. The detector is a totally active calorimeter based on liquid scintillator, and incorporates fine-grained tracking with spatial resolution similar to that of Soudan 2. The proposed design

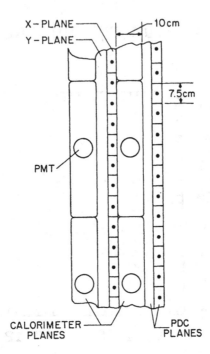

X-PLANE ————
Y-PLANE ————
———— 10cm
7.5cm
PMT
CALORIMETER ———— PLANES
PDC PLANES

Figure 7. Schematic outline of a double module of the Penn Nucleon Decay Detector (from Ref. 12). The ends of two 8-m long liquid scintillator calorimeter planes are shown, along with the crossed planes of proportional drift cells (PDC's).

is an extrapolation of a 170-ton detector for 1 GeV neutrino interactions, which is now in operation at the Brookhaven AGS [13]. Many of the properties of the nucleon decay detector can be measured directly from this "prototype". The proponents suggest that the detector be built at the recently proposed National Underground Physics Laboratory [14,15], although a definite site has not yet been selected. The experiment could be ready to operate in 1985.

Figure 7 shows a cross section of the calorimeter construction. It utilizes 6700 liquid scintillator cells, each 10 cm × 50 cm × 8 m and viewed by photomultiplier tubes at both ends. Between the planes of scintillator cells are two crossed planes of proportional drift cells, to give two high-resolution views of each event. The assembled detector would be 8 m × 8 m × 61 m and would have an average density of 0.64 g/cm^3. It contains 84,000 drift cells and 13,400 3-inch photomultipliers. Construction techniques and costs are already well understood from the Brookhaven detector experience. A $p \rightarrow e^+\pi^0$ decay would produce about 60 drift-cell hits in the detector and would be characterized by an energy resolution $\sigma_E/E \sim 10\%$. The Penn detector determines the charge of stopping pions and kaons by observing the nuclear stars produced by stopping negative hadrons and the decay products of stopping positive mesons. Energy-loss data from the liquid scintillator will give this calorimeter excellent particle-identification capabilities. Muon decay and muon polarization are easily detectable in liquid scintillator, but muon charge must be inferred from the charges of accompanying hadron decay products. The rejection of cosmic-ray-induced backgrounds is improved by timing information ($\sigma_t \sim 2$ nsec), which allows the direction of motion of most particles to be determined.

5. Issues to be Resolved for the U.S. Program

The recommendations of the DOE Technical Assessment Panel on Proton Decay reflect a disagreement among researchers on a number of issues which are essentially technical. Such uncertainties are inevitable when experiments of unprecedented size and sensitivity are proposed to search for a rare process with unknown characteristics in an environment where the important backgrounds have not yet been measured experimentally. The questions which must be resolved before the U.S. program can proceed will be discussed in depth at the workshop at Argonne this summer. A synopsis of the most important issues is given here:

(1) When should the next generation of experiments begin?

Some argue that nothing should be decided until we have results from the water Cerenkovs: if these experiments work well and do not see decays, then the technique should be extended to larger masses; if a signal is seen, we will know much better what kind of device to build to study proton decay in detail. Others argue that the physics importance of _any_ result, positive or negative, will require it to be verified by techniques which are as different as possible from the water Cerenkovs. Both the decay modes and the backgrounds are unknown, and experimental sensitivity and background rejection depend on the actual decay modes and on the detection technique. Finally,

the experiments are large and difficult. It may be several years
before we learn the real capabilities of the water Cerenkovs, and it
will certainly take years to bring any new experiment into operation.
If we want to know the answer in this decade, we'd better get started!

(2) What is the optimum balance between mass and background rejection?

The ultimate sensitivity of any experimental technique depends on
both the total mass monitored for decays and on the ability of the
detector to reject background. The largest detector will not
necessarily be the most sensitive.

(3) At what depth should the next large detector be built?

Existing sites deeper than 2000 m of water equivalent are rare in
the U.S. Excavating a new, deep site costs money which might better
be spent on detectors. If cosmic-ray muon induced backgrounds at
modest depths can be easily reduced well below the level of the depth-
independent neutrino backgrounds (as claimed in the IMB, HPW, and
Soudan proposals), then excavating deeper laboratories is unnecessary.

(4) What type of detector should be built next?

Three techniques have been extensively developed: water
Cerenkovs, fine-grained sampling calorimeters, and totally active
calorimeters. The water Cerenkovs provide the largest mass per
dollar, but their relatively poor pattern recognition for complex
topologies may limit background rejection capability. Sampling
calorimeters can have excellent pattern recognition ability, and if
ionization is measured they can usually determine the direction of a
particle's motion, perhaps as well as detectors which measure time of
flight. All techniques have some particle-identification capability,
but there are significant differences. Water Cerenkovs cannot measure
the charge of stopping tracks but can measure muon polarization; iron
calorimeters can often measure muon and hadron charges but not muon
polarization. Liquid scintillator detectors can usually determine
hadron charges and muon polarization, and their potentially excellent
energy resolution can be combined with track-chamber pattern
recognition to give them the best background rejection for two-body
decay modes with neutrinos. In general, the more information produced
for each event, the greater the cost per ton of detector.

(5)   What is the real cost of detectors, normalized for sensitivity?

In addition to the cost per ton of instrumented detector, the fiducial-to-total-mass ratio and decay detection efficiency must be taken into account. For decay modes with electrons and photons, iron calorimeters can have much higher fiducial-mass fractions than water or scintillator detectors. However, this gain is partially offset by the higher losses due to nuclear absorption of pions in iron, and is complicated by the greater "crosstalk" between channels with pions. Uncertainties about actual fabrication costs (as opposed to estimates), and the practicality and reliability of different types of experiments in the underground environment can be resolved by careful engineering and by the operation of large prototype detectors.

(6)   How important is other underground physics likely to be?

The proton decay detectors themselves have varying capabilities for the study of other physics topics, for example cosmic-ray neutrinos and muons, and monopole searches. Other important experiments also require underground laboratories (for example solar neutrino experiments), and perhaps construction costs should be shared among several efforts in addition to proton decay, for both practical and physics benefits. Some have speculated that the real discoveries which will result from the search for proton decay will be very different from anything we now imagine. Moreover, grand unified theories now suggest that the most fundamental experiments in particle physics may require underground facilities on the scale of those now provided at the accelerator laboratories. If these predictions prove to be correct, the notion of a national underground physics laboratory may be an idea whose time has come.

Acknowledgements

It is a pleasure to thank my colleagues on the Soudan experiments for their hard work in obtaining and understanding the results presented here. My insight into the interplay of physics and instrumentation issues in the search for proton decay has benefited from spirited discussions with the proponents of other experiments. This work was supported in part by the U.S. Department of Energy and the U.K. Science and Engineering Research Council.

References

[1]   "Report of the Technical Assessment Panel on Proton Decay," S.
      P. Rosen (chairman), February 1982 (unpublished).

[2]   Collaborators on the Soudan Nucleon Decay Program are: J.
      Bartelt, H. Courant, K. Heller, M. Marshak (spokesman), E.
      Peterson, K. Ruddick, and M. Shupe (University of Minnesota); D.
      Ayres, K. Coover, J. Dawson, T. Fields, N. Hill, D. Jankowski,
      E. May, and L. Price (Argonne National Laboratory); W. Allison,
      C. Brooks, J. Cobb, D. Perkins and B. Saitta (Oxford
      University).

[3]   M. A. Shupe, "Experiments at the Soudan Mine: Operating and
      Proposed," proceedings of the Second Workshop on Grand
      Unification, Ann Arbor, Michigan, edited by J. Leveille, L.
      Sulak and D. Unger (Birkhauser Boston, 1981), p. 41.

[4]   J. Bartelt et al., "Soudan 2, a 1000 Ton Tracking Calorimeter
      for Nucleon Decay," proposal to the U.S. Department of Energy
      and the U.K. Science and Engineering Research Council by the
      Minnesota-Argonne-Oxford collaboration, University of Minnesota
      report COO-1764-410, September 1981 (unpublished).

[5]   D. S. Ayres, "The Soudan Nucleon Decay Program," Argonne
      National Laboratory report ANL-HEP-CP-82-03, and proceedings of
      the GUD Workshop, Rome, 1981 (to be published).

[6]   L. E. Price et al., IEEE Trans. Nucl. Sci. NS-29, 383 (1982); L.
      E. Price, "Long Drift Techniques for Calorimeters," Proceedings
      of the International Conference on Instrumentation for Colliding
      Beam Physics, SLAC, Stanford, CA, 1982 (to be published) and
      Argonne National Laboratory report ANL-HEP-CP-82-07.

[7]   V. S. Narasimham and D. Cline, private communications.

[8]   B. Cabrera, Stanford Univ. preprint, March 1982 (submitted to
      Phys. Rev. Lett.).

[9]   J. D. Ullman, Phys. Rev. Lett. 47, 289 (1981).

[10] R. Bonarelli et al. Bologna preprint IFUB 82/1, 1982
      (unpublished).

[11] Collaborators on the Penn Design Study are: A. K. Mann
      (spokesman), E. W. Beier, L. S. Durkin, S. H. Heagy, and H. H.
      Williams (University of Pennsylvania).

[12] A. K. Mann et al., "Design Study for a Fine-Grained Multi-
      kiloton Detector for Nucleon Decay and Neutrino Studies,"
      University of Pennsylvania report, July 1981 (unpublished).

[13] Brookhaven AGS Experiment 734, being carried out by the
      Brookhaven-Brown-INS(Tokyo)-KEK-Osaka-Pennsylvania-Stony Brook
      Neutrino Collaboration.

[14] A. K. Mann, "Proposal for the Establishment of a National
      Underground Physics Laboratory," University of Pennsylvania
      report, August 1981 (unpublished).

[15] R. R. Sharp et al., "Preliminary Site Selection and Evaluation
      for a National Underground Physics Laboratory," Los Alamos
      National Laboratory report LA-UR-82-556, December 1981
      (unpublished).

RECENT DEVELOPMENTS OF THE INVISIBLE AXION

Jihn E. Kim

Department of Physics
Seoul National University
Seoul 151, Korea

1. Introduction

Although the weak interactions violate the CP invariance in various ways, the strong interactions seem to conserve it very accurately. However, it has been known that the QCD vacuum effectively adds a CP-nonconserving piece to the Lagrangian[1]

$$L_{\bar{\theta}} = \bar{\theta} \frac{g^2}{64\pi^2} \varepsilon_{\mu\nu\rho\sigma} G^{a\mu\nu} G^{a\rho\sigma} \tag{1}$$

where $\bar{\theta} = \theta + \arg \det M$. The strong CP problem has been to set $|\bar{\theta}| < 10^{-8}$–10^{-9} not to upset the observed upper limit on the neutron electric dipole moment[2].

Five years ago, a very cute solution to the strong CP problem was suggested by Peccei and Quinn by introducing a $U(1)_A$ symmetry in the quark sector so that the symmetry transformation can rotate away $L_{\bar{\theta}}$. The comprehensive study by Langacker and Pagels[3] shows that $m_u=0$ is not viable phenomenologically. Though $m_u=0$ were successful phenomenologically, some naturalists are not satisfied with this solution to the strong CP problem, since there is no reason why the coefficient of $\bar{u}_L u_R$ should be exactly zero. The global symmetry suggested by Peccei and Quinn must be broken somewhere, and we expect a Goldstone boson called axion[4]. The property of the axion is largely dependent upon the scale where the $U(1)_A$ symmetry is broken without transferring it to a lower energy scale[5].

Since the standard Peccei-Quinn-Weinberg-Wilczek axion seems not to exist[6], several attempts have been tried to make the beautiful idea of Peccei and Quinn phenomenologically successful. Tye[7] tried a heavy axion. A few years ago Kim[8], and Shifman, Vainstein and Zakharov[9]

305

constructed a very light axion (the invisible axion) model introducing a heavy quark. Last year, Dine, Fischler, and Srednicki[10] revived the interest by slightly extending the original Peceei-Quinn model, followed by applications to the grand unified theories[11-23].

In Sec. 2, the axion component is systematically discussed. In Sec. 3, the mass of the invisible axion is calculated, and then, astrophysical and cosmological bounds on the axion parameters are sketched in Sec. 4. In Sec. 5, we discuss a specific grand unification model SU(11) of Georgi which has exactly 45 chiral fields at low energy and an automatic Peccei-Quinn symmetry. Sec. 6 is a brief conclusion.

2. The Axion Component and Realization of a Goldstone Boson in Nature

The axion is a Goldstone boson. The following discussion therefore applies to any Goldstone boson. A Goldstone boson can be properly represented as a phase field a of a global symmetry $U(1)_\Gamma$. The Lagrangian is invariant under the transformation

$$U(1)_\Gamma : a \longrightarrow a + \lambda', \quad \lambda' = \text{const},$$
$$\phi_1 \longrightarrow e^{i\Gamma_1\lambda} \phi_1,$$
$$\phi_2 \longrightarrow e^{i\Gamma_2\lambda} \phi_2, \tag{2}$$
$$\text{etc.}$$

For simplicity, let us assume that ϕ_1 and ϕ_2 are $SU(2) \times U(1)$ doublet Higgs fields. Under a suitable definition of the Higgs phase, the Goldstone boson a is a linear combination of the phase fields P_1 and P_2 of Higgs fields ϕ_1 and ϕ_2. The other orthogonal combination is a kind of Higgs field which we denote as h. We represent ϕ_1 and ϕ_2 as

$$\phi_1 = (\rho_1 + v_1) \, e^{iP_1/v_1}$$
$$\phi_2 = (\rho_2 + v_2) \, e^{iP_2/v_2} \tag{3}$$

where $<\phi_1> = v_1$ and $<\phi_2> = v_2$, and

$$a \equiv \cos\theta \, P_1 + \sin\theta \, P_2$$
$$h \equiv -\sin\theta \, P_1 + \cos\theta \, P_2 \tag{4}$$

or

$$P_1 = \cos\theta \; a - \sin\theta \; h$$

$$P_2 = \sin\theta \; a + \cos\theta \; h \tag{4)'}$$

From (2), (3), and (4)', we obtain

$$\Gamma_1\lambda = \frac{\cos\theta}{v_1}\lambda', \qquad \Gamma_2\lambda = \frac{\sin\theta}{v_2}\lambda'$$

or

$$\cos\theta = \frac{v_1\Gamma_1}{(v_1^2\Gamma_1^2 + v_2^2\Gamma_2^2)^{\frac{1}{2}}}$$

$$\sin\theta = \frac{v_2\Gamma_2}{(v_1^2\Gamma_1^2 + v_2^2\Gamma_2^2)^{\frac{1}{2}}} \tag{5}$$

In general, we obtain the axion (or Goldstone) field from the phase fields of $U(1)_\Gamma$ for nontrivial Higgs fields,

$$a = \frac{1}{N^{\frac{1}{2}}} \sum_i v_i\Gamma_i P_i \tag{6}$$

where $N = \sum_i v_i^2\Gamma_i^2$ and P_i are the phase fields[5].

As an example, let us consider the Peccei-Quinn-Weinberg-Wilczek axion and the Goldstone boson absorbed into Z^0 vector boson. The U(1) quantum numbers are taken as follows,

	Γ	$Z \; (=-I_3+Y)$	
ϕ_1^0	v_2/v_1	1	(7)
ϕ_2^0	v_1/v_2	-1	

where Γ is the Peccei-Quinn generator. Therefore, we obtain

$$a_{P.Q.} = \frac{v_2 P_1 + v_1 P_2}{(v_1^2 + v_2^2)^{\frac{1}{2}}} \tag{8}$$

while $Z \propto (v_1 P_1 - v_2 P_2)$.

In general, the axion coupling to matter fields are

$$a \sum_i \frac{m_i}{N^{\frac{1}{2}}} \bar{f}_i \gamma_5 f_i (\Gamma_{iL} + \Gamma_{iR}) \tag{9}$$

where N is given in Eq.(6). How can we make the axion invisible? Certainly, $N^{\frac{1}{2}} \to \infty$ will do but this is possible only for a weak interaction singlet complex Higgs field σ^8, otherwise the SU(2) × U(1) gauge model is affected. As will be discussed in the following section, the axion mass then tends to become zero, and therefore, we have to worry about the long range force so as not to alter appreciably Einsteinian gravity. Dicus et al noted in their famous paper[24] that there is no lower bound on the axion mass due to the γ_5 coupling to the matter fields. (When I read this paper, I immediately realized the importance of their comment and wrote to them "the comment that $a\bar{u}\gamma_5 u$ coupling does not set any lower limit on the corresponding mass of the pseudoscalar a from the observed gravitational effects is very interesting and seems to be original." Then I saw the immense possibilities of a massless Goldstone boson which could be realized in Nature. Namely, the dead Goldstone boson of early 1960's is revived in early 1980's.)

Three invisible Goldstone bosons become visible in the literature:

(i) Heavy quark invisible axion[8,9] - an extra quark carries a non-vanishing Γ

(ii) Dine-Fischler-Srednicki invisible axion[10] - extra Higgs field carries a nonvanishing Γ

(iii) Chikashige-Mohapatra-Peccei Majoron[25] - extra neutrino carries a nonvanishing Γ

Cases (i) and (ii) solve the strong CP problem, since there exist quarks which carry nonvanishing and anomalous Γ quantum numbers. The Dine-Fischler-Srednicki (DFS) axion is simple and attractive since one can view that they added σ in the Peccei-Quinn model. But the physics of the DFS axion is unique only in nonsupersymmetric SU(5) or SO(10). In other grand unified models, we expect in general that all three types of physics come in together. The Majoron can be obtained from (i) by replacing the heavy quark Q to a heavy Majorana neutrino N. Since N does not carry unbroken nonabelian gauge quantum number, its mass is exactly zero. But in grand unified theories, the Majoron is is not massless any more in many cases. In supersymmetric grand unified theories, the Higgs bosons accompany Higgsinos and hence the heavy quark contribution cannot be neglected even in SU(5).

3. Mass of the Invisible Axion

To begin with, let us discuss contributions from heavy quarks.

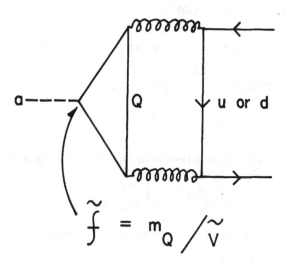

$$\tilde{f} = m_Q / \tilde{v}$$

Following Collins, Wilczek, and Zee[26], we calculate the axion coupling to light quarks through the heavy quark (carrying nonvanishing Γ) triangle graph to obtain

$$- i \, \varepsilon_q \, \bar{q} \, \gamma_5 \, q \, a, \qquad q = u \text{ or } d \qquad (10)$$

where

$$\varepsilon_q = \frac{\alpha_c^2}{\pi^2} \frac{m_Q}{\tilde{v}} \frac{m_q}{m_Q} \ln \frac{m_Q}{m_q} \qquad (11)$$

The form (11) is rather obvious. If $m_q = 0$, there is a global symmetry $q \to \exp(i\gamma_5\alpha) \, q$, and others are irrelevant. Therefore, $\bar{q}\gamma_5 q$ a cannot be generated in higher orders. Namely, ε_q must be proportional to m_q.

Following Bardeen and Tye[27], we can write an effective interaction involving a so that the U(1) symmetry be explicit

$$L_{eff} = \frac{1}{2} \partial_\mu a \, \partial^\mu a + \bar{u} \, i\slashed{\partial} \, u + \bar{d} \, i\slashed{\partial} \, d$$

$$- m_u \{\bar{u}u \cos \frac{\varepsilon_u a}{m_u} + \bar{u}i\gamma_5 u \sin \frac{\varepsilon_u a}{m_u}\}$$

$$- m_d \{\bar{d}d \cos \frac{\varepsilon_d a}{m_d} + \bar{d}i\gamma_5 d \sin \frac{\varepsilon_d a}{m_d}\} \tag{12}$$

The explicit U(1) symmetry in (12) is

$$a \longrightarrow a + \lambda, \qquad \lambda = \text{small}$$

$$u \longrightarrow (1 - i \frac{\lambda\varepsilon_u}{2m_u} \gamma_5)u \tag{13}$$

$$d \longrightarrow (1 - i \frac{\lambda\varepsilon_d}{2m_d} \gamma_5)d$$

from which we extract the conserved U(1) current up to the anomaly term

$$J_\mu^{5s} = v \partial_\mu a + \bar{u}\gamma_\mu\gamma_5 u \frac{v\varepsilon_u}{2m_u} + \bar{d}\gamma_\mu\gamma_5 d \frac{v\varepsilon_d}{2m_d} \tag{14}$$

where v is chosen as a dimensional parameter. The conserved current is

$$J_\mu^{5a} = J_\mu^{5s} - (\frac{v\varepsilon_u}{2m_u} + \frac{v\varepsilon_d}{2m_d}) \{\frac{1}{1+Z} \bar{u}\gamma_\mu\gamma_5 u + \frac{Z}{1+Z} \bar{d}\gamma_\mu\gamma_5 d\} \tag{15}$$

where $Z = m_u\langle\bar{u}u\rangle/m_d\langle\bar{d}d\rangle = m_u/m_d$ for the SU(3) invariant vacuum. Note that $\partial^\mu J_\mu^{5a}$ does not have $G\tilde{G}$ term. Following Bardeen and Tye[27], we obtain

$$m_a^Q = \frac{Z^{1/2}}{1+Z} \frac{F_\pi m_\pi}{\tilde{v}} \frac{\alpha_c^2}{\pi^2} \ln \frac{m_Q^2}{m_u m_d} \tag{16}$$

Let us note that the mass is inversely proportional to $\tilde{v}$ the scale where $U(1)_\Gamma$ is broken. It is gratifying, since we can solve the strong CP problem in supersymmetric grand unified theories too.

Secondly, let us consider the Dine-Fischler-Srednicki contribution to the axion mass. The DFS model has a relevant term in the Higgs potential

$$\varepsilon_{ij} \phi_1^i \phi_2^j \sigma\sigma,$$

and the model is invariant under the Peccei-Quinn symmetry

$$\phi_1 : \Gamma_1, \qquad \phi_2 : \Gamma_2, \qquad \sigma : -(\Gamma_1 + \Gamma_2)/2,$$

$$u_R : -\Gamma_1, \qquad u_L : 0, \tag{17}$$

$$d_R : -\Gamma_2, \qquad d_L : 0$$

Since the Z boson component is proportional to $(v_1 P_1 - v_2 P_2)$, we obtain

$$a_{DFS} = \frac{1}{N^{\frac{1}{2}}} \{v_1 v_2^2 \, P_1 + v_2 v_1^2 \, P_2 - \frac{\tilde{v}}{2} (v_1^2 + v_2^2) \, P_\sigma\} \tag{18}$$

with obvious notations for the VEV's and P's. Following Bardeen and Tye[27], we obtain

$$m_a^{DFS} = \frac{2^{\frac{1}{2}}}{\tilde{v}} F_\pi m_\pi N_g \frac{Z^{\frac{1}{2}}}{1+Z} \tag{19}$$

where N_g is the generation number.

For $N_g = 3$ and $N_Q = 3$, we obtain, assuming for the same Γ,

$$\frac{m_a^Q}{m_a^{DFS}} \simeq 0.03 \tag{20}$$

In the SU(11) model of Georgi, we have $N_Q = 59$, and $m_a^Q/m_a^{DFS} \simeq 0.6$. There is another intriguing point if some heavy quarks get masses by radiative effects, i.e. $m_Q = \alpha^n \tilde{v}$. Then the ratio m_a^Q/m_a^{DFS} is increased by a factor of $1/\alpha^n$ since the loop integration to obtain ε_q of Eq.(11) involves the mass of the quark in the loop.

4. Constraints from Astrophysics and Cosmology

Dicus et al[24] considered the invisible axion emission processes in the stars and obtained

$$m_a^Q < 0.01 \text{ eV} \tag{21}$$

from the requirement that the energy loss rate by the invisible axion emission is not important. This translates to the Peccei-Quinn symmetry breaking scale $\tilde{v} > 10^9$ GeV for the DFS axion ($\tilde{v} > 10^7$ GeV for the heavy quark invisible axion if the heavy quark is the only contribution to the axion mass.) This bound is obtained mainly from the study on red giants.

Recently, Fukugita, Watamura, and Yoshimura[28] argued that the stellar evolution in a light red giant is rather model dependent and excluded the light red giant from the study. They obtained a new bound on the axion mass

$$m_a^{DFS} < 1 \text{ eV} \qquad (22)$$

which translates to $\tilde{v} > 4 \times 10^7$ GeV.

Since the astrophysical bounds on $\tilde{v}$ is in the intermediate mass scale 10^7 GeV ~ 10^9 GeV, one may try to introduce an intermediate mass scale $\tilde{v}$ in a grand unified theory so that the bad feature of the invisibility is made visible in the stars. Indeed Reiss[29], and Mohapatra and Senjanovic[30] considered such a symmetry breaking patterns in SO(10) × U(1)$_A$ by introducing 45_H, 16_H, and 10_H where 45_H is the superheavy scale, 16_H is the intermediate scale, and 10_H is the electroweak scale. $\tilde{v}$ can be as low as 10^9 GeV without upsetting the electroweak phenomenology.[30] It is conceivable to construct other grand unified models with even lower than $\tilde{v} \doteq 10^7$ GeV. The tau neutrino might be relatively heavy, $m_{v_\tau} \doteq (20 \text{ GeV})^2/10^7$ GeV $\doteq$ 40 KeV. This rather heavy mass of v_τ does not conflict the mass density of the universe since it decays[25] by emitting an invisible axion (remember that the Majoron is the same as the invisible axion), $v_\tau \rightarrow v + a$.

Recently, Watamura and Yoshimura[31] tried to obtain the scale $\tilde{v}$ for the Peccei-Quinn symmetry violation from the baryon number generation via Higgs decay[32] (Barr-Segre-Weldon-Nanopoulos-Weinberg-Cox-Yildiz mechanism). Assuming that SU(5) × U(1)$_A$ real fermions do not contribute significantly to the B number generation, i.e. $M_F > M_{colored\ Higgs}$, they obtained $\tilde{v} \geq 10^{15}$ GeV. If $\tilde{v} << 10^{15}$ GeV, one can classify the colored Higgs interactions which preserve SU(3)$_c$ × SU(2) × U(1) × U(1)$_\Gamma$ invariance. Baryon number generation is studied for example by looking at a diagram

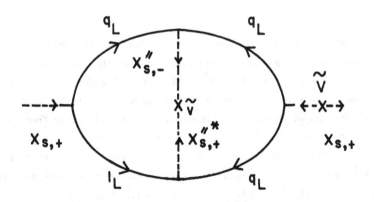

where crosses(×) denotes points where the Peccei-Quinn symmetry is broken. The arrows denote flow of $U(1)_\Gamma$ quantum numbers. If $U(1)_\Gamma$ is a good symmetry, this diagram gives a vanishing ΔB[33]. Therefore, the expected ΔB is proportional to $\tilde{v}^2$,

$$\Delta B \tilde{=} (\frac{\tilde{v}}{M_X})^2 (\Delta B)_{BSW} \tag{23}$$

Namely $\tilde{v} \geq 10^{15}$ GeV since $(\Delta B)_{BSW}$ is acceptable phenomenologically.[34]

Sickivie[35] has made an interesting remark on the remnant symmetries $Z(N)$ with N Peccei-Quinn quarks when the global Peccei-Quinn symmetry is broken by the QCD anomaly[36]. Since there is an N-fold degeneracy in the vacuum, domain walls are created between different vacua. From the restriction that energy density of the domain walls would not exceed the critical energy density of the universe, Sickivie obtains

$$F_\pi m_\pi \tilde{v} \leq 10^{-5} \; (GeV)^3 \tag{24}$$

which is not consistent for a reasonable value of $\tilde{v}$. He argues that $Z(N)$ for $N \geq 2$ must be broken softly by introducing $-\mu^3\sigma$ term in the Higgs potential with an unnaturally small $\mu \leq 10^{-3}(F_\pi^2 \, m_\pi^2/\tilde{v})^{1/3}$. Therefore, one might try to obtain a nondegenerate vacuum solution N=1 and construct an SU(5) model by introducing $5_F + 5_F^*$ fermions and a complex SU(5) singlet Higgs field σ(or a complex adjoint $\sum$),

$$5_F = \begin{pmatrix} Q \\ Q \\ Q \\ E^+ \\ E^o \end{pmatrix}_L \qquad 5_F^* = \begin{pmatrix} Q \\ Q \\ Q \\ E^+ \\ E^o \end{pmatrix}_R \tag{25}$$

For example with a complex σ, we have $SU(5) \times U(1)_A$ invariant Yukawa couplings and Higgs potential

$$L_Y = f \, \overline{5_F^*} \, \sigma \, 5_F + h.c. \tag{26}$$

$$V = -\mu^2 \sigma^* \sigma + \lambda (\sigma^* \sigma)^2 \tag{27}$$

5. An SU(11) Model with an Automatic $U(1)_A$ Symmetry

It is an attractive idea to embed automatically (from the gauge group and the representation contents) the Peccei-Quinn symmetry in grand unified theories. For model building, we pursue under the following rules:

  (i) Automatic embedding of $U(1)_A$ - This principle has been used by several people[19-22]. At low energy, models of Refs 20 and 21 cannot be distinguished, but above grand unification scale, these models are different.

 (ii) Flavor unification - It is conceivable that the Peccei-Quinn $U(1)_A$ is responsible for fermion mass hierarchy[20,17,18].

(iii) Only one global $U(1)_A$ - This is for simplicity. Then, we know that the masses of the invisible axion and the neutrinos are related since both of them arise by breaking the Peccei-Quinn symmetry.

 (iv) Simple gauge group structure.

  (v) No repetition of fermion representations.

 (vi) Gravity respects $U(1)_A$.[19]

Under the above rules, I have searched for possibilities from SU(5) to SU(11), and found that Georgi's SU(11)[37] is the only group satisfying the rules. This is a very attractive model since it has exactly 45 chiral fields at low energy due to the survival hypothesis[37]

$$(10 + 5^*)_e + (10 + 5^*)_\mu + (10 + 5^*)_\tau \tag{28}$$

from the original $561 = \psi^{\alpha\beta\gamma\delta} + \psi_{\alpha\beta\gamma} + \psi_{\alpha\beta} + \psi_\alpha$ chiral fields. We fix the Higgs structure by introducing many 165 and 482 Higgs fields (H^{abc}, H^{abcde}). Then we have the following relevant interaction terms,

$$L = f_1 \psi^{abcd} C \psi^{efgh} H^{ijk} \varepsilon_{abcdefghijk} + f_2 \psi_d C \psi^{abcd} H_{abc}$$

$$+ f_3 \psi_{ab} C \psi_c H^{abc} + f_4 \psi_{abc} C \psi_{de} H^{abcde} \tag{29}$$

$$+ M H^{abcde} H_1^{fgh} H_2^{ijk} \varepsilon_{abcdefghijk}$$

Out of six complex fields, we have five independent couplings, and obtain a $U(1)_A$ global symmetry. The full symmetry of the theory is

$$SU(11)_{gauge} \times U(1)_A^{global} \tag{30}$$

The quantum number X of the $U(1)_A$ symmetry is;

$$\psi_a \; : \; -3 \; (1 \text{ quark}) \qquad H^{abc} \; : \; -2$$

$$\psi_{ab} \; : \; 5 \; (9 \text{ quarks}) \qquad H^{abcde} \; : \; 4$$

$$\tag{31}$$

$$\psi_{abc} \; : \; -9 \; (36 \text{ quarks}) \qquad H^a_b \; : \; 0$$

$$\psi^{abcd} \; : \; 1 \; (84 \text{ quarks})$$

Note that the N-ality zero fields carry zero X quantum numbers for an automatic embedding of $U(1)_{P.Q.}$. Note also that the global current is anomalous for the quark sector, and can be used for the Peccei–Quinn symmetry.

When we break $SU(11) \times U(1)_A$ down to $SU(5)_{GG} \times U(1)_\Gamma$ by $\langle H^{abcde} \rangle$ except H^{12345}, the $U(1)_\Gamma$ charge is

$$\Gamma = 25X + \frac{1}{7} (44 \; Y_a + 20 \; Y_b) \tag{32}$$

where

$$Y_a = \text{diag.} \; (2 \; \; 2 \; \; 2 \; \; 2 \; \; 2 \; -5 \; -5 \; \; 0 \; \; 0 \; \; 0 \; \; 0) \tag{33}$$

$$Y_b = \text{diag.} \; (4 \; \; 4 \; \; 4 \; \; 4 \; \; 4 \; \; 4 \; \; 4 \; -7 \; -7 \; -7 \; -7) \tag{34}$$

The 561 chiral fermions have the Γ charges given in Table 1. Here α, β, $\cdots = 1, 2, \cdots, 5$, and A, B, $\cdots = 6, \cdots, 11$. The only $SU(5) \times U(1)_\Gamma$ real fermion pair occur from $\psi_{ABC} + \psi_{AB}$. We have to break Γ symmetry to remove $SU(5)$ real fermions, by giving VEV's to H^{ABC} and H^{12345} which carry $= -110, 220$, respectively. Excluding tree level masses we obtain

$$9\psi^{\alpha\beta AB} + 6\psi_{\alpha\beta A} + 19\psi^{\alpha ABC} + 15\psi_{\alpha AB} + 6\psi_{\alpha A} + \psi_\alpha + \text{singlets} \tag{35}$$

Table 1

No.	Fermions	Γ	No.	Fermions	Γ
1	$\psi^{\alpha\beta\gamma\delta}$	121	15	$\psi_{\alpha AB}$	-209
6	$\psi^{\alpha\beta\gamma A}$	77	20	ψ_{ABC}	-165
15	$\psi^{\alpha\beta AB}$	33	1	$\psi_{\alpha\beta}$	77
20	$\psi^{\alpha ABC}$	-11	6	$\psi_{\alpha A}$	121
15	ψ^{ABCD}	-55	15	ψ_{AB}	165
1	$\psi_{\alpha\beta\gamma}$	-297	1	ψ_{α}	-99
6	$\psi_{\alpha\beta A}$	-253	6	ψ_{A}	-55

Georgi's survival hypothesis[37] implies that all the SU(5) real fermion pairs should get masses. Indeed, they get masses at higher orders. Let us denote masses generated by the loops collectively as M_I. Then for a perturbative grand unification, we obtain

$$M_I = M_w \left(\frac{M_{P\ell}}{M_X}\right)^{\frac{31}{74}} \left(\frac{M_X}{M_w}\right)^{\frac{53}{74}} \qquad (36)$$

where M_X is the SU(5) unification scale, 5×10^{15} GeV. We obtained (36) by assuming at $M_{P\ell}$, the gauge coupling is the same as the strong coupling α_c at M_w. Thus the unification coupling becomes strong at $M_{P\ell}$. For $\sin^2\theta_w = 0.21$[38], we obtain $M_I = 0.84 \times 10^{13}$ GeV. The schematic behavior is the following.

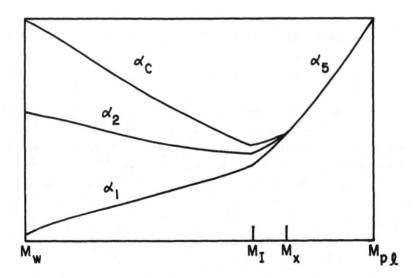

Since the Peccei-Quinn symmetry is broken at $M_{P\ell}$,

$$m_a \approx \frac{F_\pi m_\pi}{M_{P\ell}} \approx 10^{-12} \text{ eV} \qquad (37)$$

The neutrino mass is controlled by the intermediate mass scale M_I[39],

$$m_\nu \approx \frac{m_t^2}{M_I} \approx 0.05 - 1.2 \text{ eV} \qquad (38)$$

for $m_t = 20 - 100$ GeV. The result (38) is interesting, because it is within the experimentally detectable region[40]. Since the breaking of $U(1)_{P.Q.}$ leads to m_a and m_ν, we obtain in general

$$m_a = \alpha^n \frac{F_\pi m_\pi}{m_t^2} m_\nu \qquad (39)$$

where α^n is the effect of loops as pointed out by Witten[39]. On the other hand the relation obtained by Mohapatra and Senjanovic

$$m_a = \frac{F_\pi m_\pi}{m_q^2} m_\nu$$

is not general[30].

In our model, the $SU(2) \times U(1)$ symmetry must be broken by $\langle H^{\alpha ABCD}\rangle$. If $\langle H^{\alpha AB}\rangle \neq 0$, we obtain a bad relation $m_t = m_c$. Therefore

$$\langle H^{\alpha AB}\rangle = 0 \qquad (40)$$

In this case, all light fermion masses are generated by loop diagrams. For example, the diagrams shown on the next page give masses to low energy fermions.

As Georgi, Hall, and Wise commented[19], the gravitational interaction must preserve $U(1)_A$ symmetry, otherwise $U(1)_A$ breaking nonrenormalizable interactions make the idea of Peccei and Quinn useless. For consistency, therefore, we assume that the gravity is described by Zee-Adler type[41] (the so-called spontaneously generated gravity)

$$S = \int d^4x \ g^{\frac{1}{2}} \{\frac{1}{2} \epsilon \phi^2 R + \cdots\} \qquad (41)$$

For an automatic embedding of $U(1)_A$ symmetry, $\phi^2 R$ term or the gravity respects $U(1)_A$ symmetry. In $SU(11)$, there are sufficiently large

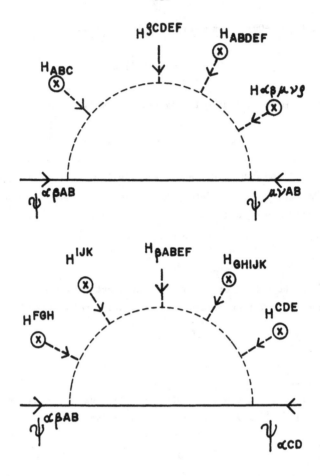

number of fermions above M_X which make all couplings (SU(11) coupling and the gravity) strong at $M_{P\ell}$.

6. Conclusion

There are several good features for having the Peccei-Quinn symmetry: $\bar{\theta} = 0$, $m_\nu \neq 0$, no unbroken global symmetry, and possibly, a solution to fermion mass hierarchy problem[18]. The unattractive features are the existence of a hidden axion and the requirement for gravity to respect $U(1)_A$ symmetry.

In grand unified theories, the invisible axion is likely to appear and the heavy quark contribution[8] and the Dine-Fischler-Srednicki contribution[10] are both important. Probably, there would not exist a "massless" Majoron. Perhaps, one must consider that the Majoron is the same as the invisible axion[42]. We have discussed the interesting SU(11 model which gives $m_a \cong 10^{-12}$ eV, $m_\nu \cong 1$ eV, and $M_I \cong 10^{13}$ GeV. Since the SU(11) unifies the flavor, the fermion masses and mixing angles are calculable in principle. SU(11) is a large group, but it has exactly the same low energy predictions as SU(5) except for the neutrino masses. Who knows how the universe started in the beginning? We, being created 15 billion years later, know only the debris of the giant unifying group, $SU(3)_c \times SU(2) \times U(1)$, 45 chiral fields, and the proton decay.

Acknowledgements

I would like to thank the organizing committee and KOSEF, and in particular Professor Howard Georgi for giving me a chance to talk about the invisible axion. I have greatly benefitted from discussions with R. Barbieri, S. Barr, A. Davidson, P. Frampton, H. Georgi, G. Harvey, K. Kang, Y.J. Ng, S.-Y. Pi, S. Raby, and D.B. Reiss.

References

1. A. Belavin et al., Phys. Lett. 59B, 85(1975); G. 't Hooft, Phys. Rev. D14, 3432(1976); R. Jackiw and C. Rebbi, Phys. Rev. Lett. 37, 172(1976); C.G. Callan, R.F. Dashen and D. Gross, Phys. Lett. 63B, 334(1976).

2. N. Ramsay quotes $d_n < 6 \times 10^{-25}$ e cm (private conversation). He expects to lower the bound to 10^{-27} e cm in two years.

3. P. Langacker and H. Pagels, Phys. Rev. D19, 2070(1979).

4. S. Weinberg, Phys. Rev. Lett. 40, 223(1978); F. Wilczek, Phys. Rev. Lett. 40, 279(1978).

5. J.E. Kim, ICTP preprint IC/81/123(1981).

6. R.D. Peccei, Proc. 19th International Conference on High Energy Physics, ed. S. Homman et al. (Phys. Society of Japan, Tokyo, 1979), p.385; C. Edwards et al., Phys. Rev. Lett. 48, 903(1982).

7. S.-H. H. Tye, Phys. Rev. Lett. 47, 1035(1981).

8. J.E. Kim, Phys. Rev. Lett. 43, 103(1979).

9. M.A. Shifman, A.I. Vainshtein, and V.I. Zakharov, Nucl. Phys. B166, 493(1980).

10. M. Dine, W. Fischler, and M. Srednicki, Phys. Lett. 104B, 199(1981).

11. M.B. Wise, H. Georgi, and S.L. Glashow, Phys. Rev. Lett. 47, 402 (1981).

12. J. Ellis, M.K. Gaillard, D.V. Nanopoulos, and S. Rudaz, Phys. Lett. 106B, 298(1981).

13. H.P. Nilles and S. Raby, SLAC-Pub-2743(1981).

14. R. Barbieri, R.N. Mohapatra, D.V. Nanopoulos, and D. Wyler, Phys. Lett. 107B, 80(1981).

15. P.H. Frampton, Phys. Rev. D25, 294(1982).

16. S.P. De Alwis and P.H. Frampton, Phys. Rev. D24, 3345(1981).

17. A. Davidson and K.C. Wali, Phys. Rev. Lett. 48, 11(1982).

18. R. Barbieri, D.V. Nanopoulos, and D. Wyler, Phys. Lett. 106B, 303 (1981).

19. H. Georgi, L.J. Hall, and M.B. Wise, Nucl. Phys. B192, 409(1981).

20. J.E. Kim, Phys. Rev. D24, 3007(1981).

21. M. Claudson, P.H. Cox, and A. Yildiz, Phys. Rev. Lett. 47, 1698 (1981).

22. P.H. Frampton and T.W. Kephart, Univ. of North Carolina Preprint UNC-164(1981); P.H. Frampton, T.W. Kephart, Y.J. Ng, and H. Van Dam, IFP 167-UNC(1981).

23. K. Kang, I-G Koh, and S. Ouvry, Brown-HET-472(1982).

24. D.A. Dicus, E.W. Kolb, V.L. Teplitz, and R.W. Wagoner, Phys. Rev. D22, 839(1980); V.L. Teplitz, private conversation at the Coral Gables meeting, Jan., 1980.

25. Y. Chikashige, R.N. Mohapatra, and R.D. Peccei, Phys. Rev. Lett. 45, 1926(1980).

26. J. Collins, F. Wilczek, and A. Zee, Phys. Rev. D18, 242(1978).

27. W.A. Bardeen and S.-H. H. Tye, Phys. Lett. 74B, 229(1978).

28. M. Fukugita, S. Watamura, and M. Yoshimura, KEK preprint KEK-TH 40 (1982).

29. D.B. Reiss, Univ. Washington preprint 40048-82-PT7 (1981).

30. R.N. Mohapatra and G. Senjanovic, CCNY and Brookhaven preprint (1982).

31. S. Watamura and M. Yoshimura, Preprint KEK-TH 36 (1981).

32. M. Yoshimura, in Proc. 4th Kyoto Summer Institute, ed. M. Konuma and T. Maskawa (World Science Publishing Co., Singapore, 1981), p.235.

33. S.M. Barr, G. Segre, and A. Weldon, Phys. Rev. D20, 2494(1979); D. V. Nanopoulos and S. Weinberg, Phys. Rev. D20, 2454(1979); A. Yildiz and P. Cox, Phys. Rev. D21, 906(1980).

34. See, also, R. Barbieri, R.N. Mohapatra, D.V. Nanopoulos, and D. Wyler, Phys. Lett. 107B, 80(1981); A. Masiero and G. Segre, to be published.

35. P. Sickivie, Phys. Rev. Lett. 48, 1156(1982).

36. G. Harvey, private conversation.

37. H. Georgi, Nucl. Phys. B156, 126(1979).

38. W. Marciano and A. Sirlin, Phys. Rev. Lett. 46 163(1981).

39. E. Witten, Phys. Lett. 90B, 81(1980).

40. D. Silverman and A. Soni, Phys. Rev. Lett. 46, 467(1981).

41. A. Zee, Phys. Rev. Lett. 42, 417(1979); S. Adler, Rev. Mod. Phys.,
    July 1 (1982).

42. J.E. Kim, Phys. Lett. 107B, 69(1981).

43. For a supersymmetric SU(11), see, S.P. De Alwis and J.E. Kim, to be
    published in Z. Phys. C. (1982).

NEUTRON-ANTINEUTRON CONVERSION EXPERIMENTS*

H. L. Anderson†

1. Introduction

A great deal of attention has been given in this Workshop to proton decay experiments. These experiments look for a violation of baryon number $\Delta B = 1$, as predicted by Grand Unified Theories. There are many experiments searching for proton decay in deep mines and tunnels, all over the world. Some are in progress, others expect to start operating soon, and although clear positive evidence is still lacking second generation experiments are being actively proposed. All are being followed with great interest for the evidence they should provide about the validity and the nature of these theories.

There is another class of experiments which bears on the same question in a different way. These also search for a violation of baryon number, but with $\Delta B = 2$. With $\Delta B = 2$ the spontaneous conversion of a neutron to an antineutron, $n \rightarrow \bar{n}$, becomes possible. In a number of unified theories the predicted rate of $n \rightarrow \bar{n}$ conversion is within the range of experimental possibility.

To my knowledge, the first reference to the $n \rightarrow \bar{n}$ process appeared in a paper by M. Gell Mann and A. Pais,[1] entitled, "Behavior of neutral particles under charge conjugation". The process is allowed under charge conjugation, but requires $\Delta B = 2$. In 1955, baryon number conservation was considered to be an inviolable property of nature so the further implications of the process were not pursued. However, the paper set the groundwork for another process involving charge conjugation, namely the process $K^0 \rightleftarrows \bar{K}^0$ which can proceed under a violation of strangeness conservation, $\Delta S = 2$, for which there were no established prejudice. The subject was revived again in 1970 by V. A. Kuz'min,[2] in a paper under the title, "CP non-invariance and the baryon asymmetry of the universe". In this paper Kuz'min actually estimated the mixing time and showed that a measurement was feasible. However, no attempt

to carry out such an experiment was made. The experimentalists got into the act after S. L. Glashow,[3] high priest of Grand Unification,
launched his campaign to emphasize the importance and the challenges of
experiments that would provide a realistic basis for these ideas. For
the $n \to \bar{n}$ process he proposed an extension of SU(5) with a 6-fermion
coupling that would allow the annihilation of 6 quarks, or alternative-
ly, the conversion of 3 quarks into 3 anti-quarks. The diagram for
this process is shown in Figure 1.

Glashow estimated an upper bound for the interaction strength of
this coupling from the lower bound on nuclear stability.[4,5] Thus,
with $\Gamma = \hbar/10^{30}y$ and m somewhat larger than the nucleon mass we have,

$$\delta m < \sqrt{\Gamma m} < 10^{-21} \text{ eV} \quad .$$

Using the relation $\delta m \simeq m^6/M^5$, appropriate to a 6-fermion coupling, the
unification mass turns out to be $M \simeq 10^6$ GeV. This is much smaller
than the value 10^{15} GeV which characterizes simple SU(5). Because of
the M^{-5} behavior of the interaction strength in the 6-fermion coupling
it is clear that if SU(5) turns out to be the correct theory the chance
for observing $n \to \bar{n}$ will be vanishingly small. However, there are
other theories of the Pati-Salam type in which a 10^6 GeV unification
mass emerges naturally. Such intermediate values for the unification
mass are also characteristic of theories based on SU(16) and SU(4)[4].
Recently, Tosa and Marshak[6] have described a theory in which the
$n \to \bar{n}$ process arises in a "natural" way. It appears that $n \to \bar{n}$ con-

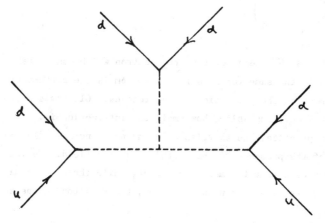

Figure 1 Annihilation of 6 quarks according to Glashow.[3]

Table 1 Theories predicting proton decay or n-n̄ conversion

Proton Decay	n → n̄	Theory	
yes	no	GUT	SU(5) type grand unified theory
yes	yes	EUT	extended unified theory
no	yes	PUT	partial unified theory
no	no	???	unified theory unknown

version is complementary to proton decay in demonstrating which of many possible unification theories nature follows. The above chart, due to Chang,[7] displays how the outcome of proton decay or n → n̄ conversion experiments can help decide which of the theories might have some validity.

2.   Neutron-Antineutron Conversion

The physics involved in the process n → n̄ with $\Delta B = 2$ is analogous to the process $K^O → \bar{K}^O$ with $\Delta S = 2$. The neutron and anitneutron may be treated as members of a two component system. The system which may start out at $t = 0$ as a pure neutron amplitude will, if unperturbed as time goes on, build up an appreciable antineutron amplitude. The time available for the conversion is limited by neutron beta decay, whose mean life is $\sim 10^3$ seconds.

In the unperturbed case, e.g., in the absence of magnetic fields, the time behavior of the system is determined by the 2 x 2 Hamiltonian operator,

$$\begin{pmatrix} E_o & \delta m \\ \delta m & E_o \end{pmatrix}$$

where E_o is the energy of the free neutron which can be taken to be its rest mass, the same for n as for n̄; and δm is the perturbation energy through which the transition n → n̄ proceeds. Glashow's estimate, $\delta m < 10^{-21}$ eV, emphasizes how small the interaction is.

In practice, the energies of n and n̄ are not exactly the same because of the presence of the earth's magnetic field. We write $\Delta E = \mu H$ and recognize that the magnetic moment, while the same in magnitude, has opposite signs for n and n̄. Thus, the Hamiltonian operator is written

$$\begin{pmatrix} E_o + \Delta E & \delta m \\ \delta m & E_o - \Delta E \end{pmatrix} \quad .$$

If an $n - \bar{n}$ mixing exists, neutrons and antineutrons are no longer eigenstates but can be expressed as a mixture of new eigenstates n_1 and n_2,

$$n = n_1 \cos\theta + n_2 \sin\theta \quad ,$$
$$\bar{n} = -n_1 \sin\theta + n_2 \cos\theta \quad ,$$

with $\tan 2\theta = \delta m / \Delta E$.

In a neutron beam the new states change phase with time according to

$$n_1(t) = n_1(0) \exp(-iE_1 t/\hbar) \quad ,$$
$$n_2(t) = n_2(0) \exp(-iE_2 t/\hbar) \quad .$$

E_1 and E_2 are the eigenvalues of energy obtained by diagonalizing the matrix,

$$E_1 = E_o + (\delta m^2 + \Delta E^2)^{1/2} \quad ,$$
$$E_2 = E_o - (\delta m^2 + \Delta E^2)^{1/2} \quad .$$

Thus,

$$n(t) = n_1(0) \exp(-iE_1 t/\hbar) \cos\theta + n_2(0) \exp(-iE_2 t/\hbar) \sin\theta$$

The probability of an $n \rightarrow \bar{n}$ conversion after a time t is obtained by calculating the overlap of $\bar{n}$ with $n(t)$ and squaring

$$P_{\bar{n}}(t) = |< \bar{n}|n(t) >|^2 \quad ,$$

with the result

$$P_{\bar{n}}(t) = \frac{\delta m^2}{\delta m^2 + \Delta E^2} \sin^2 [(\delta m^2 + \Delta E^2)^{1/2} t/\hbar] \; e^{-\lambda t} \quad ,$$

where the extra factor $e^{-\lambda t}$ has been added to take into account the beta decay of the neutron.

Table 2 Number of events in a plausible experiment

$\tau = 10^6$ sec

$\delta m = 7 \times 10^{-22}$ eV

$\phi = 10^{12}$ n/s

$T = 10^7$ sec

$\ell = 50$ meters, drift length

$v = 2500$ m/s neutron velocity

$t = 2 \times 10^{-2}$ sec, drift time

$\varepsilon = 0.5$

$\tau' = \hbar/(\delta m^2 + \Delta E^2)^{1/2}$

H gauss	ΔE eV	τ' s	ν
0.0	0	10^6	2×10^3
0.5	3×10^{-12}	2×10^{-4}	1×10^{-1}
10^{-3}	6×10^{-15}	10^{-1}	2×10^3

We refer to $\tau = \hbar/\delta m$ as the mixing time. For $\delta m = 10^{-21}$ eV, $\tau = 7 \times 10^5$ seconds. We refer to $\tau' = \hbar/(\delta m^2 + \Delta E^2)^{1/2}$ as the oscillation time. For $\Delta E = \mu H$ with $H = 0.5$ gauss, $\Delta E = 3 \times 10^{-12}$ eV and $\tau' = 2 \times 10^{-4}$ seconds.

The number of $n \rightarrow \bar{n}$ conversions detected in a time of observation T is,

$$\nu = P_{\bar{n}}(t) \; \phi T \varepsilon \quad ,$$

where ϕ is the flux, the number of neutrons/s in the beam, and ε is the detection efficiency. Plausible values give the numbers listed in Table 2.

The importance of magnetic field shielding in the experiment is shown in Figure 2. The relative conversion probability has been calculated for a thermal neutron beam having a Maxwellian distribution of velocities corresponding to a temperature of 400°K and a drift length of 10 meters. For this plot $\int B \, dl$ is an invariant so the values of B have to be divided by 5 if the drift length is 50 meters. It can be seen that if the loss in conversion is to be kept below 10%, the magnetic field intensity must be kept below 1.3×10^{-3} gauss. This requires a rather elaborate mu-metal shield together with current

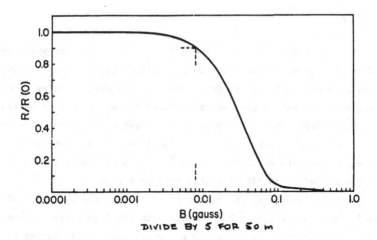

Figure 2   Effect of magnetic field on n - n̄ conversion. The curve is
calculated for a 10 meter drift length and a Maxwellian distribution of velocities at T = 400°K. Values along the
abscissa should be divided by 5 for a 50 meter drift length.

carrying coils to buck out residual fields. An advantage of this
sensitivity to magnetic field is that the effect can be turned off and
a reliable background rate established by operating with a field above
0.1 gauss.

3.   Experimental Limit

The n → n̄ process can occur spontaneously in nuclear matter and
give rise to a striking signature; the release of 2 GeV in energy when
the n̄ annihilates with another nucleon in the same nucleus. There is
an emission of 4-5 pions on average, some of which may be π^+ which
escape absorption and decay to μ^+. Such $\pi^+ \to \mu^+$ decays have been
looked for by Learned et al.[4] and more recently by Cherry et al.,[5]
in deep mine experiments. These experiments have been used to set a
limit of ∿10^{30} years for proton decay. To the extent that the branching ratio for the production of π^+ in proton decay is not very different from that in n̄ annihilation, we can take the same lifetime to apply
to either type of nucleon instability.

We can make a rough estimate of this lifetime by calculating the transition rate using Fermi's Golden Rule,

$$\Gamma = \frac{2\pi}{\hbar} |H|^2 \frac{dN}{dE} \quad ,$$

where the transition matrix element $H = \delta m$ x an overlap integral, and dN/dE is the number of final states per unit energy interval. Using plausible values, a level spacing, $(dN/dE)^{-1} \cong 1$ GeV, a mixing time of 10^6 seconds, and taking the overlap between two nucleons in a nucleus to be of the order of 10^{-2} (Mohapatra and Marshak[7] estimate), we obtain $\Gamma^{-1} = 7$ x 10^{29} years. Thus, an experiment on neutron-antineutron conversion that sets a mixing time $>10^6$ seconds will establish a new limit for nucleon stability for this type of process.

Various estimates of the mixing time based on the experimental limit for nucleon stability have appeared in the literature. Although the formulas obtained may be accurate enough, the results differ because the parameters are not so well known. These estimates are listed in Table 3.

It appears that experiments with a sensitivity $\tau = 10^7$ would be quite important. Because of the quadratic dependence $\Gamma^{-1} \propto \tau^2$, a sensitivity of 10^8-10^9 s would allow the experiment to explore the stability of matter in the range 10^{32}-10^{34} years, namely in the region of the most ambitious projects to search for proton decay.

Table 3 Estimates of mixing time

Ref.	Year	Author	Γ^{-1}	$\tau_{nn}(s)$
2	1970	V.A. Kuz'min	–	$10^4/2\pi$
3	1979	S.L. Glashow	–	10^5-10^6
7	1980	R.N. Mohapatra and R.E. Marshak	10^{30} y	10^5
8	1980	M.V. Kazarnovskii et al.	10^{30} y	2×10^7
9	1981	K.G. Chetyrkin et al.	3×10^{30} y	3×10^7
10	1980	P.G.H. Sandars	10^{30} y	1.8×10^7
11	1981	Riazzudin	10^{30} y	3×10^6
12	1980	L.N. Chang and N.P. Chang	10^{30} y	10^7
13	1980	R. Cowsik and S. Nussinov	10^{31} y	5×10^7

4. The n-n̄ Conversion Experiment

To observe the conversion n-n̄ we arrange to have a large number of neutrons moving as slowly as possible within a long evacuated drift space carefully shielded from magnetic fields. The interaction of n̄ with the target produces a spectacular signature. The annihilation reaction n̄n or n̄p occurs in the nucleus with a high cross-section and results in the emission of 4-5 pions on average, and the release of 2 GeV in energy. However, to realize the full sensitivity needed in such an experiment, the utmost care must be taken in the design of the detector so that it will distinguish, with a high degree of certainty, annihilation events from the cosmic ray background. The difficulty comes from the fact that if 1 event is found after 1 year of running, we want to be sure it is an annihilation event and not caused by one of the 10^{11} cosmic rays that have traversed the detector during this time. Moreover, the detector must function almost perfectly in the presence of a large background of the capture gamma rays that invariably accompany slow neutrons.

5. Annihilation Events

Although the annihilation events are quite striking, involving the emission of 4-5 pions on average, their energy is in the range of several MeV, appreciably lower than the multi-GeV particles that high energy experimentalists have become accustomed to. The rest mass of the charged pions accounts for a substantial fraction of the 2 GeV available, and since the probability of a nuclear interaction is high, much of the energy goes into nuclear excitation and disintegration. A good detector will make this energy visible. Multiple scattering is more serious with these low energy particles, making it more difficult to reconstruct the vertex.

The characteristics of the annihilation events are known in considerable detail from studies that have been made of p̄p annihilations at rest. The branching ratios for the different modes of annihilation are given in Table 4.[14] The difference in the case of n̄n should be minor because the same isotopic spin is involved in the initial state. In the case of n̄p annihilation a different charge and isotopic state is involved and some difference in the charge composition of the pions emitted can be expected.[15] The mean multiplicities and energies of

Table 4   Contribution of pionic states to $\bar{p}p$ annihilations at rest*

Final state	Resonant intermediate state	Percentage of all annihilations CERN	Columbia
All neutral particles		$4.1^{+0.2}_{-0.6}$	3.2 ± 0.5
$\pi^+\pi^-$		0.37 ± 0.03	0.32 ± 0.03
$\pi^+\pi^-\pi^0$		6.9 ± 0.35	7.8 ± 0.9
	$\rho\pi$	5.8 ± 0.3	4.1 ± 0.4
	$f^0\pi^0$	0.24 ± 0.07	
$\pi^+\pi^-$ MM		35.8 ± 0.8	34.5 ± 1.2
	$\eta\pi^+\pi^-$	0.8 ± 0.1	
$2\pi^+2\pi^-$		6.9 ± 0.6	5.8 ± 0.3
	$A_2^\pm \pi^\mp$ $\mathrel{\llcorner\!\!\rightarrow} \rho^0\pi^\pm$	2.0 ± 0.3	
	$\rho^0 f^0$	0.90 ± 0.2	
	$\rho^0\pi^+\pi^-$	1.50 ± 0.3	
	$\rho^0\rho^0$	0.12 ± 0.12	0.4 ± 0.3
$2\pi^+2\pi^-\pi^0$		19.6 ± 0.7	18.7 ± 0.9
	$\omega^0\pi^+\pi^-$	3.0 ± 0.3	3.3 ± 0.4
	$\omega^0\rho^0$	2.1 ± 0.2	0.7 ± 0.3
	$\omega^0 f^0$	1.7 ± 0.2	
	$\rho^0\pi^+\pi^-\pi^0$	} 13.7 ± 0.6	7.3 ± 1.7
	$\rho^\pm\pi^\mp\pi^+\pi^-$		6.4 ± 1.8
	$B^\pm \pi^\mp$ $\mathrel{\llcorner\!\!\rightarrow} \omega\pi^\pm$	0.7 ± 0.1	
	$\eta\pi^+\pi^-$	0.35 ± 0.04	0.34 ± 0.1
	$A_2^\pm \pi^\mp$ $\mathrel{\llcorner\!\!\rightarrow} \eta\pi^\pm$	0.13 ± 0.03	
$2\pi^+2\pi^-$ MM		20.8 ± 0.7	21.3 ± 1.1
	$\eta'\pi^+\pi^-$	0.11 ± 0.02	
$3\pi^+3\pi^-$		2.1 ± 0.2	1.9 ± 0.2
$3\pi^+3\pi^-\pi^0$		1.9 ± 0.2	1.6 ± 0.3
	$\omega 2\pi^+2\pi^-$	1.3 ± 0.3	
	$\eta 2\pi^+2\pi^-$	0.17 ± 0.07	
	$\eta'\pi^+\pi^-$	0.04 ± 0.01	
$3\pi^+3\pi^-$ MM			0.3 ± 0.1

*This table updates the one published by R. Armenteros and B. French in High Energy Physics, Vol. 4 (Academic Press,Inc.), New York, 1969, with data published later by the CERN-Collège de France Collaboration.   In quoting percentages for resonance production, no corrections have been made for decay modes not occurring in the given final state.

the pions for $\bar{p}p$ annihilations at rest according to a summary by
Enstrom et al.[16] are: $<\pi^{\pm,o}> = 5.02$, $<\pi^{\pm}> = 3.06$, $<\pi^-> = 1.53$, and
$<T_\pi> = 234$ MeV. The average number of charged pions according to
Table 4 is 2.95.

It is important to note in Table 4 that there are very few $\pi^+\pi^-$
back to back events, only 0.35%. This makes it plausible to trigger on
a minimum of three particles. To do this effectively, the trigger
should be sensitive to the gammas from π^o. In the case of $\bar{p}p$ ($\bar{n}n$)
annihilations, 3.6% have only π^o's. In 7.4% of the cases one π^o accom-
panies the $\pi^+\pi^-$ pair, and in 35% of the cases the number of π^o's accompanying the $\pi^+\pi^-$ pair is more than one. In all the remaining cases, the
number of charged pions is 4 or more, with and without π^o accompaniment.
In the case of $\bar{n}p$ annihilation, there will always be at least one π^+ and
the situation is more favorable. There are also processes involving K
meson emission, but we do not list them here because all of these
together have a branching ratio of only 4%.

In practice the $\bar{n}$ annihilation will take place on a nuclear target.
For this reason it is useful to consider the results obtained from a
study of 750 MeV antineutron annihilations in a heavy liquid bubble
chamber.[17] Although the chamber contained 22% by weight of Br, this
accounted for only 10% of the annihilations. Most of the annihilations
took place in the light elements, 18% in H, 58% in C, and 14% in F.
Each annihilation gave, on average, 2.8 $\pi^{\pm}$, 1.2 π^- and 2.1 p. The mean
kinetic energy of the pions was $\bar{E}_\pi = 322$ MeV, but as is shown in
Figure 3 the distribution peaks below 100 MeV. An appreciable amount
of energy (and momentum) goes to protons. The average energy is $\bar{E}_p$
= 88 MeV. Most of the protons would be missed in all of the detectors
proposed so far. These indications underline the importance of using
the lightest possible element for the target.

6. The n - $\bar{n}$ Experiments

One n - $\bar{n}$ experiment has already been completed. It was carried
out at the Grenoble reactor[18] and a report was given to this Workshop
last year. The experiment used one of the cold neutron beams that are
available at Grenoble. The conditions of the experiment were very
clean. The cold neutrons, with average velocity 150 m/s are bent
around a curved path by total reflection. They reach the detector
virtually free of capture gammas and fast neutrons. The flux,

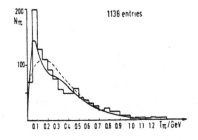

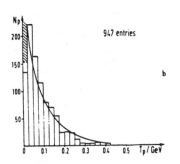

Figure 3 Kinetic energy distributions a) for pions and b) protons obtained using 750 MeV n̄ in a heavy liquid bubble chamber. The events of the n̄p type which were mainly on hydrogen were excluded from the pion plot. Solid lines: results from a model; dashed line in pion spectrum: distribution for annihilations on free nucleons.

however, is low, 10^9 neutrons/s so the sensitivity of the experiment is limited to a mixing time of about 10^6 seconds. The detector was a calorimeter, a hodoscope made of alternate layers of lead and scintillator. Although an extensive anticoincidence shield against cosmic rays was provided, all the events recorded were presumed to be due to cosmic rays, as runs with and without the magnetic field and with the reactor off showed. The event rate, about 100 per day, could not be distinguished from background and limited the determination of the mixing time to $\tau > 1.2 \times 10^5$ s with a 95% confidence level. With "zero" background, this result could have been obtained in less than one day, and a limit of 3×10^6 s could be set in 100 days.

A new run is currently under way with a greatly improved arrangement. The calorimeter is now an array of limited streamer planes, sandwiched between aluminum plates. The limited streamer chambers have good spatial resolution and allow reconstruction of the events to check that they originate from the target. The limited streamer chambers cover the region behind as well as in front of the target. Altogether, 90% of the solid angle for events originating at the target is covered. There are no results available at this time, but the report is that the discrimination against the background has been greatly improved.

A second experiment is currently being mounted in Pavia.[19] The
neutron source is a Triga reactor. Although this reactor is only
250 kw, it is being adapted exclusively for this experiment. The full
area of the thermal column is being utilized, together with a long
flight path and a large target area. As a result, a sensitivity
$\tau \simeq 10^7$ s is anticipated even though the neutrons will be at room
temperature and not cooled.

The other experiments[20] are all at the proposal stage. They
attempt to exploit neutron sources of higher intensity to extend the
sensitivity beyond 10^7 s, possibly even somewhat above 10^8 s. These
proposals require either substantial modification of the neutron source
and/or a more elaborate detector. Thus, these experiments, if they
come off at all, are some years down the road. In considering their
relative merit, it is important to judge whether the detector has sufficient power in fact to discriminate against background events well
enough to achieve the "zero' background condition that is assumed in
quoting sensitivity.

7. Detector Problems

The detector will not operate properly in the presence of the large
fluxes of fast neutrons and capture gamma rays that normally accompany
the production of thermal neutrons. The Grenoble solution, cold neutrons and a curved beam path is the most elegant but the loss of neutron intensity is a major limitation. The Grenoble experimenters hope
that a more intense beam of this type will be built and that by increasing the length of the flight path as well as the area of the target they
will be in a position to exceed 10^8 seconds in sensitivity. The other
proposals do not contemplate the use of neutrons colder than room temperature at this stage. In the experiments using a nuclear reactor the
neutrons are extracted over a large area close to the core and care must
be taken in the design of the moderator to reduce the fast neutron and
gamma ray fluxes that come from the core. This is done by a suitable
arrangement of bismuth slabs within the moderator, although it is hard
to do this without some loss in thermal neutron intensity. It helps to
place the detector as far away from the reactor as possible, i.e., a
long drift distance. It also helps to use a thin target through which
most of the neutrons and gamma rays in the beam can pass without

interacting until they reach a beam stop far beyond. The target, of course, must be thick enough to stop any of the antineutrons which may have appeared in the beam. This leads to a detector in the form of a cylinder which surrounds the target and subtends as large a solid angle around it as practical. Openings have to be left for the beam to enter and leave, to end up in the beam stop beyond. A series of baffles in the drift tube upstream of the detector is used to collimate the beam, to make it strike the target but not the inside wall of the detector around it.

In the proposal using an accelerator as a neutron source, advantage is taken of the pulsed beam. The time of flight of gamma rays and fast neutrons down the drift tube is very different from the slow thermal neutrons and allows gating the detector to respond only to the slow neutrons.

8.   Cosmic Ray Background

The problem of the cosmic ray background is more subtle. We want to be sure that the one event we may see after 200 days of running is not due to one of the 10^{11} cosmic rays that have passed through the detector in that time. Even with an effective anti-coincidence shield to protect the detector against charged particles in the cosmic rays, the uncharged particles, neutrons and wide angle bremsstrahlung, remain in sufficient number to cause plenty of mischief.

9.   Detector Criteria

The successful detector must distinguish annihilation from cosmic ray events unambiguously, but it's hard to know how elaborate (and expensive) a detector has to be to meet this requirement. It does this best if it measures as many as possible of the quantities that characterize an annihilation event, namely:

1)   Spatial reconstruction that identifies a vertex at the target.

2)   Temporal reconstruction that shows the particles to be moving from the vertex, not toward it.

3)   Identification of each particle, showing that by charge, energy, and momentum, it is a likely component of an annihilation channel.

4) The multiplicity is that of a probable annihilation channel.

5) The total energy is consistent with 2 GeV.

6) The total momentum is consistent with zero.

All detectors have limited space, time, energy, and momentum resolution. In particular, since the annihilation takes place in a nuclear target, not all the energy can be made visible, and some of the momentum will be taken up by nuclear recoil and remain unseen. Better results can be obtained at greater cost by elaborating the detector. The problem is to know what is good enough. In this the proposals differ.

We have already seen how, in the first Grenoble experiment, calorimetry plus crude spatial resolution was not enough. It remains to be seen how the new arrangement which incorporates greatly improved spatial resolution for reconstructing the vertex will work out. The other experiments deal with the problem in different ways.

10. Summary of n - n̄ Experiments

Table 5 summarizes the neutron sources used or contemplated in the experiments. Table 6 summarizes the detector characteristics. Table 7 summarizes the main parameters of each experiment, giving the "zero" background sensitivity for a 200 day run at 90% confidence level.

11. Omega West Experiment

As an example of a proposed experiment whose description has not been given before, I give an overview of the layout and design of the Los Alamos Omega West experiment. Figure 4 shows the general layout. The Omega West reactor lies at the bottom of a narrow canyon. The 50 meter drift tube extends across the canyon, with the detector building cut into the canyon wall. A useful amount of cosmic ray shielding is provided thereby. Figure 5 shows some detail at the reactor. Inside the reactor building the drift tube is 24 inches in diameter and shielded with iron plates to keep the radiation levels down in the vicinity of the reactor. Figure 6 shows how the detector building is set in the rock of the canyon wall. It also shows the layout of the detector and the beam stop. Figure 7 shows how the detector is set inside its building. The overburden on the roof of the building provides shielding against the neutrons in the cosmic rays coming from

Table 5 Neutron Sources

Grenoble:
57-MW Reactor: cold neutrons 25°K
curved beam guide 20 cm x 3 cm
low γ, low fast n
$\Phi = 10^9$ n/s $\bar{v} = 160$ m/z $\lambda = 20$ Å

Pävia:
0.25 MW Reactor: thermal neutrons
Area 1.2 m x 1.2 m Bi+Paraffin moderator
$\Phi = 2 \times 10^{11}$ n/s at target

Oak Ridge:
30 MW Reactor: thermal neutrons
Area 1400 cm^2
$\Phi = 5 \times 10^{13}$ n/s at 1 m^2 target at 20 m
$\Phi = 7 \times 10^{12}$ n/s at target with Bi+D$_2$
moderator

Los Alamos LAMPF:
Proton Linac 580 μA
Heavy Metal Beam Stop, D$_2$O moderator
$\Phi = 2\text{-}4 \times 10^{12}$ n/s thermal neutrons
(1 m target at 30 m)

Los Alamos Omega West:
8 MW Reactor: thermal neutrons
Channel tangent to core
$\Phi = 3 \times 10^{11}$ n/s (1.5 m target at 50 m)

Chalk River:
110 MW Reactor: thermal neutrons
channel tangent to core
$\Phi = 1 \times 10^{12}$ n/s (1.5 m target at 50 m)

Grenoble III:
Guide area 100 cm^2, $\lambda = 10$ Å
$\Phi = 1 \times 10^{12}$ n/s (1 m target at 30 m)

---

above. Figure 8 is a longitudinal section of the detector. A cross-sectional view of the detector near its central region is shown in Figure 9. The arrangement of the counters at the ends is shown in Figure 10. The baffling along the drift tube is shown in Figure 11. The detector surrounds a 6 foot (2 meter) diameter thin beryllium target inside the drift tube. The drift tube at this location is a 6 foot diameter evacuated aluminum tube fitted with a magnetic shield. The detector is made entirely of liquid scintillation counters copied from the design of K. Lande, et al.[21] in their Homestake Mine

Table 6   Detectors

Grenoble I:	Pb scintillator calorimeter
	760 kg 10% active. Acceptance 25% of 4π
	Thick target ^{6}LiF
Grenoble II:	Limited streamer tubes + Al plates
	Tracking calorimeter. Acceptance 95% of 4π
	Thick target ^{6}LiF
Pavia:	Pb-flash chamber tracking calorimeter
	65% of 4π acceptance
	Scintillator hodoscope + resistive plate
	chambers for trigger
	Thin target C or Be
Oak Ridge:	Pb-glass Čerenkov counter + scintillation
	counter hodoscope
	90% of 4π acceptance
LAMPF:	Time of flight and tracking chambers
	Temporal + spatial reconstruction
	93% of 4π acceptance
	No calorimetry
Los Alamos Omega West:	Time of flight + calorimetry: using liquid
	scintillators
	95% of 4π acceptance
	Checks momentum balance

---

proposal to study proton decay. The liquid scintillator is held in
polyvinylchloride (PVC) containers 26 feet long, 1 foot by 1 foot in
cross section. The scintillators close to the drift tube are 6 inches
by 6 inches in cross section. The scintillators are viewed at each end
by 5 inch hemispherical photomultipliers. These scintillators have an
attenuation length of 8 meters and a time resolution (measured) of 3 ns
FWHM. A 4 foot gap separates the inner ring of scintillators from the
innermost of the outer rings of scintillators. Thus, by time of flight
measurements it becomes possible to tell whether particles are moving
from inside out or from outside in. The full array of scintillators is
used for calorimetry. The scintillators make visible 84% of the energy
deposited. A large fraction of the charged pions are stopped within

Table 7 n → n̄ Experiment

$$\nu = (t/\tau)^2 \, \Phi T \epsilon$$

$$t = \ell/v \qquad T = 1.73 \times 10^7 \text{ s (200 d)} \qquad \tau = 10^7 \text{ s}$$

Experiment	drift length ℓ m	drift time $t=\ell/v$ s	neutron flux Φ ns^{-1}	Figure of Merit Φt^2 ns	detection efficiency ϵ	no. events for T=10^9 s ν	sensitivity at "zero" background τ_{min} 90% CL s
Grenoble II	6	3×10^{-3}	10^9	9×10^5	0.35	0.054	1.5×10^6
Pavia	16	7×10^{-3}	3×10^{11}	1.4×10^7	0.50	1.3	0.7×10^6
Oak Ridge (with Bi-D$_2$O moderator)	20	8×10^{-3}	4×10^{13}	3×10^9	0.50	215.	1×10^8
			6×10^{12}	4×10^8	0.50	29.	4×10^7
LAMPF (probable)	30	1.4×10^{-2}	2×10^{12}	4×10^8	0.50	34.	4×10^7
(possible)			4×10^{12}	8×10^8	0.50	68.	5×10^7
Omega West	50	2.3×10^{-2}	3×10^{11}	1.5×10^8	0.50	13.	2×10^7
Grenoble III	35	9×10^{-2}	10^{12}	8×10^9	0.50	700.	1.7×10^8

339

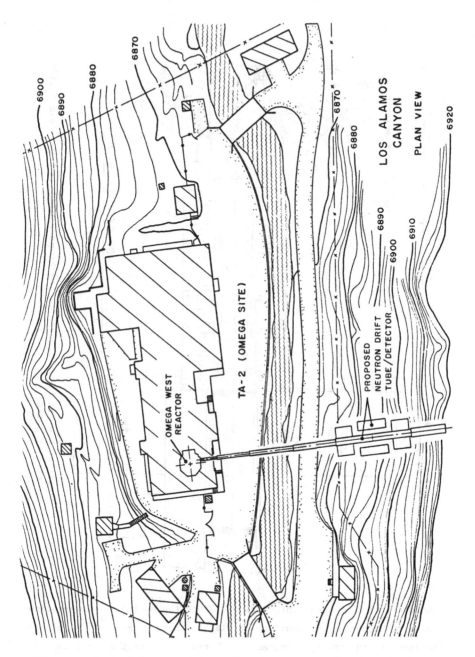

Figure 4 General layout of Omega West reactor n – n̄ conversion
experiment.

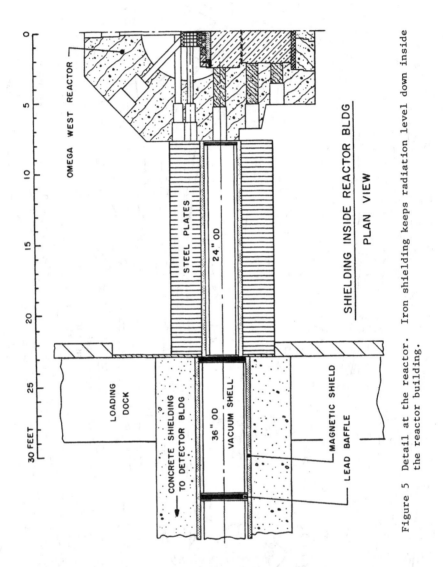

Figure 5 Detail at the reactor. Iron shielding keeps radiation level down inside the reactor building.

the detector by nuclear interaction. Since the nuclear disintegration energy is recorded by liquid scintillator, the rest mass of the pion is included in the energy measurement. Moreover, since the direction of the particle is measured by time of flight it is possible to determine the momentum from the energy measurement of each of the particles that fall within the acceptance, which is 95%. The calorimeter is made thick enough to contain a large fraction of the electromagnetic energy

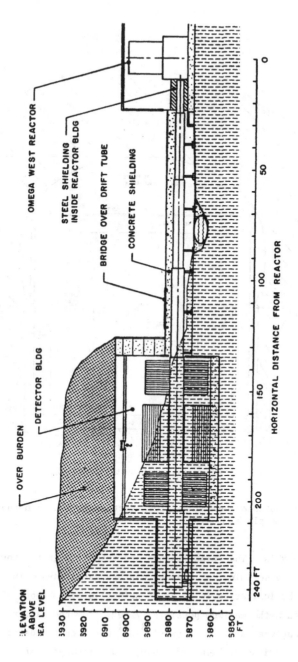

Figure 6 Detector building set in canyon wall.

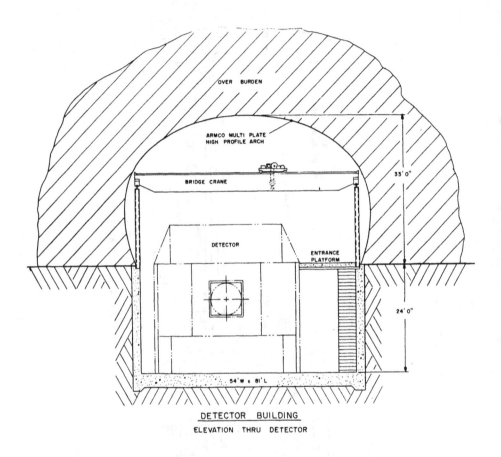

Figure 7   Detector building, cross-sectional view with the detector
in place.

from the π^o's. The scintillators are thick enough to provide a useful
means for discriminating against capture gamma rays. The energy deposited by ionization in 15 cm of scintillator is about 27 MeV, much
higher than the maximum possible, 7 MeV, from the capture gamma rays.
With a capture gamma ray flux $<10^8$ s^{-1} entering the detector, pile-up
is not a serious problem and a simple discriminator at each phototube
will make the detector rather insensitive to the capture gamma rays.
The time of flight capability is used to veto cosmic ray events.

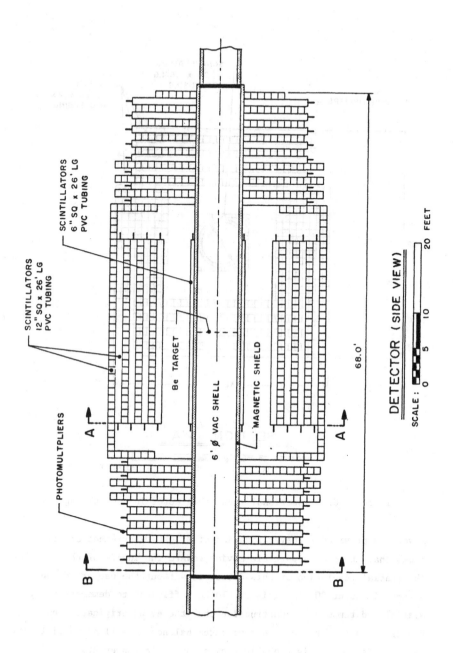

Figure 8   Detector: Longitudinal view showing disposition of
26' x 1' x 1' scintillator modules.

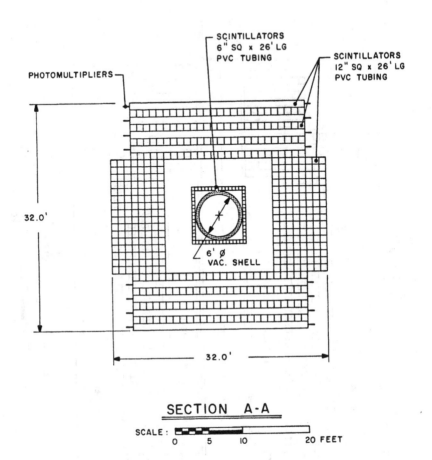

Figure 9 Cross-sectional view of detector in central region.

A veto is provided whenever the time of flight measurement clearly shows that there is a particle moving toward the inside of the detector. The spatial resolution of this detector, without the use of chamber planes, is about 30 cm. This should be sufficient to demonstrate by spatial and temporal reconstruction that the event originates near the target and that energy and momentum balance as well as multiplicity corresponds to what is probable from an annihilation event.

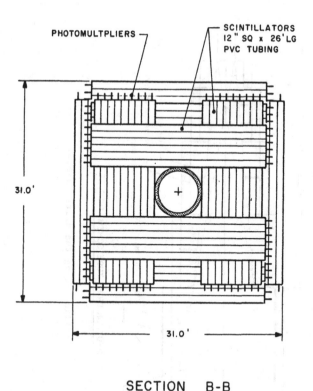

PHOTOMULTPLIERS

SCINTILLATORS
12" SQ x 26'LG
PVC TUBING

31.0'

31.0'

SECTION B-B

SCALE :

0   5   10   20 FEET

Figure 10   Cross-sectional view of end counters.

References

*Talk presented at the Third Workshop on Grand Unification, The
University of North Carolina, Chapel Hill, NC, April 11-17, 1982.

†Los Alamos National Laboratory. Work performed under the auspices
of the U.S.D.O.E.

[1] M. Gell-Mann and A. Pais, Phys. Rev. 97, 1387 (1955).

[2] V. A. Kuz'min, JETP Lett. 12, 228 (1970) [Original: Pis'ma Zh.
     Eksp. Teor. Fiz. 12, 335 (1970)].

[3] S. L. Glashow, Harvard reports HUTO-79/A040,A059 (1979).

[4] J. Learned, F. Reines, and A. Soni, Phys. Rev. Lett. 43, 907 (1979).

[5] M. L. Cherry, M. Deakyne, K. Lande, C. K. Lee, R. I. Steinberg, and
     B. Cleveland, Phys. Rev. Lett. 47, 1509 (1981).

[6] Y. Tosa and R. E. Marshak, VPI-HEP-81/10. R. N. Mohapatra and

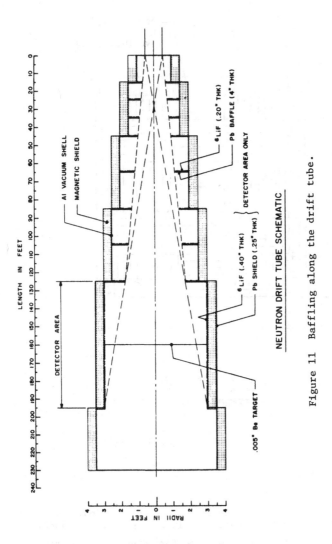

Figure 11 Baffling along the drift tube.

R. E. Marshak, Phys. Rev. Lett. <u>44</u>, 1316 (1980). R. N. Mohapatra and R. E. Marshak, Phys. Lett. <u>94B</u>, 183 (1980). G. Feinberg, M. Goldhaber and B. Steigman, Phys. Rev. D <u>18</u>, 1602 (1978).

[7] N. P. Chang, in Proc. Workshop on Weak Interactions as Probes of Unification, Blacksburg, Va., 1980 [AIP Conf. Proc. No. 72: Particles and Fields, subseries No. 23 DOE-CONF 801244 (1981)].

[8] M. V. Kazarnovskii, V. A. Kuz'min, K. G. Chetyrkin, and M. E. Shaposhnikov, JETP Lett. <u>32</u>, 82 (1980) [Original: Pis'ma Zh. Eksp. Teor. Fiz. <u>32</u>, 88 (1980)]

[9] K. G. Chetyrkin, M. V. Kazarnovskii, V. A. Kuz'min and M. E. Shaposhnikov, Phys. Lett. <u>99B</u>, 358 (1981).

[10] P. G. H. Sandars, J. Phys. G. (Nuclear Physics) $\underline{6}$, L616 (1980).

[11] Riazzudin, Virginia Polytech. Inst. report, VPI-HEP 81/4 (1981).

[12] L. N. Chang and N. P. Chang, Phys. Lett. $\underline{92B}$, 103 (1980); and Errata: Phys. Lett. $\underline{94B}$, 551 (1980).

[13] R. Cowsik and S. Nussinov, Univ. of Maryland, August 1980.

[14] ASTERIX collaboration proposal, "A Study of $\bar{p}p$ Interactions at Rest in a H_2 Gas Target at LEAR", CERN/PSCC/80-101 PSCC/P28, August 29, 1980. See also, J. Diaz et al., Nucl. Phys. $\underline{B16}$, 239 (1970), and references therein.

[15] A. Bettini et al., Il Nuovo Cimento $\underline{47A}$, 642 (1967).

[16] J. E. Enstrom, T. Ferbel, P. F. Slattery, B. L. Werner, Z. G. T. Guiragossián, Y. Sumi, and T. Yoshida, "$\overline{N}N$ and $\overline{N}D$ Interactions-- A Compilation", LBL-58, Berkeley (1972).

[17] H.-J. Besch, H. W. Eisermann, G. Nöldeke, W. Vollrath, D. Waldren, H. Kowalski, H.-J. von Eyss, and H. von der Schmitt, Z. Physik $\underline{A292}$, 197 (1979).

[18] G. Fidecaro, Neutrino '81 Conference on Neutrino Physics and Astrophysics, Maui, Hawaii, July (1981). Milla Baldo-Ceolin, Proceedings of the 15th LAMPF Users Group Meeting, November (1981). K. Green, Review of neutron-antineutron experiments, in Proc. 2nd Workshop on Grand Unification, Ann Arbor, Mich. (1981).

[19] S. Ratti, Invited talk ICOBAN, Bombay, January 1982. NADIR proposal [Pavia nuclear reactor (1981)].

[20] R. Wilson et al. (Harvard-Oak Ridge-Univ. Tennessee Collaboration), A sensitive search for neutron-antineutron transitions, ORR research proposal, 1981. R. Ellis et al. [Los Alamos-Texas (A & M)- Univ. Texas-WP/DNA-William and Mary Coll. Collaboration], LAMPF proposal No. 647, rev. 22 June 1981.

[21] K. Lande et al., University of Pennsylvania proposal to construct the Homestake tracking spectrometer, a deep 1400-ton nuclear decay detector. September 1981.

MASS OF THE t-QUARK

Sandip Pakvasa*
Department of Physics and Astronomy
University of Hawaii at Manoa
Honolulu, Hawaii 96822

ABSTRACT

In many grand unified models as well as in many horizontal symmetry schemes fermion mass relations of the form $m_c/m_t/m_{t'} \cdots = m_\mu/m_\tau/m_L \cdots = m_s/m_b/m_{b'} \cdots$ are expected. A gauge independent formulation for the effective quark mass based on the momentum substraction scheme is used to compute the naked t-threshold with the result: $2M_t = 51 \pm 1$ GeV. If the fourth charged lepton is at 30 GeV, the fourth generation quark masses are found to be $M_{b'} \sim 73 \pm 3$ GeV, $M_{t'} \sim 360 \pm 10$ GeV.

It is fair to say that to date the generation puzzle is not solved and remains one of the important unsolved problems. There are, however, a number of attempts at understanding or at least "guessing" the intergenerational pattern of fermion masses and mixings. A large number of them lead to proportionality between up, down, and leptonic mass matrices, i.e.

$$M_u \propto M_\ell \propto M_d \tag{1}$$

ignoring phases. Then the eigenvalues are also proportional and (neglecting m_e, m_u, and m_d) one expects

$$m_c/m_t/m_{t'} \cdots = m_\mu/m_\tau/m_L \cdots$$
$$= m_s/m_b/m_{b'} \cdots \tag{2}$$

Relations like this arise[1-6] in models in which some symmetry is imposed on the Higgses which give masses to fermions e.g. permutation symmetry, horizontal gauge symmetry, some versions of grand unified models SU(5), SO(10), E_6, etc.

*On behalf of K. Kanaya, H. Sugawara, S. Pakvasa, and S. F. Tuan.

Some more specific ansatzes yield the determinant formula[1,5] for three generations:

$$\begin{vmatrix} m_e^2 & m_\mu^2 & m_\tau^2 \\ m_u^2 & m_c^2 & m_t^2 \\ m_d^2 & m_s^2 & m_b^2 \end{vmatrix} = 0 \tag{3}$$

For practical purposes this is identical to $m_c/m_t = m_\mu/m_\tau$.

These relations (2) are valid at some mass scale $\mu_0 \gg m_W$; e.g. in GUTS with desert $\mu_0 \sim M$unification $\sim 10^{14}$ GeV, in global or gauged horizontal symmetry schemes μ_0 may be $\geq 10^5$ GeV. In this talk, I am going to discuss some recent work[7] on extracting information about physical masses from these mass formulae.

The problem in using mass formulae such as in Eq. (2) is how to define the mass of a quark which is "confined" and hence never on mass-shell? It seems that the most reasonable definition is due to Georgi and Politzer[8]. Their prescription is to choose a momentum subtraction scheme in which the quark propagator $S(p)$ is specified to be

$$i\, S^{-1}(p) = \not{p} - m(\mu) \tag{4}$$

at a space-like $p^2 = -\mu^2$. Then $m(\mu)$ is the running mass of the quark at $p^2 = -\mu^2$. For a heavy quark, the threshold (for pair production) is when

$$m(2M) = M. \tag{5}$$

Heavy means M such that $g(2M)$ is small. We take the threshold to refer to the naked flavor threshold e.g. for charm $2M = 3768$ MeV. I shall return to this later. Knowing $m(2M) = M$, to extrapolate to some other value $m(\mu)$, we need to know the anomalous dimension for mass:

$$\gamma_m = \frac{\mu}{m}\frac{dm}{d\mu}$$

which can be calculated from the self energy $\Sigma(p)$:

$$\gamma_m = \mu\frac{\partial}{\partial\mu}\left[-\frac{1}{4\mu^2}\, \text{Tr}\,\not{p}\,\Sigma(p) + \frac{1}{4m}\,\text{Tr}\,\Sigma(p)\right] \tag{6}$$

With the quark propagator defined in the usual way i.e.

$$S(x,y) = \langle 0| T(\psi(x)\,\bar\psi(y))|0\rangle \tag{7}$$

γ_m can be calculated and the result is

$$\gamma_m^{GP} = g^2/3\pi^2 \left[-\tfrac{3}{2}\left\{1 - \eta \, \ell n(1 + 1/\eta)\right\} - \alpha\eta\left\{1 - (\eta + \tfrac{1}{2}) \, \ell n \, (1 + 1/\eta)\right\}\right] \quad (8)$$

Where $\eta = (m/\mu)^2$ and α is the gauge parameter. Hence, as expected, this γ_m and this extrapolation of $m(\mu)$ are gauge dependent. If we were calculating S-matrix elements this would not matter since all gauge dependence would vanish in the final results. However, for the purpose at hand viz. to predict quark masses at threshold using some mass relations, the gauge dependence creates a serious arbitrariness.

To remove this gauge dependence, we introduce the path ordered phase factor (POP):

$$S_z(x,y) = \frac{1}{N_c} <0|Tq_\alpha(y) \left\{P \exp ig \int_x^y A_\mu(z)dz^\mu\right\}_{\beta\alpha} \bar{q}_\beta(x)|0> , \quad (9)$$

Such propagators were introduced in QED by Schwinger[9] long ago. Now, however, the propagator has path dependence. So we should average over all paths with an appropriate weight factor. A suitable choice for the weight factor which satisfies Lorentz invariance, connectivity, etc., is

$$W_z = \exp\left\{-im' \int_x^y \sqrt{(dz_\mu)^2}\right\} . \quad (10)$$

We had to introduce m' which is a dimensional parameter. Now the propagator is

$$S(x,y) = \frac{1}{N}\int [dz]W_z S_z(x,y) \quad (11)$$

where $N = \int [dz]W_z$. The physical meaning of this propagator becomes clearer when rewritten on a lattice. It can be shown that $S(x,y)$ is equivalent on a lattice to

$$S(x,y) = \frac{1}{N_c N} <0|T(q_\beta(y) \phi_\beta^*(y)\bar{q}_\alpha(x)\phi_\alpha(x))|0> \quad (12)$$

with $N = <0|T(\phi^*(y) \phi(x)|0>_{g=0}$ where $\phi(x)$ is a color triplet scalar field of mass m_0 given by

$$e^{m'a} = 2d + m_0^2 a^2 \quad (13)$$

where a is the lattice spacing and d is the number of space-time dimensions. Hence this propagator resembles that for a color singlet fermion (made up of $q_\alpha \phi_\alpha^*$) normalized by the free propagator for ϕ. It is perhaps natural to define the mass of a confined quark bound in a color singlet hadron this way.

To calculate the self-energy $\Sigma(p)$ and γ_m from this propagator $S(x,y)$, as it stands, is difficult. Some approximation is necessary. The approximation we make is that m' is very large. This can be justified if the colored scalar ϕ is very heavy. Then $1/m'$ which is the range of fluctuations around a straight line path is negligible and we can use a straight line path to evaluate γ_m.

To order g^2, S_z consists of the free part and three self-energy diagrams, as shown in Fig. 1.

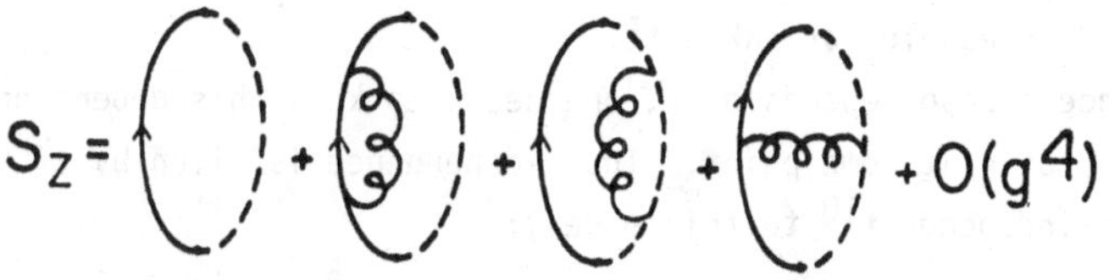

$$S_z = \left(\;\right) + \left(\!\begin{array}{c}\text{\small⊗}\end{array}\!\right) + \left(\!\begin{array}{c}\text{\small⊗⊗}\end{array}\!\right) + \left(\!\begin{array}{c}\text{\small∿}\end{array}\!\right) + O(g^4)$$

$$= S^{(0)} + S^{(1)} + S_z^{(2)} + S_z^{(3)} + O(g^4)$$

Fig. 1   Perturbation expansion of S_z. $S^{(0)}$ and $S^{(1)}$ are z-independent.

$$S_z(x,y) = S^{(0)}(x,y) + S^{(1)}(x,y) + S^{(2)}(x,y) + S^{(3)}(x,y), \qquad (14)$$

where $S^{(0)}(x,y) = \int dp^d (2\pi)^{-d} e^{ip(x-y)} i(\not{p} - m)^{-1}$. $S^{(1)}$ is the usual one loop contribution and leads to the known contribution to γ_m $\gamma_m^{(1)} = \gamma_m^{GP}$ given by Eq. (8). Since $S^{(1)} + S^{(2)} + S^{(3)}$ is gauge invariant (as checked explicitly) we can choose a gauge to work in $(\alpha = 1)$. γ_m is given by

$$\gamma_m = \gamma_m^{(1)} + \gamma_m^{(2)} + \gamma_m^{(3)} . \qquad (15)$$

The evaluation of γ_m^2 is relatively straight forward and is found to be:

$$\gamma_m^{(2)} = g^2/3\pi^2 [2n - n(1 + 2n)\ell n(1 + 1/n)] \qquad (16)$$

The calculation of $\gamma_m^{(3)}$ is much more tedious; finally we were able to show that

$$\gamma_m^{(3)} = \gamma_m^{(2)} . \qquad (17)$$

So net γ_m is

$$\gamma_m = \gamma_m^{GP}(\alpha = 1) + \gamma_m^{(2)} + \gamma_m^{(3)} = g^2/3\pi^2 [-\frac{3}{2} + 3n - 3n^2 \ell n(1 + 1/n)] \qquad (18)$$

This expression for γ_m agrees with $\gamma_m^{GP}(\alpha = 1)$ in the high energy $(\eta \to 0)$ and low energy $(1/\eta \to 0)$ limits.

Now we are ready to apply γ_m to a mass formula such as

$$m_t(\mu) = (m_\tau/m_\mu)m_c(\mu) = 16.88 \, m_c(\mu) \qquad (19)$$

which may hold at some $\mu \gg m_W$. We "know" $m_c(2M_c) = M_c$ at threshold, with $2M_c = 3.768$ GeV. We know $\gamma_m = \frac{\mu}{m}\frac{\partial m}{\partial \mu}$ and so we can extrapolate $m_c(\mu)$ from $m_c(3.768)$ to higher values of μ until at μ^* it satisfies $16.88 \, m_c(\mu^*) = \mu^*/2 = M_t = m_t(2M_t)$. This value μ^* is then the predicted threshold for naked $t\bar{t}$.

Since g also depends on μ, one needs to know this dependence and the value of g at some point. The μ-dependence is given by β^{GP} (which is gauge-independent[10] to this order):

$$\beta = \frac{\mu \partial g}{\partial \mu} = -g^3/16\pi^2 \left\{ 11 - \frac{2}{3}\sum_i \left(1 - 6\eta_i - \frac{12\eta_i^2}{\zeta_i}\ln\left|\frac{\zeta_i - 1}{\zeta_i + 1}\right|\right)\right\} \qquad (20)$$

The equations for $m(\mu)$ and $g(\mu)$ are solved until $\mu^* = 2\mu_t$ is found. The allowed range for $\alpha_s(Q^2)$ at $Q^2 = 20$ GeV^2 and 100 GeV^2 determines $g(2M_c)$ and $2M_t$ in turn is determined primarily by $g(2M_c)$ and the value used for $2M_c$. With $2M_c = 3.768$ GeV and the recent determinations[7] of α_s, we find $2M_t$ to lie between 50 and 52 GeV as in Fig. 2. The ground state topponium is expected to be about 2 GeV below this mass.

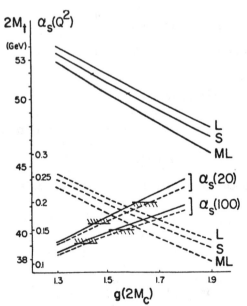

Fig. 2

Plots of top threshold $(2M_t)$ and α_s as functions of $g(2M_c)$. Solid (dashed) lines correspond to $2M_c$ = 3768 (3097) MeV. L, S, and ML are for Landau gauge, our gauge-invariant straight path, and for massless case respectively. Shadowed region on α_s lines is the region allowed by the recent experiments.

There are several uncertainties in the final value for M_t. One is that the calculation is for a space-like M^2 whereas the observed threshold is at a time-like M^2. The correction due to this is presumably small[11],[12] and would tend to <u>increase</u> M_t. Another is the choice of the value of $2M_c$ at threshold to be $2M_D$ = 3.678 GeV. As I shall show below one reason to believe this is the right choice is that the analogous choice for $2M_s$ to be at $2M_K \sim 1$ GeV leads to a reasonable value for $2M_b$. On may also ask why not define the threshold condition as $m(M) = M$ instead of $m(2M) = M$? In other words, one can equally well think of heavy quark production in the hypothetical reaction $\nu_e e^+ \to \bar{d}Q$ ($Q \equiv c$ or t) instead of $e^+e^- \to Q\bar{Q}$. We have carried out a computation of M_t with this threshold condition and find the resulting value for $2M_t$ is lowered only by about 0.5 GeV from the earlier value. Finally, ideally one should calculate the heavy quark effective mass in a meson by calculating γ_m for an object such as $<0|T \left\{ q_\beta(y)\bar{q}'_\beta(y)\bar{q}_\alpha(x)q'_\alpha(x) \right\}|0>$ i.e. replace the colored scalar ϕ in Eq. (12) by a light quark, q'.

As I mentioned above, one can calculate $2M_b$ from $m_s/m_b = m_\mu/m_\tau$ instead of the usual calculation[12] of m_b from $m_b = m_\tau$. The latter depends sensitively on the unification mass, the total number of sequential families and of course the crucial assumption of the desert. We need to know only $m_s(2M_s) = M_s$ and as with charm we take $m_s(2M_s) = M_k \cong 0.5$ GeV. Then we find the naked b threshold $2M_b$ to be between 11.2 and 12 GeV to be compared to the actual value 10.6 GeV. Considering the fact that g is quite large at 1 GeV, I consider this reasonable for $2M_b$ and a vindication of our choice for the heavy quark threshold (á posteriori justification).

The mass formulas in (2) also relate fourth generation masses to third generation masses i.e. $m_t/m_{t'} \sim m_\tau/m_L \sim m_b/m_{b'}$. But m_L, the mass of the fourth lepton, is unknown except that it is beyond PETRA or PEP range. One possibility is that lepton masses scale geometrically as speculated often[1],[13]:

$$m_L/m_\tau \sim m_\tau/m_\mu .$$

$$m_L \sim m_\tau^2/m_\mu \sim 30 \text{ GeV.} \tag{21}$$

In this case, we can compute $M_{b'}$ and $M_{t'}$. We find $M_{b'}$ to be between 70 and 76 GeV and $M_{t'}$ to be between 350 and 370 GeV. If these are indeed the fourth generation masses then:

a) Z^0 has additional decay modes $Z^0 \to L\bar{L}$, $\nu_L\bar{\nu}_L$;

b) b' being lighter than W and Z^0 can be produced in their decays such as $W^- \to b'\bar{c}$ and $Z^0 \to b'\bar{b}$ [14];

c) $M_{t'}$ saturates fermion mass bounds[15] and so this should be the last sequential family.

Returning to our predicted value for M_t of 25.5 GeV; this value is certainly consistent with bounds derived from $K_L \to \mu\bar{\mu}$ and $\delta m_{K\bar{K}}$ [16] and also is lower than the values expected in fixed point schemes (discussed at SWOGU last year)[17] and hence more encouraging for experimenters. So perhaps $m_c/m_t \sim m_\mu/m_t$ is a reasonable ansatz for m_t, our extrapolation is sensible and t is waiting to be found with $2M_t \sim 51$ GeV; if not by hadronic colliders[18], then by TRISTAN.

Acknowledgment

I would like to acknowledge collaboration and counsel by Hirotaka Sugawara, San Fu Tuan, and especially Kazuyuki Kanaya. This work was supported in part by the U.S. Department of Energy under Contract No. DE-AM03-76SF00235.

References

1. S. Pakvasa and H. Sugawara, Phys. Lett. 82B (1979) 105.

2. G. Segre, H. A. Weldon, and J. Weyers, Phys. Lett. 83B (1979) 851; D. Grosser, Phys. Lett. 83B (1979) 855.

3. T. Yanagida, Phys. Rev. D20 (1979) 2986.

4. J. D. Bjorken, SLAC-PUB-2195 (1978).

5. S. L. Glashow, Proceedings of Neutrino - 79, international conference on neutrinos, weak interactions and cosmology, Bergen, June 18 - 22, 1979, ed. A. Haatuft and C. Jarlskog, Vol. 1, p. 518; Phys. Rev. Lett. 45 (1980) 1914.

6. K. T. Mahanthappa and M. A. Sher, Phys. Lett. 86B (1979) 294; R. Barbieri and D. V. Nanopoulos, Phys. Lett. 91B (1980) 369.

7. K. Kanaya, H. Sugawara, S. Pakvasa, and S. F. Tuan, UH-511-461-82; K. Kanaya, KEK Report KEK-TH46. Details and a complete list of references will be found there.

8. H. Georgi and H. D. Politzer, Phys. Rev. D14 (1976) 1829.

9. J. Schwinger, Phys. Rev. 82 (1951) 664.

10. A nice physical discussion of the thresholds in and the gauge independence of β^{GP} can be found in W. Kummer, Phys. Lett. 105B (1981) 473.

11. R. G. Moorhouse, M. R. Pennington, and G. G. Ross, Nucl. Phys. B124 (1977) 285.

355

References (cont.)

12. A. Buras, J. Ellis, M. K. Gaillard, and D. V. Nanopoulos, Nucl. Phys. $\underline{B135}$ (1977) 66.

13. K. Tennakone and S. Pakvasa, Phys. Rev. Lett. $\underline{13}$ (1971) 757; S. Blaha, Phys. Lett. 84B (1979) 116; H. Georgi, Harvard Report HUTP-81/A057.

14. The relevance of this flavor changing mode at one loop level for producing heavy flavors in Z^0-decay was pointed out to me by T. Weiler.

15. M. Veltman, Phys. Lett. $\underline{70B}$ (1977) 252; M. Chanowitz, M. Furman, and I. Hinchliffe, Phys. Lett. 78B (1978) 285; P. Q. Hung, Phys. Rev. Lett. 42 (1979) 873; H. D. Politzer and S. Wolfram, Phys. Lett. $\underline{82B}$, (1979) 242, 421.

16. A. Buras, Phys. Rev. Lett. $\underline{46}$ (1981) 1354; V. Barger, W. Long, E. Ma, and A. Pramudita, Phys. Rev. $\underline{D25}$ (1982) 1860.

17. M. R. Pennington and G. G. Ross, Phys. Lett. $\underline{98B}$ (1981) 291; C. T. Hill, Phys. Rev. $\underline{D24}$ (1981) 691; M. Machacek and M. Vaughn, Phys. Lett. $\underline{103B}$ (1981) 427.

18. The conventional wisdom is that $t\bar{t}$ production for $M_t \sim 25$ GeV will be very difficult to detect in hadronic collisions, e.g. M. Dechantsreiter et al. Phys. Rev. $\underline{D20}$ (1979) 2862. There are some recent more optimistic estimates, however.

THE MONT-BLANC FINE-GRAIN EXPERIMENT ON NUCLEON STABILITY

G. Battistoni[1], E. Bellotti[2], G. Bologna[3], P. Campana[1], C. Castagnoli[3],
V. Chiarella[3], D. Cundy[4], B. D'Ettorre[3], E. Fiorini[2], E. Iarocci[1],
G. Mannocchi[3], G. Murtas[1], P. Negri[2], G. Nicoletti[1], L. Periale[3],
P. Picchi[3], M. Price[4], A. Pullia[2], S. Ragazzi[2], M. Rollier[2],
O. Saavedra[3], L. Trasatti[1] and L. Zanotti[2]

[1] Laboratori Nazionali dell'INFN, Frascati, Italy
[2] Istituto di Fisica dell'Università and INFN, Milan, Italy
[3] Laboratorio di Cosmogeofisica del CNR, Turin, Italy
[4] CERN, European Organization for Nuclear Research, Geneva, Switzerland

(Presented by D.C. Cundy)

The NUSEX detector (3.5 x 3.5 x 3.5 m^2) is a 150 t tracking calorimeter made up of a sandwich of 134 iron plates (1 cm thick) filled with limited streamer tubes [1] of 1 cm^2 cross section and 3.5 m length. An orthogonal pick-up strip system allows tracks to be localized in two dimensions on each tube. In principle this detector can investigate lifetimes up to 10^{31} years, but only if the background problems are well understood along with the behaviour of various particles expected in the decays.

In order to shield against atmospheric muons, the detector will be placed in a laboratory in the Mont-Blanc road tunnel at a depth of ∿ 5000 m.w.e. At this depth, only ∿ 5 muons per day will traverse the detector and we expect a negligible neutron background.

A careful calibration has been carried out in a test module for electrons and pions. The electron resolution was found to be $\sigma/E \underset{\sim}{\sim} 0.20/\sqrt{E}$ (E in GeV).

The response to 500 MeV π^- is the following:

- $\sim$ 85% pion identification,
- $\sim$ 10% π-μ ambiguity,
- $\sim$ 5% π-e confusion.

The major background in this experiment will come from atmospheric neutrino interactions which occur at a rate equivalent to a lifetime of $\sim 10^{30}$ years. In order to simulate this background, a neutrino beam was set up at CERN which had exactly the same energy spectrum up to energies of 3 GeV (fig. 1). The beam consisted of a two-interaction

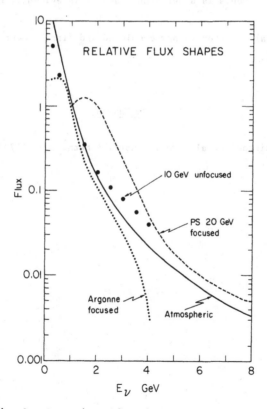

RELATIVE FLUX SHAPES

Fig. 1 - Comparison of various accelerator ν fluxes with atmospheric ν flux

length bare beryllium target followed by ∿ 10 m decay space and 12 m
of shielding. The detector module had a fiducial mass of 2 t, and
∿ 700 events have been obtained for both normal incidence and 45°
incidence of the neutrino beam. A preliminary analysis of these events
has been performed, especially with respect to events of a topology
compatible with proton decay. These events are two-prong events with
the subtended angle > 120°, and ⪆ 3-prong events with one prong backwards
with respect to the other tracks. Initial results show that this type
of event is ∿ 5% of the total neutrino interactions. If an energy cut
of 1 GeV ± 20% is applied the background can be reduced by a further
order of magnitude.

It should be stressed that these results are for neutrino events
that would be completely contained in the detector. The background for
non-contained events is under study, and it is evidently much higher.

The actual detector is now completed and first tracks were obtained
end April 82.

REFERENCE

[1]   G. Battistoni et al., Nucl. Instr. & Meth. 164 (1979) 57.

WHERE IS SUPERSYMMETRY BROKEN?[*1]

Steven Weinberg
Department of Physics
University of Texas

Supersymmetry is a wonderful toy, with which many theorists have enjoyed playing for the last six or seven years. It's a toy of whose purpose we are so far unsure, and the question what supersymmetry is good for is clouded over by the fact that, whatever this toy is good for, it is certainly broken. We would very much like to know where it is broken and what broke it. In particular, is it broken at the 300 GeV scale where the electroweak gauge symmetry is broken (surely not at a scale much lower than this, else we would already be seeing supermultiplets), or is it broken at a very high energy scale of 10^{19} GeV cr 10^{15} GeV associated with gravity or the breakdown of a grand unified symmetry, or as is increasingly popular to suppose, is it broken somewhere in between? We don't know. We also don't know whether supersymmetry is broken spontaneously, or broken intrinsically by "soft" terms in the Lagrangian, but for the most part in this talk I will just assume what seems to me the most attractive case, that it is cnly spontaneously broken. But where?

I don't have an answer to this question, and I won't even try to survey what has been said about it. I will just offer a few partly connected remarks here about the scale of supersymmetry breaking, that I hope may throw some light on the question. I will talk first about the phenomenological problems that arise if one assumes that super-symmetry is broken spontaneously at what these days are called ordinary energies, energies comparable to the 300 GeV scale of $SU(2) \times U(1)$ breaking. Then I will discuss some interesting effects of gravitation in theories of this and other types. Lastly I will come to some cosmological implications of the assumption that supersymmetry is broken at much higher energy scales.

Supersymmetry Broken at Ordinary Energies

It is sometimes said that supersymmetry must be broken at energies no higher than about 300 GeV, or else it would be of no use in solving the hierarchy problem. Be that as it may, it's well worth considering the possibility that supersymmetry survives down to energies of order 300 GeV, because this is the assumption that leads to the greatest number of observable consequences. In fact, not only does this assumption meet Popper's criterion for a scientific theory, that it be experimentally falsifiable; we will see that it is even better than that: it is not only falsifiable, it is already pretty well falsified.

A realistic supersymmetric theory would have to involve quarks and leptons, and also their scalar superpartners, squarks and sleptons, which we organize into chiral superfields Q and L. To give mass to the various quarks and leptons, we also have to include Higgs superfield doublets (H^0, H^-) and (H'^+, H'^0) [not complex conjugates] with trilinear interactions QQH, LLH, etc. Immediately we encounter some problems, that lead us to add further elements:

(a) Trilinear interactions QQQ and QQL would lead to an unsuppressed violation of baryon and lepton conservation. (This is because the color triplet squarks can couple to both quark-lepton and antiquark-antiquark channels.) This must be prevented, either by a discrete symmetry $Q \to -Q$, $L \to -L$, introduced for this purpose by Dimopoulos and Georgi, or by a new $\widetilde{U(1)}$ gauge symmetry of the sort introduced (for other reasons) by Fayet.

(b) Bilinear interactions such as HH' if allowed would revive the hierarchy problem: we would have no way of knowing why their coefficients are not of order 10^{15} GeV. Such superrenormalizable terms should therefore be naturally prohibited by some symmetry, which could be the same $\widetilde{U(1)}$ symmetry used in (a). This is not an entirely convincing conclusion, as the original theory might somehow arrange itself so that even if these terms are allowed by symmetries, they do not arise in tree approximation when we integrate out the superheavy particles, in which case they wouldn't arise in any order of perturbation theory. However, I do not know of any theory where this happens naturally. This is in contrast with the superrenormalizable "Fayet-Iliopoulos" term [proportional to the auxiliary field of a U(1) gauge superfield], which cannot arise in any finite order of perturbation

theory if the underlying grand gauge group is semi-simple, but which
is not forbidden by any symmetry of the low energy effective theory,
and may perhaps appear at ordinary energies through some sort of non-
perturbative mechanism. If superrenormalizable terms like HH' are
ruled out by symmetries, then the scale of supersymmetry (and
SU(2) x U(1)) breaking would have to be set by such Fayet-Iliopoulos
terms.

(c) The scalar partners of the quarks and leptons tend to be too
light in supersymmetric SU(3) x SU(2) x U(1) theories. As advocated by
Fayet, the squarks and sleptons can be systematically boosted in mass
by including a $\widetilde{U(1)}$ gauge group, with quantum number $\tilde{Y}$ having the same
sign for all quark and lepton superfields.

So far, the problems we have considered here have had more or less
straightforward solutions. Now we come to some other problems that may
not be so satisfactorily solved.

(d) The extra $\widetilde{U(1)}$ gauge boson gives rise to neutral current effects,
which generally would have been seen. Fayet has suggested an ingenious
way of suppressing these effects, by passing to a limit in which the
$\widetilde{U(1)}$ coupling constant tends to zero and the coefficient of the Fayet-
Iliopoulos term becomes correspondingly large. This seems to me rather
ad hoc, but perhaps not.

(e) SU(3) x SU(2) x U(1) x $\widetilde{U(1)}$ models like those of Fayet tend to have
triangle anomalies in gauge currents--for instance, there is an anomaly
in the $\widetilde{U(1)}$-SU(3)-SU(3) triangle graphs. It is not at all difficult to
add additional colored and charged superfields to cancel all these
anomalies, but this just raises another problem. As pointed out by
Alvarez-Gaumé, Claudson, and Wise, the addition of these extra superfields opens up degrees of freedom in scalar field space which allow
one to find supersymmetric minima of the potential, in which case
supersymmetry could not be spontaneously broken. As far as I know,
there are no models (realistic or not) that are free of triangle
anomalies in gauge currents and superrenormalizable germs like HH' [see
(b) above], and in which supersymmetry can be spontaneously broken.

(f) [The work of this and the next subsection was done in collaboration with Glennys Farrar.] There is a symmetry called R invariance,
which was originally introduced by Salam and Strathdee and by Fayet for

various model-building purposes, but which has remained with us whether we like it or not, like the rabbits in Australia. An R invariance is any global U(1) symmetry which is carried by the spinor superspace coordinate θ, and hence takes different values for different members of the same supermultiplet. For instance, a theory that has no superrenormalizable terms (like HH') in the superpotential is automatically invariant under an R-symmetry, R_0, which takes the values

	R_0
quarks, leptons, Higgsinos	-1/3
squarks, sleptons, Higgs bosons	+2/3
gauge bosons	0
gauginos	+1

More generally, an R symmetry is R_0 plus any ordinary global U(1) symmetry. The existence of an R symmetry leads to a variety of possible problems, but the particular problem that actually occurs depends on what we assume for the breaking of R symmetry. First, R_0 is certainly spontaneously broken, because the Higgs scalar fields carry $R_0 \neq 0$, but in SU(3) x SU(2) x U(1) x U(1) theories it may be possible to take a linear combination of R_0 and $\tilde{Y}$ which is not spontaneously broken. If this is not the case, then the spontaneous breakdown of R_0 yields a semi-weakly coupled Goldstone boson, more or less like an axion, which would probably have been seen in axion searches. If some R symmetry is not spontaneously broken by scalar vacuum expectation values of order 300 GeV, then there is no such Goldstone boson, but if this R symmetry has a QCD anomaly, then there is a Goldstone boson of a different sort.[2] The R current can be combined with the unitary singlet axial-vector current of the light quarks in such a way as to cancel QCD anomalies, forming an approximately conserved axial current whose conservation is intrinsically broken only by light quark masses. The spontaneous breakdown of this γ_5 symmetry leads to the same ninth pseudoscalar meson with mass below $\sqrt{3}\, m_\pi$ that was the crux of the old U(1) problem before 't Hooft pointed out the relevance of instantons. Finally, if the R symmetries have no QCD anomaly, we find a problem of a different sort. As noted long ago by Ferrara and Zumino, the divergence of the R current is related by supersymmetry to the trace of the energy-momentum tensor. If the R current is free of one-loop QCD triangle anomalies, then the energy-momentum tensor is free of one-loop trace anomalies; that is, the one-loop contributions to the QCD beta function cancel. (The argument breaks down beyond one loop, because the renormalization prescription which makes the R current anomaly-free to all orders is not the one which makes its divergence related to the trace of the energy-momentum tensor.) In a theory of this sort, it would be impossible for the strong and electroweak couplings to approach a common value at any energy at or below the Planck scale.

(g) Supersymmetric models tend to have a variety of light fermions, especially if there is an unbroken R symmetry. One of these which is always present is the Goldstino, with zero mass (aside from gravitational effects to be discussed below.) Another is the photon's superpartner, the photino. These two interact more or less like neutrinos (with only neutral current interactions) and hence might have escaped detection, except that, as emphasized by Dimopoulos and Turner, they contribute to the cosmic energy density and hence increase the helium abundance. A more troublesome particle is the gluino. Even if R symmetry is spontaneously broken, it is difficult to see how the gluino can get a mass above about 1 GeV, so there should be relatively light exotic hadrons, such as gluon-gluino or quark-antiquark-gluino bound states. To have escaped detection, such particles would have to be heavier than about 3 GeV.

All in all, I would say that a supersymmetry that is spontaneously broken at energies of order 300 GeV either runs counter to ideas of grand unification or is in severe conflict with experiment. There is another possibility that has been much discussed lately, and that arises naturally in certain theories: that supersymmetry may be broken at energies of order 300 GeV by soft explicit supersymmetry breaking terms in the effective Lagrangian. (This happens for instance in models like that of Dimopoulos and Raby based on Witten's "inverted hierarchy".) The problems raised above are then largely avoided but the low-energy predictions of supersymmetry in such theories are of course much less definite.

Effects of Gravitation

If supersymmetry is not broken at the energies, say of order 10^{15} GeV, where the grand gauge group is broken, and survives down to much lower energies, then in the absence of gravitation we would expect to find a number of distinct degenerate vacuum states. To see why this vacuum ambiguity arises, recall the general formula for the potential in globally supersymmetric theories:

$$V(z,z^*) = \sum_{ab} J_{ab}^{-1} F_a F_b + \frac{1}{2} \sum_A |D_A|^2$$

where

$$F_a \equiv \frac{\partial f(z)}{\partial z_a}$$

$$J_{ab} \equiv \frac{\partial^2 d(z,z^*)}{\partial z_a \partial z_b^*}$$

$$D_A \equiv \sum_{ab} \frac{\partial d(z,z^*)}{\partial z_a} (t_A)_{ab} z_b.$$

Here z_a are the complex scalar field components of chiral scalar superfields; $f(z)$ and $d(z,z^*)$ are arbitrary functions that determine the way that the superfields appear in the Lagrangian [in the so-called F-terms and D-terms, respectively] and t_A is the matrix [including coupling constant] representing the A-th gauge generator. The equations for a supersymmetric minimum (i.e., V=0) are clearly just $D_A=0$, $F_a=0$. Now, any grand gauge group G is likely to have a number of subgroups H, which are large enough so that if the z_a are constrained to be invariant under H, then all the D_A must vanish. (This is the case for instance if G is SU(N) and H is SU(N) itself, or SU(N-1) x U(1), or SU(N-M) x SU(M) x U(1), etc.) For such scalar field values, the equations for a supersymmetric minimum are that $F_a=0$. Constraining the z's to be invariant under H reduces the number of independent F's by precisely the same amount as it reduces the number of independent z's, so the number of independent complex variables is equal to the number of equations to be satisfied, and we expect to find a solution. A supersymmetric solution of this sort can be avoided by choosing $f(z)$ to have a special form, as in the O'Raiffeartaigh model, but the normal expectation and experience is that for each sufficiently large subgroup of G there is a supersymmetric vacuum solution that is invariant under that subgroup. They all have the same energy density - zero - so in these theories there is no way to distinguish the true vacuum.

Now let's consider the effects of gravitation. The potential has been calculated for a single chiral scalar superfield by Cremmer et al., and in the general case by Witten and Bagger. Their result is

$$V(z,z^*) = e^{8\pi G |d|^2} \left[\sum_{ab} J_{ab}^{-1} \tilde{F}_a \tilde{F}_b - 24\pi G |f|^2 \right] + \frac{1}{2} \sum_A |D_A|^2$$

with J and D the same as before, but now

$$\tilde{F}_a \equiv \frac{\partial f}{\partial z_a} - 8\pi Gf \frac{\partial d}{\partial z_a} \, .$$

The supersymmetric stationary points can easily be seen to be now at $D_A=0$, $\tilde{F}_a=0$. Once again, we expect a solution of these equations for each sufficiently large subgroup of the overall gauge group. However, now they do not have zero energy, but rather

$$V = -24\pi G|f|^2 \, e^{8\pi G|d|^2}$$

or to first order in G:

$$V = -24\pi G|f|_{G=0}^2 + O(G^2) .$$

The superpotentials f(z) are not generally equal at the different stationary points, so the degeneracy has been lifted. Furthermore, the splitting is quite large. In a typical grand unified theory, f at the stationary points of V will be of order $(10^{15} \text{ GeV})^3$, giving a splitting of order $(10^{13} \text{ GeV})^4$, which is much larger than any effect of spontaneous supersymmetry breaking at scales of 300 GeV.

Not only has the degeneracy been lifted, but all of the vacuum states have been negative energy densities. We can add a constant to the superpotential f in such a way as to cancel it for any one of the supersymmetric vacua, so that one solution corresponds to zero energy density and flat spacetime, but all the others will then have negative-definite energy densities, and hence an "anti-de Sitter" spacetime metric. Our own spacetime is pretty flat, so it must be the one of zero energy, and hence apparently unstable against decay into any of the other vacua.

At least, so I thought until a recent conversation with Sidney Coleman. He pointed out that in an earlier paper with de Luccia, they had found that gravity could stabilize a "false" vacuum of zero energy density, even when there is a deeper vacuum state with energy density $-\varepsilon$, provided ε is no larger than a certain critical energy density ε_c, which is of first order in G. The energy density differences found in supergravity are of first order in G, so there was at least a chance that gravity could stabilize flat space in these theories.

In fact, I should have known from the start that this stabilization must occur. There is a general theorem of Deser and Teitelboim, to the effect that in supergravity any state in which the energy density vanishes and the space becomes flat at large distances must have positive energy. Hence even though there may be some sense in

which our universe is not the state of lowest energy, if you form a
bubble of negative-energy density space in our nearly flat space the
energy gained from the volume of the bubble will always be more than
overbalanced by the surface tension energy required to make the bubble
walls. This is borne out by detailed calculation. By applying the
results of Coleman and de Luccia to the case of supergravity, one can
explicitly calculate the critical energy density ε_c to first order in
G. It comes out to be just equal to $24\pi G|f|^2$. But according to our
previous remarks, this is precisely the first-order value for the
energy density difference ε! I do not know why it turns out that
$\varepsilon=\varepsilon_c$ instead of $\varepsilon<\varepsilon_c$, but either way our universe is saved from decay
into anti-de Sitter space.

It would be of great interest to reconsider all this at finite
temperature, in the context of cosmology. Srednicki and Nanopoulos and
Tamvakis have considered how thermal effects lift the degeneracy among
supersymmetric vacua: different vacuum solutions with different
unbroken gauge subgroups have different numbers of light particle
species, yielding different T^4 terms in the free energy. I will just
remark here that once the temperature falls below about 10^{13} GeV these
thermal effects become much smaller than the gravitational vacuum
splitting, so the universe had better be in the right state by then.

Gravitino Masses

Now I want to consider some cosmological consequences of the
assumption that supergravity is spontaneously broken not at 300 GeV
but at very much higher energy scales. Whatever the energy scale $\sqrt{F}$
of supersymmetry breaking (F is normalized so that in the limit $G \to 0$
the vacuum energy is $F^2/2$) the spin 3/2 partner of the graviton, the
gravitino, will get a mass. This was calculated some time ago by Deser
and Zumino to be

$$m_g = \sqrt{\frac{4\pi}{3}} \frac{F}{m_{PL}}$$

with $m_{PL} = 1.2 \times 10^{19}$ GeV the Planck mass. At some time in the history
of the universe these gravitinos must have been as common as any other
particles, and if they have all survived to the present then as
pointed out by Pagels and Primack their mass must be less than about
1 keV, or else they would dominate the mass density of the universe,
so $\sqrt{F} < 10^6$ GeV. If this were really a firm conclusion it would
answer the question raised at the beginning of this talk: we would
know that supersymmetry is broken at relatively low energies, not at

the energy of grand unification.

However, as Pagels and Primack themselves realized, all this depends on the gravitinos surviving to the present. You may recall that for neutrinos, there are two possibilities: either they are light enough to survive to the present, in which case their mass must be below about ·100 eV, or else they are heavier than about 2 GeV, in which case they would have annihilated long before now. Well, you can easily see that the gravitino-gravitino annihilation rate is much too low for them to have disappeared like heavy neutrinos, so the only way to get rid of them is through their decay. The question is, how fast do they decay?

The gravitino decay rate is proportional to $G = 1/m_{PL}^2$, and hence on dimensional grounds also to m_g^3. We may write it as

$$\Gamma_g = \alpha_g m_g^3 / m_{PL}^2$$

with α_g a dimensionless number, which may be of order unity, or very much less. Gravitinos definitely couple with full ($\sqrt{G}$) strength to such channels as quark + squark, gauge boson + gaugino, Higgs boson + Higgsino, etc., so if any squark or slepton or gaugino or Higgsino is lighter than the gravitino then gravitino decay will occur with $\alpha_g \approx 1$. In particular, this is the case if $\sqrt{F} > 10^{16}$ GeV, because then the photino and gluino masses will be less than $(\alpha/2\pi)\sqrt{F}$, while the Goldstino mass will be greater than $10^{-3} \sqrt{F}$. Even if there is no exotic particle lighter than the gravitino, the gravitino may still decay into ordinary particles if a certain symmetry ("R-parity") is spontaneously broken, but in this case $\alpha_g \ll 1$.

The energy density ρ of the universe is dominated when the gravitinos decay by their rest mass, so $\rho \approx m_g (kT)^3$. The decay occurs when the corresponding cosmic expansion rate $H = \sqrt{8\pi\rho G/3}$ drops to the decay rate Γ_g, i.e., at a temperature

$$kT_d \approx \Gamma_g^{2/3} m_{PL}^{2/3} / m_g^{1/3} \approx \alpha_g^{2/3} m_g^{5/3} / m_{PL}^{2/3}.$$

After the decay products' energy is thermalized, the temperature of the universe will have risen to

$$kT_d' \approx \rho^{1/4} \approx \alpha_g^{1/2} m_g^{3/2} / m_{PL}^{1/2}.$$

We must require that this all happens <u>before</u> cosmic nucleosynthesis, so as not to interfere with the predicted helium and deuterium abundances. This implies that $kT_d' > 0.4$ MeV, and therefore

$$m_g > \alpha_g^{-1/3} \times 10^4 \text{ GeV}$$

corresponding to

$$\sqrt{F} > \alpha_g^{-1/6} \times 10^{11} \text{ GeV}.$$

The conclusion then is that the scale $\sqrt{F}$ of supersymmetry breaking may be less than 10^6 GeV, in which case gravitinos are too light to be a problem for cosmology, or they may be above a model-dependent lower bound, which ensures that they decay early enough so as not to be a problem either. This lower bound is above about 10^{11} GeV [because $\alpha_g \lesssim 1$] and below about 10^{16} GeV [because $\alpha_g \approx 1$ if $\sqrt{F} \gtrsim 10^{16}$ GeV.] Thus supersymmetry breaking scales above about 10^{16} GeV are cosmologically allowed in all models, and scales between about 10^6 GeV and 10^{11} GeV are allowed in no models. This is certainly not a complete answer to the question of where supersymmetry is broken, but for the present it's the best I can offer.

[*]Research supported in part by the Robert A. Welch Foundation.
[1]This talk is based in part on four recent papers: S. Weinberg, Phys. Rev. D 26, 287 (1982); G. Farrar and S. Weinberg, paper in preparation; S. Weinberg, Phys. Rev. Letters 48, 1303, 1776 (1982).
[2]This was noticed independently by both Farrar and Georgi.

THIRD WORKSHOP ON GRAND UNIFICATION

ORGANIZING COMMITTEE

Paul Frampton (Chairman)

Howard Georgi

Sheldon Glashow

Gordon Kane

Paul Langacker

Frederick Reines

Gino Segre

Lawrence Sulak

Hendrik van Dam

Asim Yildiz

1.	Francisco del Aguila	University of Florida
2.	Carl Albright	Northern Illinois University
3.	Herbert L. Anderson	Los Alamos National Lab
4.	C. S. Aulokh	City College New York
5.	Frank T. Avignone, III	University of South Carolina
6.	David S. Ayres	Argonne National Laboratory
7.	Joseph Ballam	SLAC
8.	Ricardo Barbieri	Scuola Normale Superiore, Pisa, Italy
9.	Maurice V. Barnhill, III	University of Delaware
10.	Stephen M. Barr	University of Washington,Seattle
11.	Lawrence C. Biedenharn	Duke University
12.	Jeffrey M. Bowen	Bucknell University
13.	Clyde B. Bratton	Cleveland State University
14.	Ronald Bryan	Texas A&M University
15.	Andrzej Buras	Fermilab
16.	Blas Cabrera	Stanford University
17.	Kevin Cahill	University of New Mexico
18.	Gustavo Castelo-Branco	Virginia Polytechnic Institute and State University
19.	Jayprokas Chakrabarti	University of Rochester
20.	Herbert H. Chen	University of California-Irvine
21.	Swee-Ping Chia	University of Texas, Austin/University of Malaya, Kuala Lumpur Malaysia
22.	Latif Choudhury	Elizabeth City State University
23.	Steven M. Christensen	University of North Carolina, Chapel Hill
24.	Donald C. Cundy	CERN, Geneva, Switzerland
25.	Aharon Davidson	The Weizmann Institute, Rehovot, Israel
26.	Sally Dawson	Fermilab
27.	Savas Dimopoulos	Harvard University
28.	Minh Dinh Tran	Brookhaven National Laboratory
29.	John F. Donoghue	University of Massachusetts
30.	Minh Duong-Van	Rice University
31.	Martin B. Einhorn	University of Michigan
32.	Mark Fischler	University of Pittsburgh
33.	Guy Fogleman	Indiana University
34.	Paul H. Frampton	University of North Carolina, Chapel Hill
35.	María Belén Gavela	Brandeis University/LPTHE, Orsey, France
36.	Howard Georgi	Harvard University
37.	Sheldon L. Glashow	Harvard University
38.	Hyman Goldberg	Northeastern University
39.	Maurice Goldhaber	Brookhaven National Laboratory
40.	Zeno D. Greenwood	University of South Carolina
41.	Howard Haber	University of Pennsylvania

42.	Moo-Young Han	Duke University
43.	Jeff Harvey	Princeton University
44.	Archibald W. Hendry	Indiana University
45.	John P. Hernandez	University of North Carolina, Chapel Hill
46.	Yutaka Hosotani	University of Pennsylvania
47.	Prabaham K. Kabir	University of Virginia
48.	Kyungsik Kang	Brown University
49.	Gabriel Karl	University of Guelph, Canada
50.	Boris Kayser	National Science Foundation
51.	Thomas W. Kephart	University of North Carolina, Chapel Hill
52.	Jihn E. Kim	Seoul National University, Korea
53.	William R. Kropp	University of California, Irvine
54.	T. K. Kuo	Purdue University
55.	Paul G. Langacker	University of Pennsylvania
56.	Ching-Yieh Lee	Virginia Polytechnic Institute and State University
57.	Gary Lipton	University of North Carolina, Chapel Hill
58.	Ernest Ma	University of Hawaii
59.	Marie Machacek	Northeastern University
60.	Kalyana T. Mahanthappa	University of Colorado
61.	H. S. Mani	Indian Institute of Technology, Kanpur, India
62.	Philip David Mannheim	University of Connecticut
63.	Takayuki Matsuki	Ohio State University
64.	Eugen Merzbacher	University of North Carolina, Chapel Hill
65.	Hans Meyer	University of Wuppertal, West Germany
66.	Thomas Meyer	Texas A&M University
67.	S. Mikamo	National Laboratory for High Energy Physics (KEK), Tsukuba, Ibaraki-ken, Japan
68.	Kimball A. Milton	Oklahoma State University
69.	S. Miyake	Institute for Cosmic Ray Research University of Tokyo, Japan
70.	Robert Morse	University of Wisconsin
71.	Ramana Murthy	University of Michigan
72.	Norio Nakagawa	Purdue University
73.	Satyanarayan Nandi	The University of Texas, Austin
74.	V. S. Narasimham	Tata Institute, Bombay, India
75.	Charles A. Nelson	SUNY at Binghamton
76.	Y. Jack Ng	University of North Carolina, Chapel Hill
77.	Jan S. Nilsson	Institute of Theoretical Physics, Göteborg, Sweden
78.	Burt A. Ovrut	Princeton University
79.	Stephan Ouvry	Brown University
80.	Sandip Pakvasa	University of Hawaii
81.	Elain Pasierb	University of California, Irvine

83.	Robert E. Peterkin	University of North Carolina, Chapel Hill
84.	So-Young Pi	Harvard University
85.	Milorad Popovic	City College New York
86.	Lawrence Price	Argonne National Laboratory
87.	Gordon D. Pusch	Virginia Polytechnic Institute and State University
88.	Stuart Raby	Los Alamos National Laboratory
89.	Joel D. Rauber	University of North Carolina, Chapel Hill
90.	David B. Reiss	University of Washington
91.	Jung S. Rno	University of Cincinnati
92.	S. Peter Rosen	NSF/ Purdue University
93.	Serge Rudaz	University of Minnesota
94.	Alvaro De Rújula	CERN, Geneva, Switzerland
95.	Joseph Scanio	University of Cincinnati
96.	Bert Schellekens	Fermilab
97.	Howard J. Schnitzer	Brandeis University
98.	Herman M. Schwartz	University of North Carolina, Chapel Hill
99.	Gino Segré	University of Pennsylvania
100.	Qaisar Shafi	University of Maryland
101.	Stephen Shafroth	University of North Carolina, Chapel Hill
102.	Marc Sher	University of California, Santa Cruz
103.	Marvin Silver	University of North Carolina, Chapel Hill
104.	J. J. Simpson	University of Guelph, Canada
105.	Henry W. Sobel	University of California, Irvine
106.	Pierre Sokolsky	University of Utah
107.	James Stasheff	University of North Carolina, Chapel Hill
108.	Richard Steinberg	University of Pennsylvania
109.	T. Suda	Institute for Cosmic Ray Research, University of Tokyo, Japan
110.	Lawrence R. Sulak	University of Michigan
111.	Peter Suranyi	University of Cincinnati
112.	Kasuke Takahashi	National Laboratory for High Energy Physics (KEK), Tsukuba, Ibaraki-ken, Japan
113.	Kunihiko Terasaki	Research Institute for Theoretical Physics, Hiroshima University, Japan
114.	Robert L. Thews	U. S. Department of Energy
115.	Yasunari Tosa	Virginia Polytechnic Institute and State University
116.	Henry Tye	Cornell University
117.	David G. Unger	Carnegie-Mellon University
118.	Hendrik van Dam	University of North Carolina, Chapel Hill
119.	Jack Vander Velde	University of Michigan
120.	Michael T. Vaughn	Northeastern University
121.	Petr Vogel	Caltech

122.	Walter W. Wada	Ohio State University
123.	Kameshwar C. Wali	Syracuse University
124.	Robert C. Webb	Texas A&M University
125.	Thomas J. Weiler	SLAC
126.	Steven Weinberg	University of Texas, Austin
127.	Pei-You Xue	Virginia Polytechnic Institute and State University
128.	Mieko Yamawaki	The University of Rochester
129.	James W. York, Jr.	University of North Carolina, Chapel Hill
130.	Beecher R. Znkayn	Proto-Electronics, Inc., Raleigh
131.	Guido De Zorzi	University of Rome, Italy

SYNOPSIS OF PROGRAM FOR THIRD WORKSHOP
ON GRAND UNIFICATION

Thursday (April 15, 1982)

Morning   Afternoon

Session Chairman: Merzbacher   Session Chairman: Vander Velde

Glashow   Takahashi
De Zorzi   Kang
Frampton   Steinberg
Sobel   Dimopoulos
Ovrut   Miyake and Narasimham

Friday (April 16, 1982)

Morning   Afternoon

Session Chairman: Rosen   Session Chairman: Wali

Cabrera   Langacker
Georgi   Kropp
Morse   Vaughn
Rudaz   Simpson
Chen   De Rújula
Pi

Saturday (April 17, 1982)

Session Chairman: De Rújula

Ayres
Kim
Anderson
Pakvasa
Cundy
Weinberg

Progress in Mathematics
Edited by J. Coates and S. Helgason

Progress in Physics
Edited by A. Jaffe and D. Ruelle

- A collection of research-oriented monographs, reports, notes arising from lectures or seminars
- Quickly published concurrent with research
- Easily accessible through international distribution facilities
- Reasonably priced
- Reporting research developments combining original results with an expository treatment of the particular subject area
- A contribution to the international scientific community: for colleagues and for graduate students who are seeking current information and directions in their graduate and post-graduate work.

Manuscripts

Manuscripts should be no less than 100 and preferably no more than 500 pages in length.

They are reproduced by a photographic process and therefore must be typed with extreme care. Symbols not on the typewriter should be inserted by hand in indelible black ink. Corrections to the typescript should be made by pasting in the new text or painting out errors with white correction fluid.

The typescript is reduced slightly (75%) in size during reproduction; best results will not be obtained unless the text on any one page is kept within the overall limit of 6x9½ in (16x24 cm). On request, the publisher will supply special paper with the typing area outlined.

Manuscripts should be sent to the editors or directly to: Birkhäuser Boston, Inc., P.O. Box 2007, Cambridge, Massachusetts 02139

PROGRESS IN MATHEMATICS
Already published

PM 1 Quadratic Forms in Infinite-Dimensional Vector Spaces
Herbert Gross
ISBN 3-7643-1111-8, 432 pages,   paperback

PM 2 Singularités des systèmes différentiels de Gauss-Manin
Frédéric Pham
ISBN 3-7643-3002-3, 346 pages,   paperback

PM 3 Vector Bundles on Complex Projective Spaces
C. Okonek, M. Schneider, H. Spindler
ISBN 3-7643-3000-7, 396 pages,   paperback

PM 4 Complex Approximation, Proceedings, Quebec, Canada,
July 3-8, 1978
Edited by Bernard Aupetit
ISBN 3-7643-3004-X, 128 pages,   paperback

PM 5 The Radon Transform
Sigurdur Helgason
ISBN 3-7643-3006-6, 202 pages,   paperback

PM 6 The Weil Representation, Maslov Index and Theta Series
Gérard Lion, Michèle Vergne
ISBN 3-7643-3007-4, 348 pages,   paperback

PM 7 Vector Bundles and Differential Equations
Proceedings, Nice, France, June 12-17, 1979
Edited by André Hirschowitz
ISBN 3-7643-3022-8, 256 pages,   paperback

PM 8 Dynamical Systems, C.I.M.E. Lectures, Bressanone, Italy,
June 1978
John Guckenheimer, Jürgen Moser, Sheldon E. Newhouse
ISBN 3-7643-3024-4, 298 pages,   paperback

PM 9 Linear Algebraic Groups
T. A. Springer
ISBN 3-7643-3029-5, 314 pages,   hardcover

PM10 Ergodic Theory and Dynamical Systems I
A. Katok
ISBN 3-7643-3036-8, 346 pages,   hardcover

PM11 18th Scandinavian Congress of Mathematicians, Aarhus,
Denmark, 1980
Edited by Erik Balslev
ISBN 3-7643-3040-6, 526 pages,   hardcover

PM12 Séminaire de Théorie des Nombres, Paris 1979-80
      Edited by Marie-José Bertin
      ISBN 3-7643-3035-X, 404 pages,   hardcover

PM13 Topics in Harmonic Analysis on Homogeneous Spaces
      Sigurdur Helgason
      ISBN 3-7643-3051-1, 152 pages,   hardcover

PM14 Manifolds and Lie Groups, Papers in Honor of Yozô Matsushima
      *Edited by J. Hano, A. Marimoto, S. Murakami, K. Okamoto, and
      H. Ozeki*
      ISBN 3-7643-3053-8, 476 pages,   hardcover

PM15 Representations of Real Reductive Lie Groups
      David A. Vogan, Jr.
      ISBN 3-7643-3037-6, 776 pages,   hardcover

PM16 Rational Homotopy Theory and Differential Forms
      Phillip A. Griffiths, John W. Morgan
      ISBN 3-7643-3041-4, 258 pages,   hardcover

PM17 Triangular Products of Group Representations and
      their Applications
      S.M. Vovsi
      ISBN 3-7643-3062-7, 142 pages,   hardcover

PM18 Géométrie Analytique Rigide et Applications
      Jean Fresnel, Marius van der Put
      ISBN 3-7643-3069-4, 232 pages,   hardcover

PM19 Periods of Hilbert Modular Surfaces
      Takayuki Oda
      ISBN 3-7643-3084-8, 144 pages,   hardcover

PM20 Arithmetic on Modular Curves
      Glenn Stevens
      ISBN 3-7643-3088-0, 236 pages,   hardcover

PM21 Ergodic Theory and Dynamical Systems II
      A. Katok, editor
      ISBN 3-7643-3096-1, 226 pages,   hardcover

PM22 Séminaire de Théorie des Nombres, Paris 1980-81
      Marie-José Bertin, editor
      ISBN 3-7643-3066-X, 374 pages,   hardcover

PM23 Adeles and Algebraic Groups
      A. Weil
      ISBN 3-7643-3092-9, 138 pages,   hardcover

PM24 Ennumerative Geometry and Classical Algebraic Geometry
      Patrick Le Barz, Yves Hervier, editors
      ISBN 3-7643-3106-2.   hardcover

PROGRESS IN PHYSICS

Already published

PPh1 Iterated Maps on the Interval as Dynamical Systems
Pierre Collet and Jean-Pierre Eckmann
ISBN 3-7643-3026-0, 256 pages,   hardcover

PPh2 Vortices and Monopoles, Structure of Static Gauge Theories
Arthur Jaffe and Clifford Taubes
ISBN 3-7643-3025-2, 294 pages,   hardcover

PPh3 Mathematics and Physics
Yu. I. Manin
ISBN 3-7643-3027-9, 112 pages,   hardcover

PPh4 Lectures on Lepton Nucleon Scattering and Quantum
Chromodynamics
W.B. Atwood, J.D. Bjorken, S.J. Brodsky, and R. Stroynowski
ISBN 3-7643-3079-1, 574 pages,   hardcover

PPh5 Gauge Theories: Fundamental Interactions and Rigorous Results
P. Dita, V. Georgescu, R. Purice, editors
ISBN 3-7643-3095-3, 406 pages,   hardcover